D0848546

Quantum Methods
with *Mathematica®*

JAMES M. FEAGIN

Quantum Methods
with *Mathematica*®

Includes diskette

THE
ELECTRONIC
LIBRARY
OF
SCIENCE

Springer-Verlag

James M. Feagin
Department of Physics
California State University–Fullerton
Fullerton, CA 92634 USA

Published by TELOS, The Electronic Library of Science, Santa Clara, CA.
Publisher: Allan M. Wylde
Publishing Associate: Cindy Peterson
TELOS Production Manager: Sue Purdy Pelosi
Springer-Verlag Production Manager: Karen Phillips
LaTeX conversion and page layout: Joe Kaiping
Cover Designer: André Kuzniarek
Manufacturing Supervisor: Vincent Scelta

Library of Congress Cataloging-in-Publication Data
Feagin, J.M. (James M.)
        Quantum methods with mathematica / J.M. Feagin.
            p. cm.
        Includes bibliographical references and index.
        ISBN 0-387-97973-5. -- ISBN 3-540-97973-5
        1. Quantum theory--Mathematics. I. Title.
    QC174.17.M35F4 1993   93-29028
    530.1′2--dc20

Printed on acid-free paper.

Photocomposed pages prepared from the author's *Mathematica* files using LaTeX.
Printed and bound by Hamilton Printing Co., Rensselaer, NY.
Printed in the United States of America.

9 8 7 6 5 4 3 2 1

ISBN 0-387-97973-5 Springer-Verlag New York Berlin Heidelberg
ISBN 3-540-97973-5 Springer-Verlag Berlin Heidelberg New York

THE ELECTRONIC LIBRARY OF SCIENCE

TELOS, The Electronic Library of Science, is an imprint of Springer-Verlag New York, with publishing facilities in Santa Clara, California. Its publishing domain encompasses the natural and physical sciences, computer science, mathematics, and engineering. All TELOS publications have a computational orientation to them, as TELOS' primary publishing strategy is to wed the traditional print medium with the emerging new electronic media in order to provide the reader with a truly interactive multimedia information environment. To achieve this, every TELOS publication delivered on paper has an associated electronic component. This can take the form of book/diskette combinations, book/CD-ROM packages, books delivered via networks, electronic journals, newsletters, plus a multitude of other exciting possibilities. Since TELOS is not committed to any one technology, any delivery medium can be considered.

The range of TELOS publications extends from research level reference works through textbook materials for the higher education audience, practical handbooks for working professionals, as well as more broadly accessible science, computer science, and high technology trade publications. Many TELOS publications are interdisciplinary in nature, and most are targeted for the individual buyer, which dictates that TELOS publications be priced accordingly.

Of the numerous definitions of the Greek word "telos," the one most representative of our publishing philosophy is "to turn," or "turning point." We perceive the establishment of the TELOS publishing program to be a significant step towards attaining a new plateau of high quality information packaging and dissemination in the interactive learning environment of the future. TELOS welcomes you to join us in the exploration and development of this frontier as a reader and user, an author, editor, consultant, strategic partner, or in whatever other capacity might be appropriate.

TELOS, The Electronic Library of Science
Springer-Verlag Publishers
3600 Pruneridge Avenue, Suite 200
Santa Clara, CA 95051

Hi David, Jenna, Michael, Birgit!

Foreword

The teaching and learning of nonrelativistic quantum mechanics have in recent years been profoundly affected by two developments. First, advances in experimental technique, especially in quantum optics and electronics and in neutron interferometry, have turned many quantum phenomena and "thought experiments," which previously had only been the subject of hypothetical inferences, into laboratory reality. The other line of progress has been, after much premature advertising, the advent of computational physics as a tool of the trade that now belongs, and fits, into every student's bag.

Quantum Methods with Mathematica® places itself comfortably four-square in our contemporary culture by using the full technology of *Mathematica* as its pedagogic environment. I confess that I approached the manuscript of this text with apprehension, fearing that, with no previous exposure to *Mathematica* and well past the nimble age, I might quickly come to grief and frustration. As I write this, I have survived the course, and I have learned the wonders (and some of the foibles) of *Mathematica*, as well as some new things about quantum mechanics. And, under Jim Feagin's deft guidance, the experience has been fun!

"DO quantum mechanics" might have been the title of this book, which exploits the symbolic manipulation, numerical computing, and graphical potential of *Mathematica*. Discussion of the concepts is kept to the minimum necessary for working through the exercises and problems on a computer. This strategy will appeal to those of us who prefer to convey to beginning students of quantum mechanics their own understanding of the ideas and implications of the theory. Without cutting corners, the payoff is freedom from the tedium of long-winded and distracting blackboard derivations of the mathematical apparatus and the possibility of solving problems in settings that are more realistic and interesting than the standard schematic textbook examples. Meticulous debugging of the programs we write is, of course, the price we must pay.

Jim Feagin has resisted the temptation to cover the waterfront and limited his topics to a selection that could be treated in depth and that makes for a coherent interrelated story. It all culminates in a remarkably fresh analysis of the hydrogen atom, with touches of insight gained from the author's own research on the quantum mechanics of simple atoms and molecules, with or without external fields.

Surely, as computers become integrated into physics education, this book will serve as a model for further curricular innovation. Those who follow Jim Feagin's lead will be guided by his infectious enthusiasm and sparkling style as much as by his important message.

Eugen Merzbacher
Chapel Hill, North Carolina
July 1993

Preface

This book is brought to you by the evolution of "computer algebra systems" on personal computers and our expanding notion of computational physics. Its objective is to show you that computer quantum physics need not merely mean finding numerical solutions of the Schrödinger equation. Using the computer, you can calculate quantities like matrix elements symbolically as well as compute them on a grid numerically. This book is intended both for new students learning quantum mechanics and for experienced practitioners wanting to automate their efforts with modern machines. To help in the transition to the computer, I present several standard topics in a conventional way, for example series solution of the harmonic oscillator Schrödinger equation. I also consider, however, problems that tend to be too bulky to be much fun by hand like the time development of free-particle and harmonic oscillator wavepackets. The computer is simply our tool for the task at hand, and I have chosen *Mathematica* as the working medium over other notable alternatives because of its flexibility and elegance. This is therefore also a book about *Mathematica*.

The physics in this book is appropriate for upper-division undergraduate and first-year graduate courses in non-relativistic quantum mechanics. I have tried to avoid drafting another quantum mechanics text, and this book should be used in conjunction with lecture notes or with a favorite textbook. At the same time, I don't attempt to compete with conventional texts on computers in physics that take a purely numerical approach. Nevertheless, I have sought to make the book self-contained and to connect applications.

The book is divided into two parts: *Systems in One Dimension* and *Quantum Dynamics*. Part I emphasizes topics from a first year course on quantum mechanics, while Part II includes more advanced topics. Part II, however, is certainly not off limits to undergraduates. On the contrary, commutators, angular momentum, and the hydrogen atom are essential elements of undergraduate training. In order to keep the applications throughout the book accessible to undergraduates, I have focussed on specific examples and refrained from generality. For example, I discuss the matrix representation of angular momentum by working explicitly with a spin-one particle. I have taught most of Part I and selected topics from Part II in a one semester senior quantum mechanics course. We spent about half our time outside lecture in the computer laboratory working interactively.

The early sections of the book emphasize *Mathematica* usage and syntax in order to help beginners jump right in with the physics. Lengthy or repeated applications of *Mathematica* are backed up by appendices designed for reference or for a more systematic approach if someone is just learning about *Mathematica*, indeed computers. Appendix V gives an introduction to vector calculus in curvilinear coordinates and sets the stage for the operator formalism presented in Part II, especially chapters 18 and beyond. There are many problems and exercises integrated into the book. The exercises focus on material from the text and have shorter solutions, while the problems provide topics for further study and generally require more effort. In either case, I give hints for setting up solutions, particularly the *Mathematica*. I sometimes suggest paper and pencil be used to solve certain problems not particularly well suited for the computer, usually because they can be solved by simple formal manipulations, but contribute to the book's overall completeness.

I find the *Mathematica notebook* front-end to be indispensable for minimizing the effort required to set up and work through problems on the computer. New users almost always prefer it over other front-ends for such tasks, even when better machines are available, and I have used it for the format of this book. Even *Mathematica* (version 2.2) notebooks, however, do not currently support symbols and greek letters in the input and output. For example, integration or summation is performed by typing in the built-in functions **Integrate** or **Sum** explicitly, and a wavefunction like $\psi(x)$ must be spelled out, for example **psi[x]**. Thus, to help in the translation between the discussion in the text and the *Mathematica* input and output, I have mostly employed *Mathematica* syntax throughout and minimized word processing of symbols and formulas.

Despite the (undoubtedly temporary) lack of greek letters, typing in words for mathematical operations is still generally more efficient than word processing conventional formulas. Although students don't seem to have difficulty with this, I often find that physicists used to paper and pencil tend to view the lack of conventional formulas as a limitation. Users should consider, however, that *Mathematica* can be fairly adaptable to individual needs if full advantage of its pattern-matching resources is taken. Thus, much of the *Mathematica* in this book involves pattern matching together with built-in functions like **Expand** and **Factor**, which do not have symbolic representations in conventional mathematics. Functions like **Integrate** and **Sum** which do are seldom needed. Moreover, *Mathematica* achieves much of its elegance when operations are nested, a procedure that would be awkward and at best cumbersome if conventional symbols like integrals and sums were used. Although algebra with pattern matching takes getting used to, the result can be compact and effective. You should be encouraged by the fact that the derivations in this book exhibit all of the steps since this is required by the computer. Thus, unlike conventional textbooks, you can break down a derivation step-by-step and examine each step separately, of course, only on the computer.

This book has benefited from thorough readings and criticisms of several colleagues well acquainted with the challenges of teaching physics with a computer. I am particularly indebted to David Cook, Paul Abbott, Alec Schramm, Eugen Merzbacher, and Markus van Almsick for their detailed comments and suggestions. I also appreciate the contributions and long-term support of Volker Engel, Ladislav Kocbach and Richard Stevens. Several students worked through early versions of the manuscript and suggested numerous improvements, and I would especially like to thank Karen Poelstra, Alex Tosheff and Richard Belansky. Alex Tosheff also helped to prepare figures. I happily acknowledge Joe Kaiping for translating *Mathematica* notebooks into LATEX and skillfully contributing to the book's format. I also would like to thank Paul Visinger for numerous suggestions regarding the book's organization and for creating the book's diskette.

I much appreciate the interest of Eugen Merzbacher, John Briggs, Willis Lamb, David Park, Wolfgang Christian, Patrick Tam, and Bill Thompson in this project. Finally, I have enjoyed the encouragement and greatly benefited from the support of my colleagues here at home, especially Heidi Fearn.

Jim Feagin
Fullerton, Fall 1993

Using This Book Interactively

The 3.5" diskette included with this book contains all of the text's *Mathematica* input and supporting packages for working through the book on the computer. Both *Mathematica* notebooks and ASCII text for each chapter are provided. Since the book was developed under *Mathematica* version 2.0, the diskette also provides some minor changes for running under version 2.2. Additional details can be found in the *readme.1st* file on the diskette.

The supporting packages (see also Appendix IV of the book) are located in the diskette's **Quantum** folder or directory. In order for the calls throughout the book to these packages (**Get** and **Needs** commands) to function properly, this folder must be placed in the *Mathematica* Packages folder or directory. Several of the underlying rule sets are developed in Appendix III of the book.

The diskette has been tested on Macintosh®, IBM®-compatible, and UNIX® computers, and we have found no discrepancies in the *Mathematica* output across any of these platforms. To give you an idea of how long steps in this book take to evaluate, time estimates are given in the text for tasks requiring roughly more than two minutes of run (CPU) time on a 16 MHz, 8 Mb (Macintosh 68030) computer. As indicated in the text, however, time estimates in Chapter 12 and in some of the figure captions refer to a 33 MHz, 16 Mb (NeXT 68040) computer.

Contents

PART I

Systems in One Dimension

Systems in One Dimension

In the first six chapters, we adopt two problems inherent to introductory quantum mechanics, a particle in a rectangular well and the harmonic oscillator, and emphasize translation onto the computer and into *Mathematica*. Although a significant portion of the physics is relegated to the problems and exercises, the discussion and references in the text are designed to guide you through the underlying physics. Chapters 7, 9, 11 and 12 introduce additional basic tools including approximation methods and the matrix, momentum and lattice representations. The remaining chapters in Part I provide applications and extensions of these tools.

If you're new to *Mathematica*, indeed to computers, you should work through Appendices I and II and the first nine exercises in Appendix III. Then as you work through the book, you can turn to the remaining exercises in Appendix III, when referenced in the text. Keep in mind that almost all of the calculations given in the text show all of the steps, since this is required for the computer. As long as you are working on a computer, which is what this book is intended for, you can dissect any expression and examine what *Mathematica* does with each command in turn from the full sequence of commands given in the text. Remember the maxim: *Divide and rule.*

For emphasis and reference, certain mathematical relations are highlighted and numbered such as the Schrödinger equation. These expressions, while generally not appropriate as *Mathematica* input, are intended to indicate *Mathematica* syntax and help bridge the gap between conventional mathematical notation and *Mathematica* input.

We shall find ourselves repeating certain operations and wanting procedures which help automate them. Several such functions are developed in Section 2 of Appendix III and are included in the packages in the **Quantum** directory. For example, we often require an extension of the built-in function **Conjugate** for symbolic complex arguments, which we define by entering

```
<<Quantum`QuickReIm`
```

This package also extends the built-in functions **Re** and **Im** to symbolic arguments.

Whenever possible, *Mathematica* commands and operators will be applied from the right-hand side of an expression using the postfix syntax, e.g. *expr* **//Expand** rather than **Expand** [*expr*] (see Exercise 1.8, Appendix III). This helps us to emphasize the physics and minimize the entanglement with *Mathematica*. If we need to include an option, we can still use the postfix form by introducing a *pure function*, as for example *expr* **//Expand [#, Trig -> True] &**, where **#** is a slot to place the expression properly in the operator's argument list and **&** signals the end of the operator. For compactness, the usual *Mathematica* **In** and **Out** labels have been removed from the text. Instead, input cells are shown in bold face and indented to distinguish them from ordinary text cells, while output cells are shown in plain face and are further indented relative to input cells. Also, most *Mathematica* warning messages have been removed.

Finally, we use **h** to represent Planck's constant divided by **2Pi**, i.e. $\hbar$, rather than a longer alternative like **hBar**. Although this might take some getting used to, a simple **h** is well justified by the compactness it affords in the *Mathematica* input and output.

Basic Wave Mechanics 1

We begin by entering the quantum mechanical hamiltonian describing a particle of mass **m** moving along the x axis in a one-dimensional *(1D)* potential. It is convenient to set this up for an arbitrary potential energy **V_** and to act on an arbitrary wavefunction **psi_** (see Exercise 1.4, Appendix III):

```
hamiltonian[V_] @ psi_ := -h^2/(2m) D[psi,{x,2}] + V psi
```

The *second-order* derivative **D** term on the right-hand side represents the kinetic energy of the particle. Since the hamiltonian is properly thought of as an operator which operates on the wavefunction, we use the prefix form **Q @ psi = Q[psi]** to apply the **hamiltonian** to **psi** (see Exercise 1.8, Appendix III).

Consider, for example, a planewave function **E^(I k x)** and a constant potential energy **Vo**. Operating with the hamiltonian and simplifying (see Exercise 2.1, Appendix III)

```
hamiltonian[Vo] @ (E^(I k x)) //Collect[#,E^(I k x)]&
```

$$E^{I k x} \left(\frac{h^2 k^2}{2 m} + Vo \right)$$

we see we get back a constant times the planewave. Now whenever a differential operator **Q** operating on a function **f** is equivalent to just a constant **q** times the function, **Q @ f = q f**, the function is said to be an *eigenfunction* and the constant its corresponding *eigenvalue*. Hence, the planewave is an eigenfunction of the hamiltonian describing a free particle (constant potential). If we identify **2Pi/k** as the particle's *deBroglie wavelength* and **k** as its *wavenumber*, so that **h k** is the particle's momentum and **(h k)^2/(2m)** its kinetic energy, we find the eigenvalue of the hamiltonian to be the particle's total energy. As we will see shortly, we've just found a solution of the time-independent *Schrödinger equation* describing a free particle.

We can easily verify that the hamiltonian is a *linear* operator, an important property fundamental to quantum mechanics. We need merely note, given any two functions **f[x]** and **g[x]** and two arbitrary complex numbers **a** and **b**, that

```
hamiltonian[V] @ (a f[x] + b g[x]) ==
    a hamiltonian[V] @ f[x] + b hamiltonian[V] @ g[x] //
        ExpandAll
```

> True

Here the symbol **==** is *logical* **Equal** and is used to check if the *patterns* on the left- and right-hand sides of the equation are identical, in which case it returns **True**. Otherwise, it returns the equation. (Refer to Exercise 1.7 in Appendix III.)

1.1   Equations of Motion

The *time-dependent Schrödinger equation* or *wave equation* takes the form

$$I\ h\ D[psi[x,t],t]\ ==\ hamiltonian[V]\ @\ psi[x,t] \tag{1.1-r1}$$

The wavefunction **psi[x,t]** is assumed to contain all information about the state of the system at time **t**, and the terms *wavefunction* and *state* are used interchangeably.

The explicit appearance of **I == Sqrt[-1]** means that solutions **psi[x, t]** of the wave equation are in general *complex* functions. A basic assumption of quantum mechanics is that the square of the wavefunction **Conjugate[psi]psi == Abs[psi]^2** represents the *probability density* of finding the particle at **x** when the function is normalized in the sense that

$$Integrate[Conjugate[psi]\ psi,\ \{x,-Infinity,Infinity\}]\ ==\ 1$$
$$\tag{1.1-r2}$$

The wavefunction **psi** is thus also referred to as a *probability amplitude*. Here **Conjugate** means *complex conjugate* and can be evaluated symbolically using the rule from the package **Quantum`QuickReIm`** developed in Exercise 2.4, Appendix III. We'll work with it shortly.

Relations such as (1.1-r1) and (1.1-r2) are for reference and for indicating *Mathematica* syntax and are not generally appropriate for *Mathematica* input. Their form is intended instead to aid the translation of conventional mathematical formulas into *Mathematica* expressions.

A probability density which is a function of **x** means for example that the average position of the particle is given by the *expectation value* of **x** defined by

> *xExp == Integrate[x Conjugate[psi] psi,{x,-Infinity,Infinity}]*
> (1.1-r3)

This quantity can be compared with the classical position of the particle. Note that the expectation value of a quantity **Q** is usually expressed as **<Q>**, a notation we will often use in the text but one which is awkward as *Mathematica* input. (Try it for yourself.)

In fact, it is a basic assumption of quantum mechanics that expectation values of physically observable quantities are to satisfy the laws of classical mechanics. At the very least, they are assumed to be real values, i.e. **<Q> == Conjugate[<Q>]**. In order for this to be generally the case, however, we must represent a physical quantity by an operator, which operates on the wavefunction, and compute the expectation value by integrating **Conjugate[psi]Q @ psi**. An important example is the momentum along the *x* axis that we enter as

> *p @ psi_ := -I h D[psi, x]*

We thus ensure that the expectation value of **p**

> *pExp ==*
> *Integrate[Conjugate[psi] p @ psi,{x,-Infinity,Infinity}]*
> (1.1-r4)

represents the classical momentum of the particle and is real. The appearance of **-I** in the definition of **p** is noteworthy.

An operator which satisfies **<Q> == Conjugate[<Q>]** is said to be *hermitian* (see also Problem 1.1-1) and we therefore require that physically observable quantities be represented in quantum mechanics by hermitian operators. It is also evident that hermitian operators have real eigenvalues. For the time being, we will not need to concern ourselves much more with quantum operators beyond computing their expectation values, and we shall postpone a detailed investigation until Part II.

The form of the momentum operator also justifies the form of the kinetic energy operator in **hamiltonian[V]**. Since the kinetic energy is classically **K = p^2/(2m)**, we express it quantum mechanically by

> *1/(2m) p @ p @ psi[x]*

```
          2
    -(h  psi''[x])
    ─────────────────
          2 m
```

which we can compare with **hamiltonian[V]** for **V = 0**:

```
hamiltonian[0] @ psi[x] == %
```

```
True
```

One easily verifies that both the kinetic energy and the hamiltonian (as long as the potential energy is real) are hermitian operators.

Exercise 1.1-1

Show that the *free-particle planewave functions* `E^(I k x)` are eigenfunctions of the momentum and the kinetic energy with eigenvalues `h k` and `(hk)^2/(2m)`, respectively. Hence, `E^(I k x)` represents a traveling wave along the positive x axis and `E^(-I k x)` represents one along the negative x axis. Show that `Sin[k x]` and `Cos[k x]` are also eigenfunctions of the kinetic energy but not of the momentum. (See also Exercise 18.1-1.)

Exercise 1.1-2

(a) Show that the momentum **p** is a linear operator, a property shared by most operators relevant to quantum mechanics. (An important example of a non-linear operator is `Conjugate`.)

(b) A fundamental quantity in quantum mechanics is the *commutator* of **x** and **p** usually denoted `[x,p]` and formally defined by `[x,p] = x p - p x`. Verify that `[x,p]` is equivalent to `I h`. Thus, enter for example

```
x p @ psi[x] - p @ (x psi[x]) == I h psi[x] //ExpandAll
```

Refer to Chapter 18 for more details.

Problem 1.1-1

Here are two problems too important to leave out, but perhaps better done with paper and pencil than on the computer.

(a) Verify that `<p> == Conjugate[<p>]` through integration by parts and assuming that `psi -> 0` as `x -> ±Infinity`. This condition on the wavefunction means physically that the probability of finding the particle at infinity vanishes. Verify that the kinetic energy and the hamiltonian are also hermitian.

(b) Show generally that

```
Integrate[Conjugate[f] Q @ g,{x,-Infinity,Infinity}] ==
    Conjugate[
        Integrate[Conjugate[g] Q @ f,{x,-Infinity,Infinity}]
    ]
```

$$(1.1\text{-r5})$$

for a Hermitian operator `Q` and any two functions `f[x]` and `g[x]` which vanish at infinity. Hint: Form the linear combination `f + c g` with `c` a complex constant.

Stationary States   1.2

When the potential energy is conservative, the **Energy** of the system, defined by the expectation value of **hamiltonian[V]**, is a constant independent of time as in classical mechanics. Since the hamiltonian is hermitian, the **Energy** is guaranteed to be real. One then easily separates variables **x** and **t** in the wave equation (1.1-r1) and derives (see Exercise 1.2-1) the time-independent *Schrödinger equation*

$$\text{hamiltonian[V] @ psi[x] == Energy psi[x]} \qquad \text{(1.2-r1)}$$

for **psi[x]** the *time-independent* wavefunction. At the same time, one finds that **psi[x,t] == E^(-I Energy t/h) psi[x]** is the corresponding solution of the wave equation. Hence, the probability density **Conjugate[psi[x,t]] psi[x,t] == Conjugate[psi[x]] psi[x]** is constant in time, and solutions of the Schrödinger equation are therefore also referred to as *stationary states.*

Comparing relation (1.2-r1) with our earlier example, we thus see that planewaves **E^(I k x)** are solutions of the free-particle (constant potential energy) Schrödinger equation with **Energy = (h k)^2/(2m)**, the particle's kinetic energy.

Exercise 1.2-1

Assume a wavefunction of the form **psi[x,t] == f[t] psi[x]** and perform a separation of variables on the wave equation. Show that **f[t] = E^(-I w t)** where **h w** is the separation constant. Try the built-in function **DSolve**. Equate **h w** to the **Energy** by evaluating the expectation value of **hamiltonian[V]** in the state **psi[x,t]**.

A Well-Posed Problem   1.3

For a given potential energy, we shall seek solutions of the wave equation subject to certain *boundary conditions* on the wavefunction. This is the essential mission of non-relativistic quantum mechanics. The resulting Schrödinger equation is an eigenvalue equation, and solutions which are consistent with the boundary conditions are called *energy eigenfunctions* and the corresponding eigenvalues the *eigenenergies*. In a conservative potential, this defines the *energy levels* or the *energy level spectrum* of the system.

As we shall see, when the system is bounded the energy spectrum is discrete, and when the system is unbounded the spectrum is continuous. The wavefunction of a bounded system vanishes outside the boundaries so that the probability of finding the particle there is zero. Eigenfunctions belonging to the discrete portion of the energy spectrum thus describe *bound states*, while those belonging to the continuous portion unbound or *continuum states*.

The wavefunction must be everywhere finite, since infinite probabilities are physically unreasonable. (Actually, we only require that physical observables be represented by hermitian operators, and therefore in some instances we might tolerate mild singularities as long as all expectation values remain defined. See discussions in Park [50], Chapter 6, and Exercise 20.2-1.) We shall refer to such solutions as *well-behaved*. Moreover, the wavefunction must be continuous and have a continuous first derivative in order to be a solution of the Schrödinger equation. This is because the kinetic energy involves a second derivative. Hence, if the value or the slope of the wavefunction were discontinuous at a point where the potential energy is finite, the wavefunction would not satisfy the Schrödinger equation at that point.

It will be convenient for what follows to define the differential operator

```
schroedingerD[V_] @ psi_ :=
    hamiltonian[V] @ psi - Energy psi
```

which when equated to zero gives the Schrödinger equation:

```
schroedingerD[V[x]] @ psi[x] == 0
```

$$-(\text{Energy psi}[x]) + \text{psi}[x] \ V[x] - \frac{h^2 \ \text{psi}''[x]}{2 \ m} == 0$$

Both the wave equation and the Schrödinger equation are linear differential equations, so that any linear combination of solutions of one of these equations is also a solution of that equation. We shall refer to this as the *principle of superposition*. Moreover, since the Schrödinger equation is a real differential equation (as long as the potential energy is real), a solution `psi[x]` can always be chosen to be real by a judicious choice of phase, although the corresponding solution of the wave equation `psi[x,t]` is in general complex.

1.4 Time-Development Operator

Given an arbitrary initial wavefunction `psi[x,0]`, one finds that if the hamiltonian is time-independent the wave equation has a simple formal solution given by

```
    psi[x,t] == E^(-I hamiltonian[V] t/h) @ psi[x,0]        (1.4-r1)
```

This is a useful result, readily proven by differentiating both sides with respect to `t`, that defines the *time-development operator* `E^(-I hamiltonian[V] t/h)`. Although the appearance of an operator in an exponential may at first seem somewhat mysterious, we can easily define it in terms of something we can in principle calculate, namely, an exponential power series of the form

```
E^Q == 1 + Q/(1!) + Q @ Q/(2!) + Q @ Q @ Q/(3!) + ...
```
$$(1.4\text{-}r2)$$

Then, for example, the time-development operator is clearly equivalent to the phase `E^(-I Energy t/h)` from Exercise 1.2-1 when `psi[x,0] = psi[x]` is a solution of the Schrödinger equation (since `(hamiltonian[V] @ psi[x])^n == (Energy @ psi[x])^n`). Moreover, the time-development operator is evidently a linear operator since the hamiltonian is itself.

In any case, the wavefunction is determined for all **t** by its value (for all **x**) at any given time, e.g. by `psi[x,0]` at **t = 0**, a consequence of the wave equation being a *first-order* differential equation in **t**.

Extra Dimensions 1.5

Although we will work primarily with one-dimensional models in Part I, one can easily generalize the hamiltonian and the equations of motion to higher dimensions by introducing the laplacian and rewriting the kinetic energy. For example, if we generalize to three dimensions, the kinetic energy becomes `-h^2/(2m)laplacian[psi]`. This is the form needed in order to define a hermitian operator in curvilinear coordinates, such as spherical polar coordinates. However, if the potential is spherically symmetric, the system is effectively one dimensional since the non-trivial part of the problem only involves the radial coordinate. We shall loosely refer to such systems as one-dimensional and investigate the problem more systematically in Part II (see also Sections 13.0 and 14.4).

Particle in a Box

2

As a simple example, consider a particle moving in a one-dimensional (*1D*) rectangular box of length **L**. For maximum simplicity, assume that the potential energy inside the box is a constant equal to zero and at the walls of the box is infinite, so that the box has infinitely high and hard walls. Then our boundary conditions are that the wavefunction vanishes at the walls, **psi[0] = psi[L] = 0**.

Analytical Eigenfunctions   2.1

We now show that the energy eigenfunctions which satisfy these boundary conditions are of the form

```
phi[n_,x_] := norm[n] Sin[n Pi x/L]
```

where **n** is a number to be determined and **norm[n]** is a normalization constant (see also Exercise 2.1-1 below). First of all, we see that these are solutions of the Schrödinger equation if

```
schroedingerD[0] @ phi[n,x]/phi[n,x] == 0 //ExpandAll

                 2  2    2
                h  n   Pi
   -Energy + ------------- == 0
                  2
               2 L  m
```

This expression relates the eigenenergies **Energy -> e[n]** to the number **n**. Thus,

```
e[n_] = Energy /. Solve[ %, Energy ][[1]]
```

$$\frac{h^2\ n^2\ Pi^2}{2\ L^2\ m}$$

Here the subscript `[[1]]` picks out the inner list from the list of lists returned by **Solve** so that **e[n]** is not a *{list}* (see Exercise 1.3, Appendix III). Clearly at the left-hand boundary, **phi[n,0] = 0**. At the right-hand boundary, we also have **phi[n,L] = 0** if **n** is an integer. However, **n = 0** gives the trivial solution **phi[0,x] = 0** for all **x**, and **n < 0** gives nothing extra since **phi[-n,x] == -phi[n,x]**. We conclude that **n** is a positive integer and that **n = 1** corresponds to the ground state and **n > 1** to the excited states.

We thus say that the system and its energies are *quantized* and speak of a *spectrum of discrete energy levels* for the infinite box. The values of **Energy** or **n** which solve the Schrödinger equation and the boundary conditions are called *eigenvalues* and the corresponding solutions *eigenfunctions*. The eigenvalue **n** is also called a *quantum number*. The energy levels are discrete because the system is bounded for all energies. If we let the length **L** of the box increase, the energy levels become more and more closely spaced or *quasi-continuous*. In the limit of an infinite box, the particle is completely free and the energy level spectrum is a *continuum*.

A notable feature of this idealized system common to all quantum mechanical systems is the *finite* minimum eigenenergy relative to the bottom of the potential. That is, the ground-state energy of any potential is always greater than the potential minimum, even if the system is held near absolute-zero temperature. This so-called *zero-point motion* is a consequence of the *Heisenberg uncertainty principle* (see Chapter 5 and refer to Merzbacher [45], Chapter 5).

Exercise 2.1-1

Starting with a linear superposition **A E^(I k x) + B E^(-I k x)** of independent plane waves, where **A** and **B** are constants, verify the box eigenfunctions and eigenenergies given above. Thus, show that this superposition is a solution of the Schrödinger equation and, by invoking the boundary conditions, that **k -> n Pi/L**. Use the built-in function **ComplexExpand** or the function **ComplexToTrig** from the package **Algebra`Trigonometry`** to transform **E^(±I k x)** to **Cos[k x] ± I Sin[k x]** (see Exercise 2.4, Appendix III, and Section 2.5).

It's interesting to note in passing that the *1D* box eigenfunctions are also classically the eigenfunctions of a taut string. However, whereas the quantum mechanical energies scale as **n^2**, the classical eigenfrequencies of the string's normal modes are linear in **n**. This is a consequence of the classical wave equation being *second order* in time in contrast to the quantum wave equation being *first order*.

We evaluate the normalization constant by requiring that the probability of finding the particle somewhere inside the box be one. Since these eigenfunctions are real, we have that

```
norm[n_] = norm[n] /.
   Solve[
      Integrate[ phi[n,x]^2, {x,0,L} ] == 1 /.
            Sin[m_Integer n Pi] -> 0,
      norm[n]
   ][[1]]
 Sqrt[2]
 ───────
 Sqrt[L]
```

where we have used the fact that **Sin[m n Pi]** vanishes for integer **m** and **n**.

The first three box eigenfunctions and eigenenergies are shown in Figure 2.1-1 using **h = m = L = 1** for convenience. The **n^2** scaling of the energies is evident. Also, one sees that **n-1** determines the number of nodes in the wavefunction inside the well starting with the ground state, *which has no node*. This is a general feature of Schrödinger wavefunctions, and can be used for example as a criterion for generating and ranking wavefunctions numerically.

Exercise 2.1-2

Reproduce Figure 2.1-1 but try adding a few more eigenfunctions. Note these plots are schematic in that the eigenfunctions have been scaled arbitrarily and shifted by their eigenenergy. Make similar plots of the probability distributions **phi[n,x]^2**.

Exercise 2.1-3

Calculate the expectation values **<x>** and **<p>** of the position and momentum of a particle in an eigenstate of the box. Interpret your results classically. (Refer to Section 2.5 below.)

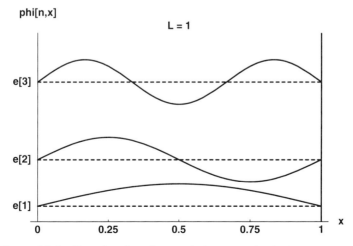

Figure 2.1-1. Box eigenfunctions and eigenenergies for **h = m = L = 1**.

2.2 Numerical Eigenfunctions

For a one-dimensional system, the most direct and in general the most efficient way of *computing* the wavefunction is to integrate the Schrödinger equation numerically. Let's examine how this works by generating a box eigenfunction using the built-in numerical differential equation solver **NDSolve**
(see Appendix II).

In order to do this, we need to scale the Schrödinger equation or otherwise
provide numerical values for all of the parameters. Let's simply continue to
take **L = h = m = 1**, so that

```
schroedingerD[0] @ phi[x] == 0 /.{h -> 1, m -> 1}
```

$$-(\text{Energy phi}[x]) - \frac{\text{phi}''[x]}{2} == 0$$

Since this is a *second-order* differential equation, **NDSolve** needs starting
values for **phi** and its slope **phi'** at some point inside the box. Therefore,
we will integrate out from the origin at **x = 0** to the right-hand wall of the
box at **x = L** and specify that **phi[0] == 0** and **phi'[0] ==** *constant*. Since
the wavefunction has to be normalized anyway, i.e. multiplied by a constant,
we can take *constant* to be unity.

Now all we need to do is to provide a value for **Energy**. Our choice,
however, will determine where the integration will end up when it reaches the
right-hand wall. If we guess the correct ground-state energy, the numerical
wavefunction will have no nodes and vanish when it hits the right-hand wall,
as required by the boundary condition **phi[L] == 0**. Of course, unless we're
lucky, our guess will not be the correct value, and the wavefunction will miss
the right-hand boundary condition, although in general the quality of our
guess will determine how close we come to hitting our mark.

Let's start with the value **Energy -> 4.0**, which is somewhat below the
ground-state energy **e[1]**, and ask **NDSolve** to integrate to just beyond the
wall of the box:

```
NDSolve[
    {schroedingerD[0] @ phi[x] == 0,
        phi[0] == 0, phi'[0] == 1} /.
            {h -> 1, m -> 1, Energy -> 4.0},
    phi, {x,0,1.1}
]
```

```
{{phi -> InterpolatingFunction[{0., 1.1}, <>]}}
```

The result is a replacement rule for evaluating **phi** on the interval **{x, 0,
1.1}**. Although the numerical integration generates a list of discrete data
points, **NDSolve** automatically interpolates the points for us (see Appendix

```
Plot[
    phi[x] /.%, {x,0,1.1},
    PlotRange -> {0,0.5},
    AxesLabel -> {" x","phi[e, x]"},
    PlotLabel -> "e = 4.0",
    Ticks ->
        {{{0,0},{.25,0.25},{.5,0.5},{.75,0.75},{1,1}},
         {{0,0},{.25,0.25},{.5,0.5}}}},
    Epilog -> {{Line[{{1,0},{1,2}}]}}
];
```

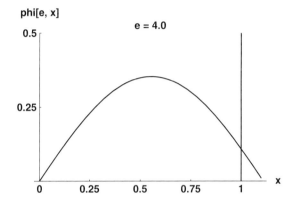

Figure 2.2-1.   Numerical estimate of the ground-state wavefunction of the box.

II and also Exercise 3.5, Appendix III). Hence, we can plot our numerical wavefunction as in Figure 2.2-1 just like any other continuous function of **x**.

Clearly, our guess for the eigenenergy was off the mark since the wave-function does not vanish at the right-hand wall. Thus we need to make a new guess and integrate again, and, unless we're satisfied with the result, repeat the process until we are. In this way, we converge on the eigenenergy to some desired number of significant digits, assuming of course that **NDSolve** is reliable. (In general, we may have to adjust the options **AccuracyGoal** and **MaxSteps** to obtain a desired precision. See **??NDSolve** and also Section 14.1.) This trial and error numerical technique for computing eigen-functions of the *1D* Schrödinger equation is commonly employed and is known as the *shooting method*. Although many refinements are possible to ensure accuracy and reliability, its direct application outlined here will serve us throughout most of this book. Readers seeking more generality should refer to Koonin [37], Chapter 3, and to the popular reference by Johnson [35].

Exercise 2.2-1

Use the *shooting method* to check the first five eigenenergies of the box to four significant figures. For convenience use **h = m = L = 1**.

2.3 Two Basic Properties

Returning to the analytic solutions, we see that the box eigenfunctions are also *orthogonal* in the sense that if **m** $\neq$ **n** then the *overlap integral*

```
Integrate[ phi[m,x] phi[n,x], {x,0,L} ]
```

$$\frac{2\ n\ Cos[n\ Pi]\ Sin[m\ Pi]}{(m^2 - n^2)\ Pi} + \frac{2\ m\ Cos[m\ Pi]\ Sin[n\ Pi]}{(-m^2 + n^2)\ Pi}$$

vanishes for integer **m** and **n** since **Sin[m Pi]** and **Sin[n Pi]** vanish. Orthogonality is actually a direct consequence of the hermiticity of the hamiltonian, and a set of eigenfunctions which are both normalized and orthogonal is said to be *orthonormal*.

Exercise 2.3-1

Show with paper and pencil that any two eigenfunctions **phi[m,x]** and **phi[n,x]** of a hermitian operator **Q** corresponding to distinct eigenvalues **q[m]** and **q[n]** on the interval **{x,a,b}** are *orthogonal* in the sense that

```
Integrate[ Conjugate[phi[m,x]] phi[n,x], {x,a,b} ] == 0
```

Note that in general **Conjugate** is required, although the box eigenfunctions are real.

Another important property of the eigenfunctions is that they form a *complete set*. This means that any well-behaved wavefunction can always be expanded to arbitrary accuracy in a series of eigenfunctions. In the present example, such an expansion is of course a *Fourier series*. Therefore we can use the *partial sum* (cf. Exercise 1.4, Appendix III)

```
psi[x_][nmax_] := Sum[ c[n] phi[n,x], {n,1,nmax} ]
```

to represent a wavefunction to an accuracy determined by the number of terms **nmax** in the sum. As we shall discuss later on, this *completeness* is also a general consequence of the Schrödinger equation, but in the present case follows from theorems on Fourier series. (Refer to the *Dirichlet conditions* in Arfken [4], Chapter 14, and in Boas [9], Chapter 7.)

The expansion coefficients **c[n]** are determined from the orthogonality of the eigenfunctions. Multiplying both sides of the partial sum by **phi[m,x]** and integrating (see Exercise 2.3-1), we find that

```
c[n_] := Integrate[ phi[n,x] psi[x], {x,0,L} ]
```                    (2.3-r1)

Rectangular Wave 2.4

Keeping our example simple, consider the expansion of a function `psi[x]` which is a *constant* on a short interval inside the box, say for `L/4` $\leq$ `x` $\leq$ `3L/4`, and vanishes otherwise. Normalization determines that the *constant* = `Sqrt[2/L]`. Such a spatially localized wave is referred to as a *wavepacket*, and this function is depicted in Figure 2.4-1.

The calculation of the Fourier expansion coefficients from relation (2.3-r1) is straightforward. We thus have

```
c[n_] = Integrate[Sqrt[2/L] phi[n,x],{x,L/4,3L/4}]
```

$$
\frac{2 \ \mathrm{Cos}[\dfrac{n \ Pi}{4}]}{n \ Pi} \ - \ \frac{2 \ \mathrm{Cos}[\dfrac{3 \ n \ Pi}{4}]}{n \ Pi}
$$

We can visualize these coefficients as a **BarChart** (from the package **Graphics`Graphics`**) as a function of **n**. Thus Figure 2.4-2 makes clear, for example, that the even-**n** coefficients vanish and that the series alternates in *sign* with every other pair of terms after **c[1]**.

We now easily calculate the first few terms in the Fourier series:

```
psi[x][8]
```

$$
\frac{4 \ \mathrm{Sin}[\dfrac{Pi \ x}{L}]}{Sqrt[L] \ Pi} \ - \ \frac{4 \ \mathrm{Sin}[\dfrac{3 \ Pi \ x}{L}]}{3 \ Sqrt[L] \ Pi} \ - \ \frac{4 \ \mathrm{Sin}[\dfrac{5 \ Pi \ x}{L}]}{5 \ Sqrt[L] \ Pi} \ + \ \frac{4 \ \mathrm{Sin}[\dfrac{7 \ Pi \ x}{L}]}{7 \ Sqrt[L] \ Pi}
$$

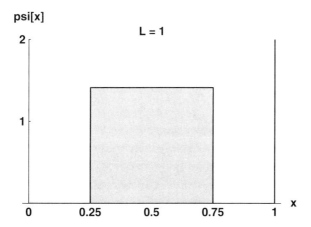

Figure 2.4-1. A normalized rectangular function in the box.

```
<<Graphics'Graphics'

BarChart[
    Table[c[n], {n,14}],
    PlotRange -> {-.5,1},
    AxesLabel -> {" n", "c[n]"},
    Ticks -> {Automatic,{{-1,-1},{-.5,-0.5},{.5,0.5},{1,1}}}
];
```

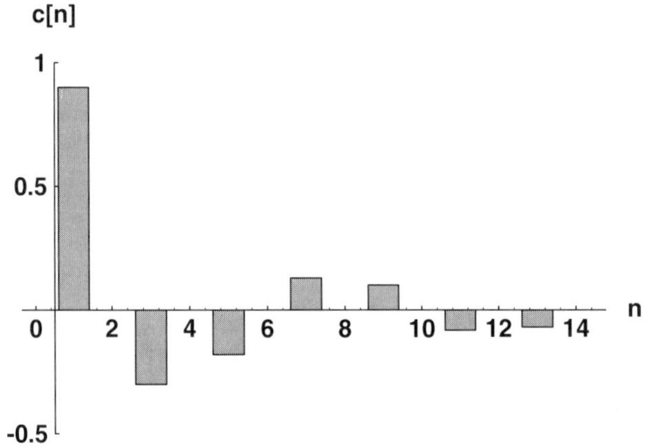

Figure 2.4-2.   Expansion coefficients of the rectangular wave in the box.

Exercise 2.4-1

Explain on the basis of symmetry of the eigenfunctions why the even terms are missing. Explain qualitatively why `c[1]` is by far the largest coefficient. (Refer to Figure 2.4-1 and the next Exercise 2.4-2.)

Although the rectangular wave is instructional from the point of view of Fourier series, we shouldn't consider it to be a representative state of the particle since it contains discontinuities. It would mean, for example, that the probability is discontinuous, which is unusual. The chance of finding a particle at some point inside a smooth box can generally be expected to be a smooth function of the particle's position inside the box. In any case, it is not a solution of the Schrödinger equation at the discontinuities (see Section 1.3). On the other hand, every partial sum in the Fourier series gives a wavefunction with a similar form which is everywhere continuous and a solution of the Schrödinger equation and the box boundary conditions.

We conveniently investigate convergence of the Fourier series by plotting partial sums for different values of **nmax** and comparing them with an exact plot of the rectangular wave. In order to compare with Figure 2.4-1, we again take the box length **L = 1**.

We can use **GraphicsArray** to display the various plots. We generate in this way a static representation of an animation, and many plots in this

book will follow this format. Thus, we define a function **psiSeriesPlot**
to plot and label **psi[x][nmax]** for a given **nmax**. Then we make a table
of **psiSeriesPlot** which we give first to **GraphicsArray** and then
pass the result on to **Show**. In order not to display each plot twice, we
suppress plotting in **psiSeriesPlot** with the option **DisplayFunction
-> Identity**, and then include the option **DisplayFunction -> $Dis-
playFunction** in **Show**.

The rectangular wave is drawn in the **Epilog** of **psiSeriesPlot** simply using **Line** to connect the corners of the rectangle. Plot labels with
the appropriate values of **nmax** are generated via the function **Text** in the
Epilog using **ToString[nmax]** and the operator **<>** (**StringJoin**).

The improved convergence of the partial sums to the square wave is evident in Figure 2.4-3 as **nmax** increases. The "wiggles" in these series approximations, especially around the corners or discontinuities of the square wave,
is a familiar feature of Fourier series. In fact, these overshoots of the discontinuities don't go away as **nmax** increases. They simply become narrower
and narrower with a height equal to about 9% of the jump. This artifact of
Fourier series is known as *Gibbs phenomenon,* and the overshoots are sometimes referred to as "Gibbs horns." The interested practitioner might note
that Gibbs horns can be dramatically reduced and the overall convergence
of Fourier series greatly improved with the introduction of so-called *Lanczos
sigma-factors.* See Problem 2.4-3 below.

Often the Fourier expansion coefficients cannot be conveniently calculated,
or calculated at all, analytically. In such cases, one might resort to numerical
integration, an approach we investigate in Problem 2.4-2 below. If we can
content ourselves with sampling the wavefunction a finite number of times on
a discrete grid or lattice of points, the coefficients can be efficiently estimated
using *fast Fourier transform (FFT)* routines, for example, using the built-in
function **Fourier**. We shall examine this technique later on in Section 12.3
and in Problems 12.3-1 and 12.3-2.

Exercise 2.4-2

Using the box eigenfunctions, generate a series for another constant but off-center
wavefunction which is nonvanishing on **L/4** $\leq$ **x** $\leq$ **L/2**. Note the appearance of
both even and odd terms in the series. Make a plot similar to Figure 2.4-3. This
wavepacket is more peaked than the one used in Figure 2.4-1. How does this affect
the convergence of the series?

Problem 2.4-1

With box eigenfunctions, expand a triangular wavefunction with a peak at the center of
the box. Determine the peak height by requiring that the wavefunction be normalized.
Show that the expansion coefficients **c[n]** are given by

```
psiSeriesPlot[nmax_] :=
    Plot[
        Evaluate[ psi[x][nmax] /. L -> 1 //N ],
        {x,0,1},
        PlotRange -> {-.2,2},
        AxesLabel -> {" x",""}, PlotPoints -> 50,
        Ticks -> {{{.25,0.25},{.5,0.5},{.75,0.75}},
                    {{0,0},{1,1},{2,2}}},
        Epilog ->
            {Text[
                "nmax = " <> ToString[nmax],
                {0.35,1.9},{-1,0}
             ],
             {Dashing[{.01,.01}],
             Line[{{0.25,0},{0.25,1.414},
                    {0.75,1.414},{0.75,0}}]},
             {Line[{{1,0},{1,2}}]}}},
        DisplayFunction -> Identity
    ];

Show[
    GraphicsArray[
        Table[
            {psiSeriesPlot[n], psiSeriesPlot[n+10]},
            {n,10,30,20}
        ]
    ],
    GraphicsSpacing -> {0.1,0.3},
    DisplayFunction -> $DisplayFunction
];
```

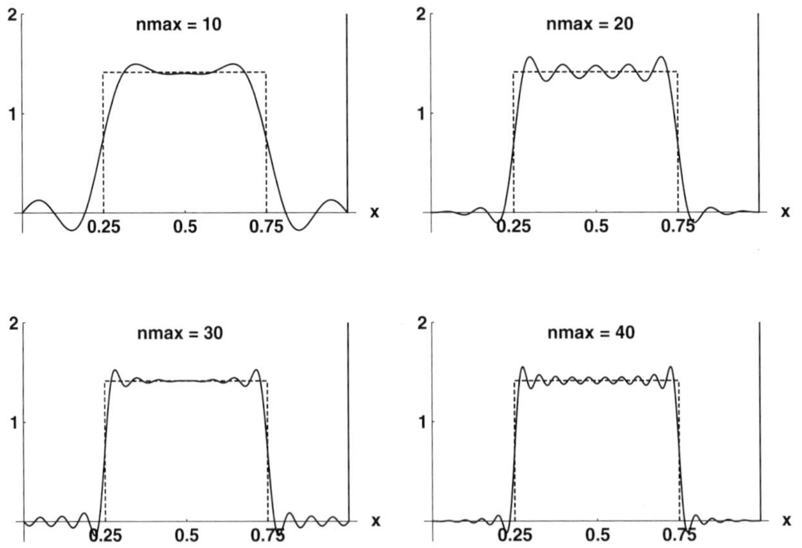

Figure 2.4-3. Fourier partial sums `psi[x][nmax]` converging to a rectangular wave.

$$\frac{\mathtt{Sqrt[96]\ Sin[\dfrac{n\ Pi}{2}]}}{\mathtt{n^2\ Pi^2}}$$

independent of **L**. Make a plot similar to Figure 2.4-3. This is the quantum analog of a classical "plucked string."

Problem 2.4-2

Expand the gaussian wavefunction

```
psi[x_] = (w/Pi)^(1/4) E^(-w (x-L/2)^2/2)
```

in box eigenfunctions for a box of length **L = 1**. Take **w = 40** and show using the built-in function **NIntegrate** that **psi[x]** is normalized to five significant figures within the confines of the box. (**psi[x]** only approximately vanishes at the walls of the box.) Use **NIntegrate** to compute the expansion coefficients.

Problem 2.4-3

Lanczos [40], Chapter 4, has shown that the oscillations of Gibbs phenomenon can be almost eliminated with the introduction of certain "focusing" factors which can be derived by a local averaging over the oscillations. It turns out, however, that even when there are no discontinuities involved the method will increase not only the slow convergence of a Fourier series but also change a divergent Fourier series into a convergent one. See also Hamming [31].

The procedure is quite simple. Given a Fourier partial sum with **nmax** terms, say for example

```
a[0]/2 + Sum[a[n] Cos[n z] + b[n] Sin[n z],{n,1,nmax}]
```

where **a[n]** and **b[n]** are the expansion coefficients, we are instructed simply to multiply each coefficient in the sum by the *Lanczos sigma factor* defined by

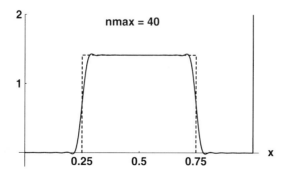

Figure 2.4-4.   A Fourier partial sum including Lanczos sigma factors converging to a rectangular wave.

```
lanczos[0] := 1
lanczos[n_] := Sin[n Pi/nmax]/(n Pi/nmax)
```

and therefore replace the sum by

```
a[0]/2 +
    Sum[lanczos[n] (a[n] Cos[n z] + b[n] Sin[n z]),{n,1,nmax}]
```

Introduce the sigma factors into our expansion of the rectangular wave in box eigenfunctions and plot your results as in Figure 2.4-3. For **nmax = 40**, compare your plot with Figure 2.4-4.

2.5 Quantum Rattle

Let us investigate the time development of a simple wavepacket. Consider at **t = 0** that the particle in the box is in a state described by a linear superposition of just the ground state and the first excited state with equal weights. We obtain the wavefunction at a later time by applying the time development operator (1.4-r1). Being a linear operator, it acts on each component eigenfunction **phi[n,x]** and generates for each component a time-dependent phase factor determined by the corresponding eigenenergy **e[n]** (see Exercise 2.5-1 below):

```
a[n_] := 1/Sqrt[2]
a[n_,t_] := a[n] E^(-I e[n] t/h)

psi[x_,t_] = Sum[a[n,t] phi[n,x],{n,1,2}]
```

$$
\frac{E^{(-I/2\ h\ Pi^2\ t)/(L^2\ m)}\ Sin[\frac{Pi\ x}{L}]}{Sqrt[L]}\ +
$$

$$
\frac{E^{(-2\ I\ h\ Pi^2\ t)/(L^2\ m)}\ Sin[\frac{2\ Pi\ x}{L}]}{Sqrt[L]}
$$

The particular choice of constant **1/Sqrt[2]** in defining **a[n]** ensures that **psi[x,t]** is normalized when the **phi[n,x]** are, as we will show in a moment. We easily verify that this wavefunction is a solution of the wave equation (1.1-r1)

```
I h D[psi[x,t], t] == hamiltonian[0] @ psi[x,t] //
    ExpandAll

    True
```

and satisfies the box boundary conditions at all times

```
psi[0,t] == psi[L,t] == 0
```

```
    True
```

Since this wavefunction is not an eigenstate of the box, the probability density is *time dependent*:

```
Needs["Quantum`QuickReIm`"]
```

```
psisq[x_,t_] = Conjugate[psi[x,t]] psi[x,t] //
               Expand //ComplexExpand
```

$$
\frac{\sin\left[\dfrac{Pi\ x\ 2}{L}\right]}{L} + \frac{2\,\cos\left[\dfrac{3\ h\ Pi^2\ t}{2\ L^2\ m}\right]\,\sin\left[\dfrac{Pi\ x}{L}\right]\,\sin\left[\dfrac{2\ Pi\ x}{L}\right]}{L} + \frac{\sin\left[\dfrac{2\ Pi\ x\ 2}{L}\right]}{L}
$$

Here we have computed the **Conjugate** *symbolically* using the rule from the package **Quantum`QuickReIm`** and we have used the built-in function **ComplexExpand** to transform the complex exponentials from the time development to **Sin** and **Cos** (see Exercise 2.4, Appendix III).

Exercise 2.5-1

The time-development phase factors can also be obtained by a separation of variables of the wave equation, similar to what we did in Exercise 1.2-1. Thus, substitute a general linear combination of the form

```
psi == Sum[a[n,t] phi[n,x],{n,nmin,nmax}]
```

into the wave equation and solve for the time-dependent coefficients `a[n,t]`. Refer to Exercise 1.2-1, but this time use the orthogonality of the eigenfunctions `phi[n,x]` and the fact that they are solutions of the Schrödinger equation to introduce the eigenenergies `e[n]`.

The wavefunction is normalized for all time. Hence, the probability of finding the particle somewhere inside the box is unity:

```
Integrate[ psisq[x,t], {x,0,L} ]
```

```
    1
```

Owing to the orthogonality of the eigenfunctions **phi[n,x]**, this imposes the restriction

```
Sum[Conjugate[a[n,t]] a[n,t],{n,1,2}]
```

```
    1
```

which expresses overall conservation of probability. The probability density oscillates, nevertheless, between the ground state and the first-excited state.

(Keep in mind that `psi[x,t]` is a superposition of just the ground and first-excited states.) Thus, for example, the probability of finding the particle inside just one half of the box

```
Integrate[ psisq[x,t], {x,0,L/2} ]
```

$$\frac{1}{2} + \frac{4 \; Cos[\frac{3 \; h \; Pi^2 \; t}{2 \; L^2 \; m}]}{3 \; Pi}$$

equals $^1/_2$ when *averaged over time*. We also have that the expectation value of the hamiltonian in the state `psi[x,t]` is a weighted sum over the eigenenergies:

```
Integrate[
    Conjugate[psi[x,t]] hamiltonian[0] @ psi[x,t],
    {x,0,L}
] ==
    Sum[Conjugate[a[n,t]] a[n,t] e[n],{n,1,2}] ==
        (e[1] + e[2])/2

    True
```

This follows from the orthogonality of the eigenfunctions `phi[n,x]`. And because the time dependence enters in as phase factors, the weights are time independent and therefore the sum is as well:

```
Sum[Conjugate[a[n,t]] a[n,t] e[n],{n,1,2}] ==
    Sum[a[n]^2 e[n],{n,1,2}]

    True
```

Hence, generally, the average total energy is also time independent and conserved. Of course, the energy expectation value is real, as are all expectation values of physical quantities. Note that in general, however, the expansion coefficients will be complex and we have to replace `a[n]^2` by `Conjugate[a[n]]a[n] == Abs[a[n]]^2`.

Exercise 2.5-2

Show that the frequency of oscillation is given by `(e[2] - e[1])/h`. Hence, show that if `L = h = m = 1` the *period* is close to 0.42. See Figure 2.5-1 below.

We can examine the motion of the wavepacket by generating a series of plots of the probability density as a function of time. Working on the computer we might make an animation, but we can also take the approach we took in generating Figure 2.4-3 above. Thus, we use `L = h = m = 1` and plot in Figure 2.5-1 for about one half period (see the previous Exercise 2.5-2).

```
psisqPlot[t_] :=
    Plot[
        Evaluate[psisq[x,t] /.{h -> 1, L -> 1, m -> 1} //N],
        {x,0,1},
        PlotRange -> {0,4},
        AxesLabel -> {" x",""},
        Ticks -> {{{0,0},{.5,0.5},{1,1}}, {{0,0},{2,2},{4,4}}},
        Epilog -> {Text["t = " <> ToString[t],{.5,3.7},{-1,0}],
                    {Line[{{1,0},{1,4}}]}},
        DisplayFunction -> Identity
    ];

Show[
    GraphicsArray[
        Table[{psisqPlot[t],psisqPlot[t+0.04]},{t,0,0.2,0.08}]
    ],
    GraphicsSpacing -> {0., 0.2},
    DisplayFunction -> $DisplayFunction
];
```

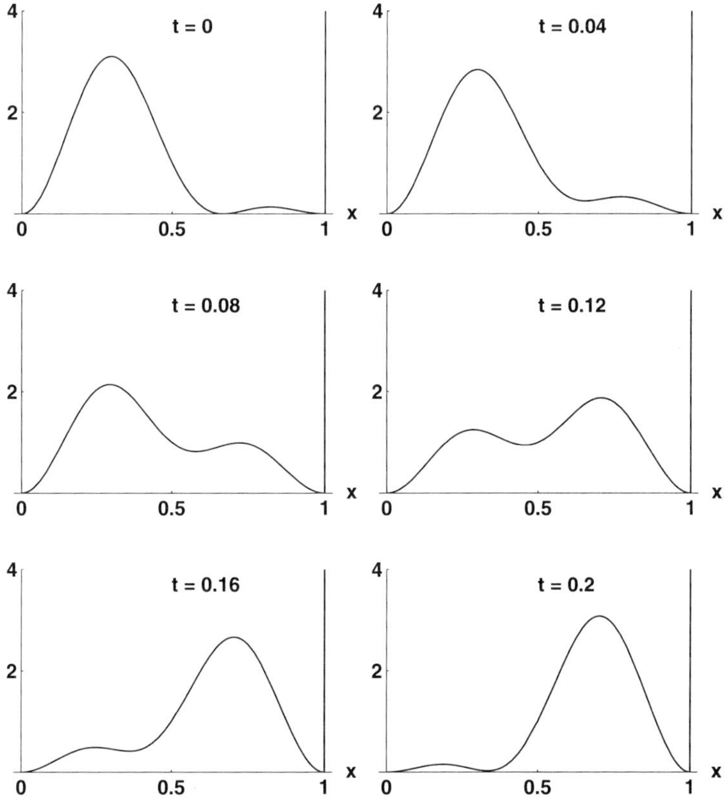

Figure 2.5-1.   Probability density `psisq[x,t]` of the two-state wavepacket as a function of time.

The position of the centroid (approximately the peak) of the wavepacket is given by the expectation value of **x**:

```
xExp[t_] = Integrate[ x psisq[x,t], {x,0,L} ]
```

$$\frac{L}{2} - \frac{16\ L\ \text{Cos}\left[\dfrac{3\ h\ \text{Pi}^2\ t}{2\ L^2\ m}\right]}{9\ \text{Pi}^2}$$

The width of the wavepacket is roughly determined by the expectation value of **x^2**:

```
xsqExp[t_] = Integrate[ x^2 psisq[x,t], {x,0,L} ]
```

$$\frac{L^2}{3} - \frac{5\ L^2}{16\ \text{Pi}^2} - \frac{16\ L^2\ \text{Cos}^2\left[\dfrac{3\ h\ \text{Pi}^2\ t}{2\ L^2\ m}\right]}{9\ \text{Pi}^2}$$

We will use it in the next chapter to define the uncertainty in the position. The momentum of the centroid of the wavepacket is given by the expectation value of the momentum operator **p**:

```
p @ psi_ := -I h D[psi, x]
pExp[t_] =
    Integrate[ Conjugate[psi[x,t]] p @ psi[x,t], {x,0,L}] //
       Expand //ComplexExpand
```

$$\frac{8\ h\ \text{Sin}\left[\dfrac{3\ h\ \text{Pi}^2\ t}{2\ L^2\ m}\right]}{3\ L}$$

Classically, the time derivative of the coordinate **x** gives the velocity of the particle or the momentum divided by the mass **p/m**. Quantum mechanically, we expect an analogous relation to hold for the expectation values. Thus, we find that

```
D[ xExp[t], t ] == pExp[t]/m
```

```
True
```

which is an example of *Ehrenfest's theorem* (see Problem 18.2-2). A similar result normally also ensures that Newton's second law be satisfied in the sense that **D[pExp,t] == <-D[V,x]>**. The box with infinite walls, however, is too pathological to satisfy the theorem's conditions. The free-particle wavepacket, which we will construct in Chapter 4, will prove to be a better example in this regard. See also Sections 8.3 and 11.11.

The expectation value of `p^2` is related to the kinetic energy

```
psqExp[t_] =
    Integrate[Conjugate[psi[x,t]] p @ p @ psi[x,t],{x,0,L}] //
        Expand //ComplexExpand
```

$$\frac{5\ h^2\ Pi^2}{2\ L^2}$$

and therefore the total energy of the wavepacket, since the potential energy is zero inside the box. That is,

```
psqExp[t]/(2m) == (e[1] + e[2])/2
```

```
    True
```

Measurements 2.6

Evidently, we can calculate the expectation value of any hermitian operator `Q` in an arbitrary state `psi[x]` simply by expanding `psi[x]` in the eigenfunctions `phi[n,x]` of `Q` and using the resulting expansion coefficients squared `Abs[a[n]]^2` in a weighted sum over all the eigenvalues `q[n]` of `Q`:

$$QExp\ ==\ Sum[\ Abs[a[n]]^2\ q[n],\ \{n,nmin,nmax\}\ ] \qquad (2.6\text{-}r1)$$

The wavepacket energy we calculated in the previous section is an example.

This simple rule also brings us to a fundamental interpretation of quantum mechanics. Namely, *the eigenvalues* `q[n]` *are the only values which can be obtained in a measurement of the physical quantity represented by* `Q`. (A measurement, for example, could be the determination of the energy of a particle by deflection from an adjustable potential barrier.) It follows that the expansion coefficient squared `Abs[a[n]]^2`, determined by the state `psi[x,t]`, gives the probability of obtaining the value `q[n]` when `Q` is measured. (Refer to Park [50], Chapter 5, and to Merzbacher [45], Chapter 8.) If the state `psi[x,t]` is already one of the eigenstates `phi[n,x]` of `Q`, then `Abs[a[n]]^2 == 1` and `Abs[a[m≠n]]^2 == 0`. In any case, immediately after a measurement the state of the system is the eigenstate `phi[n,x]` of `Q` corresponding to the eigenvalue `q[n]` obtained. This change of the wavefunction from `psi[x,t]` to an eigenstate `phi[n,x]` by the measurement is referred to as the *collapse of the wavefunction*.

We might point out in passing that many conscientious objectors who closely monitor the foundations of quantum mechanics find the interpretation in the previous paragraph too idealistic. The root of the difficulty is the concept of a quantum measurement. Readers of this book will find the paper by Fearn and Lamb [21] an interesting introduction to the vast literature on this open subject. They have investigated a model problem and the so-called *quantum Zeno effect* with the wavepacket propagation scheme we shall study in Chapter 12. See Problem 12.6-10 and also Home and Whitaker [33] and Fearn and Lamb [22].

Exercise 2.6-1

(a) Prove relation (2.6-r1) with paper and pencil.

(b) Show that the two-state expectation values **xExp[t]** and **pExp[t]** for the box can also be evaluated as sums over **n** and **np** of integrals of the form

```
Integrate[phi[n,x]] x   phi[np,x],{x,0,L}]
Integrate[phi[n,x]] p @ phi[np,x],{x,0,L}]
```

called *matrix elements* of **x** and **p**, respectively (see Chapter 9).

3
Uncertainty
Principle

A particle's wavefunction has an inherent width which naturally introduces an *uncertainty* in not only the position of the particle but in general any observable quantity Q, as for example the momentum. It is appropriate to define the *uncertainty* ΔQ by the root-mean-square deviation from the mean, where the mean is the expectation value `<Q>`. Specifically, ΔQ = `Sqrt[<(Q - <Q>)^2>] = Sqrt[<Q^2> - <Q>^2]`. Uncertainties have a special role in quantum mechanics because of the *Heisenberg uncertainty relations*, which require for example that the product of the position and the momentum uncertainties be greater than a minimum value determined by Planck's constant. In symbols, $\Delta x \, \Delta p \geq h/2$.

Physically, this means of course that **x** and **p** cannot be simultaneously measured with certainty, in sharp contrast with the classical description. See also Section 15.1 and refer to Park [50], Sec. 3.5, and Merzbacher [45], Sec. 8.6.

In the next exercise, we evaluate the position and momentum uncertainties, Δx = `delx[t]` and Δp = `delp[t]`, respectively, for the time-dependent wavepacket `psi[x,t]` we set up in Section 2.5. We find that the uncertainty product `delx[t]delp[t]` ==

```
                                           2
                                 3 h Pi  t 2
                          256 Cos[────────]
                                      2
       1       5                   2 L  m
 h Sqrt[(── - ───── - ───────────────────)
        12       2                   4
             16 Pi               81 Pi

                              2
                    3 h Pi  t 2
            64 Sin[────────]
                       2
      2            2 L  m
   5 Pi
  (───── - ───────────────)]
     2              9
```

In Figure 3.0-1, we display this quantity as a function of time for the case **L = h = m = 1** using **FilledPlot** (from the package **Graphics `FilledPlot`**) to highlight the area between the uncertainty product and the minimum uncertainty **h/2 -> 1/2**.

```
<<Graphics`FilledPlot`

FilledPlot[
    Evaluate[
        {0.5, delx[t] delp[t]} /.
        {h->1, L->1, m->1} //N}
    ],
    {t,0,2}, PlotRange -> {0,1}, PlotPoints -> 100,
    AxesLabel -> {" t","delx[t] delp[t]"}
];
```

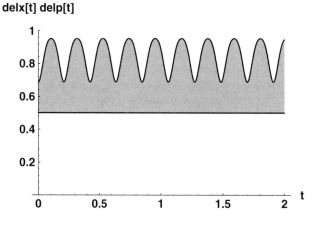

Figure 3.0-1. Uncertainty product $\Delta x\,\Delta p =$ **delx[t]delp[t]** as a function of time for the two-state wavepacket **psi[x,t]**.

Exercise 3.0-1

Evaluate the root-mean-square deviations from the mean Δ**x** = **delx[t]** and Δ**p** = **delp[t]** of the position and of the momentum of the particle, respectively, for the two-state wavepacket **psi[x,t]**. Derive the uncertainty product given in the text and displayed in Figure 3.0-1. Redo the figure for some other box lengths **L**.

We can understand how the uncertainty relation arises by a simple example. You probably know that the superposition of two planewaves **E^(I k x)** with wavenumbers **kmin** and **kmax** which are only slightly different gives rise to *beats*. The beats are characterized by minima in the superposition with a regular separation Δ**x** given by **2Pi/**Δ**k** with Δ**k** = **kmax** - **kmin**. That is, Δ**x** Δ**k** = **2Pi** or, since the momentum of each wave is **p** = **h k**, Δ**x** Δ**p** = **2Pi h > h/2**. (We will show shortly that a *gaussian* wavepacket gives rise to the minimum uncertainty product allowed, i.e. Δ**x** Δ**p** = **h/2**.)

If we now include a third planewave with a wavenumber between **kmin** and **kmax**, something remarkable happens. The beats develop peaks separated by pronounced minima but with the widths Δ**x** of the peaks still given approximately by **2Pi/**Δ**k** with Δ**k** = **kmax** - **kmin**. If we continue adding in more and more waves all between **kmin** and **kmax**, the peaks remain roughly the same but move further and further apart until in the limit of a very large number of waves we have a single-peak wavepacket of width Δ**x** at the origin. This comes about because everywhere but near the origin the phases of the component waves are almost uniformly distributed between **0** and **2Pi** and the waves destructively interfere. You might recall that a similar effect is observed in multislit interference of light as the number of slits is increased.

Let's illustrate this discussion by plotting superpositions of **n** planewaves with equal weights and with wavenumbers equally spaced in say the range **2.4** $\leq$ **k** $\leq$ **2.6**. Thus, consider

```
psiPW[n_] := Sum[E^(I k x),{k,2.4,2.6,(2.6-2.4)/(n-1)}]/n
```

which for **n** = 2 and **3** becomes

```
{psiPW[2],psiPW[3]}
```

$$\left\{ \frac{E^{2.4\,I\,x} + E^{2.6\,I\,x}}{2}, \frac{E^{2.4\,I\,x} + E^{2.5\,I\,x} + E^{2.6\,I\,x}}{3} \right\}$$

The factor **1/n** normalizes the maxima to unity, which is convenient for plotting. Figure 3.0-2 shows **Abs[psiPW[n]]^2** as a function of **x** for increasing **n**. The beats for **n** = 2 and the tendency to form a single-peak wavepacket as **n** increases are evident. Here, Δ**k** = **0.2** so that Δ**x** $\approx$ **30**.

The set of weights or amplitudes assigned to the planewave components as a function of **k** is known as the *spectral distribution* of the wavepacket, and it is clearly related to the wavepacket's *momentum distribution* since **p** = **h k**.

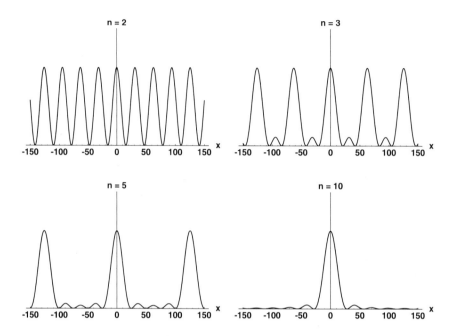

Figure 3.0-2. Superpositions of **n** planewaves **E^(I k x)** with equal weights **1/n** and with equally spaced wavenumbers in the range **2.4** ≤ **k** ≤ **2.6**. Here **Abs[psiPW[n]]^2** is plotted.

In the present example, we have taken the discrete and flat distribution **1/n** for simplicity. We examine a continuous *gaussian* distribution in the next chapter, and you can experiment with a discrete gaussian distribution in Problem 4.2-3 below.

Exercise 3.0-2

Plot as in Figure 3.0-2 the real and imaginary parts of **psiPW[n]**.

4
Free-Particle Wavepacket

Having explored the superposition of a few planewaves, we will now generalize to a superposition of planewaves $E^{(I k x)}$ over a continuous range of wavenumbers **k**. As a convenient but instructive example, consider a *gaussian* spectral distribution defined by

```
phi[k_] = (Pi w)^(-1/4) E^(-k^2/(2w))
```

$$\frac{1}{E^{k^2/(2w)} \ (Pi \ w)^{1/4}}$$

for all **{k,-Infinity,Infinity}**. This function is normalized in the sense that

```
<<Quantum`integGauss`
```

```
integGauss[phi[k]^2,{k,-Infinity,Infinity}]
```

```
    1
```

Here the function **integGauss** introduces a short list of our own rules specialized for performing integrals involving gaussians and powers. Its usage is summarized by

```
?integGauss
```

```
    integGauss[integrand,{x,-Infinity,Infinity}] integrates linear
       combinations of patterns of the form x^n E^(-a x^2 + b x + c)
       for Re[a] > 0 and integer n >= 0. WARNING: The requirements
       Re[a] > 0 and integer n >= 0 must be enforced by the user.
```

The rules are developed in Exercise 2.7, Appendix III.

4.1 Stationary Wavepacket

We therefore set up and evaluate a linear superposition of planewaves according to

```
Needs["Quantum`PowerTools`"]

psi[x_] = 1/Sqrt[2Pi] *
        integGauss[phi[k] E^(I k x),{k,-Infinity,Infinity}] //
            PowerExpand //PowerContract
```

$$\frac{\left(\dfrac{w}{Pi}\right)^{1/4}}{E^{(w\ x^2)/2}}$$

In order to obtain a somewhat simpler expression, we have introduced the built-in function **PowerExpand** and the function **PowerContract** from the package **Quantum`PowerTools`** as an inverse to **PowerExpand**. (**PowerContract** is developed in Exercise 2.8, Appendix III.) The factor **1/Sqrt[2Pi]** ensures that **psi[x]** is normalized in the usual sense that

```
integGauss[psi[x]^2,{x,-Infinity,Infinity}] //PowerExpand
```

```
1
```

We thus see that a continuous spectral distribution of finite width in **k** leads to a single-peak wavepacket of finite width in **x**. We will calculate these widths in a moment. Since each planewave component is a free particle solution of the Schrödinger equation, the superposition **psi[x]** represents a free particle. The fact that **psi[x]** is also a gaussian is a property somewhat peculiar to gaussian functions.

You might recognize **psi[x]** as the Fourier transform of **phi[k]**. In fact, these functions are *Fourier transform pairs*, a connection we will establish in Chapter 11 in order to set up the *momentum representation*. Mathematically, the superposition is just another example of expanding a function in a complete set of eigenfunctions, in this case free-particle planewaves. Because **psi[x]** extends over all space, the necessary superposition is an integral over all values of the continuous wavenumber **k**, a Fourier integral. If **psi[x]** is restricted to a finite region, such as by the walls of a box, the wavenumber **k** takes on discrete values and the integral is replaced by a sum, a Fourier series.

The fact that **psi[x]** is normalized when **phi[k]** is also leads us to the assumption that **Conjugate[phi[k]] phi[k]** is the probability density of finding the particle with wavenumber **k**, just as **Conjugate[psi[x]] psi[x]** is the probability density of finding the particle at position **x**. Therefore since **p = h k** is the particle's momentum, **Conjugate[phi[p/h]]** `phi[p/h]/h` is the probability density of finding the particle with momentum **p**. (The overall factor `1/h` ensures normalization. See Exercise 4.1-1.) It follows that we can calculate expectation values of powers of **k** or **p** as integrals over `phi[k]^2` or `phi[p/h]^2/h`, since in the present example we have chosen `phi[k]` to be real. The consummate connection of `phi[k]` with a probability amplitude, however, is that we can also calculate expectation values of powers of **p** in the usual way, namely, by applying the differential operator `p @ psi = -I h D[psi,x]` repeatedly to the wavefunction `psi[x]`, multiplying by `Conjugate[psi[x]]`, and integrating over all **x**. This property (as well as the fact that the superposition `psi[x]` is normalized simultaneously with `phi[k]`) follows from the connection of `psi[x]` and `phi[k]` as Fourier transform pairs, as we will see later on (Section 11.2).

Exercise 4.1-1

Show that the *momentum-space* amplitude `phi[p/h]/Sqrt[h]` is normalized.

We thus see that the expectation value of **k** vanishes since `phi[k]` is an even function of **k**, just as the expectation value of **x** vanishes (see the discussion at the end of the next chapter on parity). Hence, `psi[x]` represents a free-particle at rest at the origin. The uncertainties in **x** and **k** are therefore given by

```
{ delx = Sqrt[integGauss[x^2 psi[x]^2,{x,-Infinity,Infinity}]] //
            PowerExpand //PowerContract,
  delk = Sqrt[integGauss[k^2 phi[k]^2,{k,-Infinity,Infinity}]] }
```

$$\{ \frac{1}{\text{Sqrt}[2]\ \text{Sqrt}[w]}, \frac{\text{Sqrt}[w]}{\text{Sqrt}[2]} \}$$

respectively. Thus, the position-momentum uncertainty product is found to be

```
delx (h delk)
```

$$\frac{h}{2}$$

the minimum allowed by the Heisenberg uncertainty relation, a result which can be shown to be unique to gaussian wavepackets (see e.g. Merzbacher [45], Section 8.6).

Exercise 4.1-2

Check the expectation values of **p** and `p^2` by applying the differential operator `p @ psi = -I h D[psi,x]` to the coordinate wavefunction `psi[x]`.

4.2 Moving Wavepacket

Finally, let us examine the time development of **psi[x]**. First, note that we can shift the peak of the wavepacket from the origin to **x = xP** by phase shifting each planewave component:

```
integGauss[
    phi[k] E^(I k (x-xP))/Sqrt[2Pi],{k,-Infinity,Infinity}
] //Simplify //PowerExpand //PowerContract
```

$$\frac{\left(\dfrac{w}{Pi}\right)^{1/4}}{E^{(w\ (-x\ +\ xP)^2)/2}}$$

which is of course **psi[x-xP]**. Likewise, we can shift the momentum of the wavepacket from **k = 0** to **k = kP** by shifting the peak of **phi[k]**. Let's call this the *initial* wavepacket **psi[x,0]**:

```
psi[x_,0] = 1/Sqrt[2Pi] *
    integGauss[
        phi[k-kP] E^(I k x),{k,-Infinity,Infinity}
    ] /. E^m_ :> E^Expand[m] //
        PowerExpand //PowerContract
```

$$E^{I\ kP\ x\ -\ (w\ x^2)/2}\ \left(\frac{w}{Pi}\right)^{1/4}$$

which we recognize as **E^(I kP x) psi[x]**. We show in Exercise 4.2-1 that it describes the same wavepacket **psi[x]** located at the origin, in particular a minimum uncertainty wavepacket, but one now moving to the right with momentum **h kP**. The planewave factor **E^(I kP x)** is thus referred to as a *momentum boost*.

A replacement rule like **E^m_ :> E^Expand[m]** introduced here to simplify **psi[x,0]** is a handy way of gaining more control over the algebra and is usually faster than say **ExpandAll** or **Simplify**. Any command such as **Together** or **Factor** can be applied in this way. Note, however, that **RuleDelayed :>** is required, rather than **Rule ->**. With **Rule**, the system would apply **Expand** before the exponent is substituted for **m** and nothing would happen. (See Exercise 2.2, Appendix III.)

Exercise 4.2-1

(a) For the wavepacket **E^(I kP x) psi[x]** just derived, show that the expectation value of the momentum is **h kP**. Show, however, that the expectation value of the position as well as the uncertainties in position and momentum are the same as those calculated with **psi[x]** and that the expectation value of the kinetic energy is

$$\frac{h^2 kP^2}{2 m} + \frac{h^2 w^2}{4 m}$$

Calculate the necessary expectation values of momentum two ways: first as integrals over **h k**, and then as integrals over **x** using the momentum operator **p @ psi_ := -I h D[psi,x]**. (Load the package **Quantum`QuickReIm`** and use our symbolic **Conjugate** rule from Exercise 2.4, Appendix III.)

(b) Generalize the conclusions in Part **(a)** to an arbitrary wavepacket **E^(I kP x) psi[x]**.

We generate a moving wavepacket **psi[x,t]** by applying the time-development operator to the initial wavepacket **psi[x,0]**. This introduces in the usual way time-dependent phases **E^(-I e[k] t/h)** which multiply each planewave component, where **e[k] = (h k)^2/(2m)** is the corresponding planewave (free-particle) eigenenergy. Thus,

```
phi[k_,t_] := phi[k-kP] E^(-I h k^2/(2m) t)
```

```
psi[x_,t_] =
    integGauss[
        phi[k,t] E^(I k x)/Sqrt[2Pi],{k,-Infinity,Infinity}
    ] //ExpandAll //MapAll[Together,#]& //
        PowerExpand //PowerContract
```

$$\text{Power}\left[E, \frac{-(h\ kP^2\ t)}{2\ (-I\ m + h\ t\ w)} + \frac{kP\ m\ x}{-I\ m + h\ t\ w} + \frac{-\frac{I}{2}\ m\ w\ x^2}{-I\ m + h\ t\ w}\right]\ \left(\frac{w}{Pi}\right)^{1/4}$$

$$\text{Sqrt}\left[\frac{m}{m + I\ h\ t\ w}\right]$$

The corresponding time-dependent probability density **psisq[x,t]** is therefore (recall that the package **Quantum`QuickReIm`** introduces our symbolic **Conjugate** rule from Exercise 2.4, Appendix III)

```
Needs["Quantum`QuickReIm`"]
```

```
psisq[x_,t_] = Conjugate[psi[x,t]] psi[x,t] /.
                E^m_ :> E^Together[m] //
                PowerContract //ExpandAll //
                MapAll[Factor,#]&
```

$$\frac{\text{Sqrt}\left[\frac{m^2\ w}{Pi\ (m^2 + h^2\ t^2\ w^2)}\right]}{E^{(w\ (-(h\ kP\ t) + m\ x)^2)/(m^2 + h^2\ t^2\ w^2)}}$$

This function describes another gaussian wavepacket but one whose peak moves steadily to the right with the classical velocity **h kP/m** (i.e. the peak position continually shifts by **h kP t/m**). This wavepacket thus illustrates the quantum mechanical analogue of Newton's laws of force-free motion, viz. the Ehrenfest's theorem mentioned earlier in connection with the two-component box wavepacket (see also Problem 4.2-2 below and Section 8.3).

Momentum and energy are clearly conserved, since the momentum distribution is independent of time, giving

```
Conjugate[phi[k,t]] phi[k,t] == phi[k-kP]^2
```

    True

Therefore, the expectation values of **k** and **(h k)^2/(2 m)** are independent of time and equal to their initial values from Exercise 4.2-1.

Likewise, the momentum uncertainty **delp** is independent of time. The coordinate width of the wavepacket, however, increases with time and the wavepacket broadens as it moves. We thus find in Problem 4.2-2 that the position-momentum uncertainty product **delx delp** increases with time according to

$$\frac{h \; \text{Sqrt}[1 + \dfrac{h^2 \; t^2 \; w^2}{m^2}]}{2}$$

so that for **t > 0** the particle is no longer described by a minimum uncertainty wavepacket. The motion of the wavepacket is depicted in Figure 4.2-1 by a table of plots displayed with the function **StackGraphics** from the package **Graphics`Graphics3D`**.

The spreading of the wavepacket is an inevitable consequence of the fact that each planewave component **E^(I(k x - e[k]t/h))** has a *phase velocity* **h k/(2m)** different from the *group velocity* **h kP/m** of the wavepacket (cf. Park [50], Chapter 2). Consequently, the original coherence of the wavefunction is lost and the wavepacket broadens. This phenomenon is not restricted to quantum mechanics, rather it can be observed in the propagation of any classical (e.g. electromagnetic or sound) wavepulse in a dispersive medium for which the frequency is not simply proportional to wavenumber (refer to Griffiths [29], Chapter 8, and to Jackson [34], Chapter 7). In the case of a free-particle wavepacket, the wave frequency is given by **e[k]/h** and is therefore proportional to **k^2**.

In Chapter 8 and Section 11.11, we will examine a so-called *coherent-state* wavepacket which oscillates in a harmonic oscillator potential without any change of shape.

Problem 4.2-2

(a) Show that the expectation value of position **xExp** and the position uncertainty **delx** predicted by the free-particle packet **psisq[x,t]** are given by

```
Needs["Graphics`Graphics3D`"]
Show[
    StackGraphics[
        Table[
            Plot[
                Evaluate[psisq[x,t] /.{h->1,m->1,w->1,kP->2}],
                {x,-5,35},
                DisplayFunction -> Identity
            ],
            {t,0,11,1}
        ]
    ],
    PlotRange -> {0,0.6},
    BoxRatios -> {1,1,0.5}, Boxed -> False,
    Axes -> {Automatic,None,None}, AxesLabel -> {"x","",""}
];
```

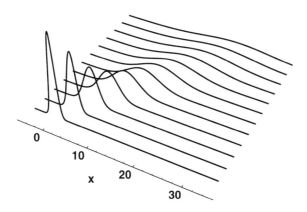

Figure 4.2-1.   Time development of the free-particle wavepacket `psisq[x,t]` with h = m = w = 1 and kP = 2. Time increases from front to back starting with t = 0 in steps dt = 1. Since the wavepacket remains symmetric, its steadily moving peak indicates the position of the corresponding classical particle.

$$\left\{ \frac{h\ kP\ t}{m}, \frac{Sqrt[1 + \frac{h^2\ t^2\ w^2}{m^2}]}{Sqrt[2]\ Sqrt[w]} \right\}$$

respectively. Hence, show that the motion of the peak of the wavepacket satisfies Newton's laws.

(b) Verify that the expectation value of momentum `pExp = h kP`. Then show that the packet's position-momentum uncertainty product `delx delp` is given by

$$\frac{h\ \mathrm{Sqrt}[1 + \dfrac{h^2\ t^2\ w^2}{m^2}]}{2}$$

Of course, the easiest way to calculate `pExp` and `delp` is to evaluate the necessary expectation values of `k` using `phi[k-kP]^2`. If you're up to it, you should be able to derive the same results by applying the momentum operator to the coordinate wavefunction `psi[x,t]`. Note that you can use your results to verify conservation of energy with the coordinate wavefunction.

Hint: You might find replacement rules such as

```
1/(a_ + I b_) -> (a - I b)/(a^2 + b^2)
Sqrt[a_] :> Sqrt[Expand[a]]
```

helpful. Compare with the analysis in Chapter 8.

Problem 4.2-3

The small submaxima between the peaks in Figure 3.0-2 arise because we superposed a finite number of component waves with equal amplitudes. For a given `n`, we can strongly attenuate these "ripples" by assigning smaller amplitudes to components near `kmin` and `kmax` than to those near `kP = (kmin + kmax)/2`.

(a) Examine this effect by assigning to each component of `psiPW[n]` a coefficient determined by a gaussian spectral distribution defined by

```
f[k_] = E^(-(k-kP)^2/(2w))/nrm
```

with `kP = 2.5` and with the width parameter `w = 0.002`. Here `nrm` is a normalization constant, which we can calculate for convenience according to

```
nrm = Sum[f[k]^2,{k,2.4,2.6,(2.6-2.4)/(n-1)}]
```

Redo Figure 3.0-2 with the resulting partial sum `phi[n]`.

(b) Redo Part **(a)** but reduce the widths in **x** approximately by a factor of five.

Problem 4.2-4

Using the rectangular-wave partial sums plotted in Figure 2.4-3 as a set of initial wavepackets `psi[x,0]` inside the box, apply the time development operator and calculate the corresponding time-dependent wavepackets `psisq[x,t]`. Compute the expectation values of position and momentum and verify the uncertainty relation for each wavepacket. Make a series of plots as a function of time for each wavepacket and animate them.

Problem 4.2-5

Analyze the motion of a free-particle wavepacket whose spectral distribution is proportional to

```
             2  2                   2  2
  -((k - kP)  L )/2      -((k + kP)  L )/2
E                    + E
```

Show that the initial wavefunction is proportional to the gaussian-modulated wave

```
  Cos[kP x]
  ─────────
    2     2
   x /(2 L )
E
```

and that the time development is described by two pulses traveling away from the origin in opposite directions. Animate the motion.

5
Parity

The recognition of a system's symmetries at the outset can be very advantageous when constructing the solution of a problem in physics. As an introduction to symmetry in quantum mechanics, let's look at some of the consequences of a *reflection* about the origin in a one-dimensional system. This is the analogue of a *coordinate inversion* in three dimensions.

We investigate another simple symmetry in Exercise 5.0-6, namely, *time reversal*.

A system which has reflection symmetry is one whose potential energy function is an even function of **x** about the origin, that is **vEven[-x] = vEven[x]**. For example, the potential of a harmonic oscillator proportional to **x^2** is clearly an even function of **x**, and when plotted as in Figure 5.0-1 appears symmetric about **x = 0**. (The plot function **FilledPlot** was introduced in Figure 3.0-1.) We can express this property generally by the following rule

```
vEven[a_. x_?Negative] := vEven[-a x]
```

which ensures that

```
vEven[-x] == vEven[x]
```

    True

Here the syntax **?Negative** constrains the blank variable **x_** to negative values only (see Exercise 1.11, Appendix III). The operation of changing **x -> -x** (and in three dimensions **rvec = {x,y,z} -> -rvec**, see Section 20.3) is known as a *parity transformation*, and an even function of **x** (or of **rvec**) is said to be *invariant under parity* or simply to have *even parity*. An odd function satisfies **fOdd[-x] == -fOdd[x]** and is said to have *odd parity*. An

```
FilledPlot[ x^2/2, {x,-5,5}, AxesLabel -> {" x","x^2/2"} ];
```

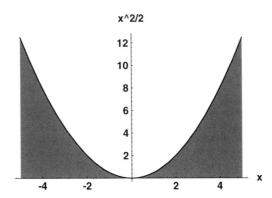

Figure 5.0-1. Harmonic oscillator potential energy with spring constant k = 1.

odd function is also said to be *antisymmetric* about **x = 0**. Note it follows that **fOdd[0] = 0**, whereas the derivative **fEven'[0] = 0**.

Exercise 5.0-1

Verify the above statements with examples and plots of the most common trigonometric and hyperbolic functions. Note that a function doesn't have to have a definite parity. Give some examples and plot them.

The significance of symmetry is that it helps us generally to characterize and classify solutions of the Schrödinger equation even when the solutions are difficult or impossible to obtain. For example, consider an even-parity potential energy and a *nondegenerate* energy level corresponding to a single eigenfunction **psi[x]**. (A *degenerate* energy level corresponds to two or more independent eigenfunctions with the same energy.) If we change the sign of **x** in the wave equation, we find that **psi[-x]** is a solution of the same equation

```
schroedingerD[vEven[-x]] @ psi[-x] == 0
```

$$-(\text{Energy psi[-x])} + \text{psi[-x] vEven[x]} - \frac{h^2\ \text{psi''[-x]}}{2\ m} == 0$$

with the same eigenenergy **Energy**.

Since the level is nondegenerate, we conclude that **psi[-x]** must be proportional to **psi[x]**. It follows that **psi[-x] = ±psi[x]** and therefore that every nondegenerate wavefunction has either even or odd parity.

Exercise 5.0-2

Prove the above statement.

This argument breaks down when the energy levels are degenerate, but it is still always possible to classify states according to their parity, as we'll see in a moment.

Also note that we have shown that the kinetic energy, which involves a second derivative, is an even-parity operator

```
(D[psi[-x],{x,2}] /.-x -> x) == D[psi[x],{x,2}]
```

    True

Exercise 5.0-3

Show that the momentum `px @ psi == -I h D[psi,x]`, however, is an *odd* parity operator.

It is sometimes convenient to have an operator which performs the parity transformation on coordinate wavefunctions. Thus, consider

```
Parity @ psi_ := psi /.{x->-x, y->-y, z->-z} //ExpandAll
```

so that for example

```
{Parity @ psi[x], Parity @ vEven[x], Parity @ x^2}
```

                        2
     {psi[-x], vEven[x], x}

We have seen that a degenerate wavefunction does not have to have a definite parity. However, given an arbitrary solution **psi[x]** of the wave equation, we can always construct a pair of independent solutions of definite parity from the even and odd parts of **psi[x]**, defined by

```
psiEven[x_] := (psi[x] + psi[-x])/2
psiOdd[ x_] := (psi[x] - psi[-x])/2
```

That these are also solutions follows from the linearity of the Schrödinger equation. It is clear that this pair is degenerate. We thus have that

```
{Parity @ psiEven[x] ==  psiEven[x],
  Parity @ psiOdd[x]  == -psiOdd[x]} //ExpandAll
```

     {True, True}

Hence, **psiEven** and **psiOdd** are also eigenfunctions of the parity operator with eigenvalues ± 1, respectively.

Finally, parity arguments can be used to predict if certain integrals vanish without having to integrate explicitly. The idea is one from first year calculus: an odd integrand **g[x]** on a symmetric interval **{x,-xo,xo}** encloses equal amounts of positive and negative area, and therefore its integral over the interval vanishes. Consider the expectation value of **x^n** with a function **f[x]** of definite parity on a symmetric interval. A change of variables **x -> -x** gives

$$Integrate[x\char`^n\ f[x]\char`^2,\{x,-xo,xo\}]\ ==$$
$$(-1)\char`^n\ Integrate[x\char`^n\ f[x]\char`^2,\{x,-xo,xo\}]\qquad\text{(5.0-r1)}$$

Therefore, the expectation value of **x^n** *on a symmetric interval* vanishes if **n** = *odd*. More generally, we see that **Integrate[x^n f[x] g[x], {x,-xo,xo}]** vanishes when **n** = *odd* and **f** and **g** have the same parities or when **n** = *even* and **f** and **g** have opposite parities.

Exercise 5.0-4

(a)Apply **Parity** to your examples from Exercise 5.0-1 above.

(b)Use the parity operator to extract the even and odd parts of the "skewed lorentzian" function **(1+x)/(1+x^2)**. Compare plots of the resulting functions.

Exercise 5.0-5

Prove generally that the expectation value of momentum in a state of definite parity vanishes. (See Exercise 5.0-3 above.)

Exercise 5.0-6

There is another symmetry which always holds as long as the potential is real. (Complex potentials can actually be of interest in modeling problems in which sources or sinks of probability arise; for example, as in exponential decay.) Show that if **psi[x,t]** is a solution of the wave equation and **V** is real, then **Conjugate[psi[x,-t]]** is also a solution, called the *time-reversed* solution. One says that the wave equation is *invariant under time reversal* or that there is *symmetry under time reversal*.

This means that if **psi[e,x]** is a stationary state of the Schrödinger equation, then **Conjugate[psi[e,x]]** is also one with the same energy **e**. What can you conclude if **psi[e,x]** is a *nondegenerate* stationary state?

As an example, consider the time-reversed state corresponding to the *moving* free-particle wavepacket **psi[x,t]** we set up in Section 4.2. A comparision of an animation of the time-reversed state with one of **psi[x,t]** on a symmetric interval **{t,-to,to}** about the initial state **psi[x,0]** will make the time-reversal symmetry clear. Also, set up a time-reversed wavepacket with initial state **Conjugate[psi[to > 0]]** from the future of **psi[x,t]** and animate its motion. (Hint: Examine the time-development operators and consider a time shift by **to**. Also, use our symbolic **Conjugate** rule from Exercise 2.4, Appendix III, and defined by the package **Quantum`QuickReIm`**.)

Problem 5.0-1

Reconsider the problem of a particle in an infinite box or rectangular well from Chapter 2. Shift the origin of coordinates to the middle of the box $-L/2 \leq x \leq L/2$ so that the potential energy now has even parity. Show that the eigenfunctions of definite parity are of the form **Cos[k x]** and **Sin[kp x]**. Apply the boundary conditions that the wavefunctions vanish at the walls of the box to show that **k -> (2n+1)Pi/L** and **kp -> 2n Pi/L**, where **n** is a positive integer or zero in the case of **k** and a nonzero, positive integer in the case of **kp**. Show that the eigenenergy spectrum is the same as before.

The ground state is now proportional to `Cos[Pi x/L]` and the first-excited state to `Sin[2Pi x/L]`. Hence, the ground state with zero nodes has even parity and the first excited state with one node has odd parity. The parity continues to alternate like this for all the higher excited states. Therefore, the box states are now also eigenfunctions of `Parity` with eigenvalues ± 1. These are general features of non-degenerate states of definite parity.

Harmonic Oscillator

We turn now to one of the most pervasive problems in all of physics: the response of a system to small displacements about a stable equilibrium. Consider the potential energy $V[x]$ of a particle of mass m moving in one-dimension (*1D*) and assume there is a relative minimum Vo in the potential at $x = xo$. This minimum is appropriately described by the Taylor expansion

```
Series[V[x], {x,xo,2}] /.
    {V[xo] -> Vo, V'[xo] -> 0, V''[xo] -> k}
                   2
         k (x - xo)                3
  Vo + ------------- + O[x - xo]
              2
```

where $V'[xo] \to 0$ defines an extremum and $V''[xo] \to k > 0$ a minimum. Hence, for small enough displacements about an equilibrium, we can almost always approximate the force on a system by a linear restoring force $-k(x - xo)$, where the *spring constant* k is defined by the (upward) curvature $V''[xo]$ of the potential at the minimum. (Note we assume here that $V''[xo] > 0$, which isn't always the case. Consider the anharmonic oscillator in Problem 7.3-2 below.)

The spring constant determines the classical frequency of oscillation $w = Sqrt[k/m]$ so that $k = m\,w^2$. Without loss of generality, it is convenient and usual to shift the origin of coordinates and the minimum of energy to the potential minimum and work instead with the potential energy (see Figure 5.0-1)

```
VHO = % /.{xo -> 0, Vo -> 0, k -> m w^2} //Normal
       2  2
   m  w   x
   ---------
       2
```

Here the built-in function **Normal** has been applied to convert the series to an ordinary algebraic expression.

Solutions of the Schrödinger equation for this potential thus find application in almost all areas of physics. Although the quantum harmonic oscillator (HO) has some pecularities, most notably its evenly spaced energy levels, it affords a full solution which is easy to work with. In particular, the HO provides a convenient starting point for investigating more complicated systems. We shall treat the HO as a paradigm of non-relativistic quantum theory and demonstrate with it many of the nuts and bolts of being a good quantum mechanic. Our approach will be mostly conventional, but nevertheless directed towards implementation on a computer. We will begin with a power series solution of the Schrödinger equation

6.1 Scaled Schrödinger Equation

We begin with the Schrödinger equation for the harmonic oscillator,

```
schroedingerD[V_] @ psi_ :=
    -h^2/(2m) D[psi,{x,2}] + V psi - Energy psi

schroedingerD[VHO] @ psi[x] == 0
```

$$-(Energy \; psi[x]) + \frac{m \; w \; x \; psi[x]}{2}^{2 \; 2} - \frac{h \; psi''[x]}{2 \; m}^{2} == 0$$

where the **schroedingerD** operator was introduced in Section 1.3. We can put this into a simpler form by scaling the coordinate **x** and the energy. Thus, we introduce a dimensionless coordinate **z = a x** and look for a convenient choice for the constant **a**:

```
-2m/(a h)^2 schroedingerD[VHO] @ psi[a x] == 0 //
    ExpandAll
```

$$\frac{2 \; Energy \; m \; psi[a \; x]}{a \; h}^{2 \; 2} - \frac{m \; w \; x \; psi[a \; x]}{a \; h}^{2 \; 2 \; 2} + psi''[a \; x] == 0$$

Using the second term as a guide, it is clear that the value **a = Sqrt[m w/h]** will simplify things:

```
% /. x -> z/a /. a -> Sqrt[m w/h]
```

$$\frac{2 \; Energy \; psi[z]}{h \; w} - z \; psi[z] + psi''[z] == 0^{2}$$

Since **h w** has units of energy, we can reduce this equation to dimensionless form by setting **Energy = h w (n + 1/2)**, where **n** is a number to be determined. The form **n + 1/2** further simplifies our results, since **n** will turn out to be an integer. Thus, the HO Schrödinger equation reduces to

```
(%[[1]] /. Energy -> h w (n + 1/2) //
    Expand //Collect[#,psi[z]]&) == 0
         2
(1 + 2 n - z ) psi[z] + psi''[z] == 0
```

where **Collect** has been applied to factor the first term in this result. (See Exercise 2.1, Appendix III.) We refer to this equation as the *scaled* HO Schrödinger equation.

Method of Solution 6.2

Power series solution of a differential equation is a particularly useful method when the coefficients of the function and its derivatives in the differential equation are polynomials, although it is applicable to other cases as well. Thus, we look for a solution of the scaled Schrödinger equation of the form

$$psi[z] == Sum[c[k] z^k, \{k,0,Infinity\}] \tag{6.2-r1}$$

where the expansion coefficients **c[k]** are independent of **z**. Note we exclude negative powers of **z** since **psi[z]** as a probability amplitude must be finite when **z = 0** (see Section 1.3).

It's always a good idea to examine first the *asymptotic* form of the Schrödinger equation for large **x** away from the center of force. In general, the potential can be ignored in this region and the Schrödinger equation solved for the asymptotic form of the wavefunction. How large **x** has to be depends on the range of the potential. In the case of the oscillator, the potential in fact becomes infinite in the regions **z -> ±Infinity** but the Schrödinger equation, nevertheless, simplifies. For large positive or negative **z**, the **z^2 psi** term dominates the scaled HO Schrödinger equation (from the previous section) and the terms **(1 + 2n) psi** can be ignored. We easily verify that **psi -> E^(-z^2/2)** is then a solution:

```
-z^2 psi + Dt[psi,{z,2}] /.psi -> E^(-z^2/2) //Factor
        2
     -z /2
   -E
```

in the limit **z -> ±Infinity**. You probably notice that this analysis doesn't exclude a solution of the form **E^(+z^2/2)**. We reject it, however, since an infinite probability is again not physically reasonable.

Returning to the solution for all **z**, we thus write the wavefunction in the form **psi[z] -> E^(-z^2/2)H[z]** and transform the wave equation to one in **H[z]**. As we shall see shortly, this has the additional advantage that the coefficients in a power series for **H[z]** satisfy a two-term *recursion relation*, whereas series coefficients for `psi[z]` satisfy a three-term recursion relation (see Exercise 6.3-1 below).

Apology to the reader. The author hopes you are not otherwise too frustrated by certain substitutions and transformations which will appear from time to time throughout this book, as for example in solving differential equations. These clever tricks are often just that, but are usually the result of someone else's effort through much trial and error. To prepare you any more would greatly lengthen the discussion and be only a poor substitute for what you can find elsewhere, as for example in the quantum books listed in the references. Refer also to Arfken [4] or to Boas [9] and at a more advanced level to Morse and Feshbach [49].

Problem 6.2-1

Use the *shooting method* from Section 2.2 to compute the bound states of a *finite* rectangular well. Take the potential to be `-Vo` for `-L < x < L` and zero otherwise, and use `h = m = L = 1` and `Vo -> 15`.

It can be shown that the bound (`Energy < 0`) levels are given by the roots of

```
k Tan[k L]  ==   K
k Cot[k L]  ==  -K
```

Here `h k = Sqrt[2m(Energy + Vo)]` and `h K = Sqrt[-2m Energy]`. Use the built-in function `FindRoot` to solve these transcendental equations and show that the well supports *four* bound levels. (Refer to `?FindRoot`.)

(a) Show that the proper asymptotic solutions are of the form `E^(-K Abs[x])`, and therefore that the wavefunctions penetrate the walls of the well. The above transcendental equations for the eigenenergies are straightforward to derive by matching the asymptotic solution and its derivative at the walls of the well to the inside solutions of definite parity of the form `Cos[k x]` and `Sin[k x]` (see Section 1.3 and Problem 5.0-1). (Refer to Park [50], Chapter 4, and to Merzbacher [45], Chapter 6.)

(b) For the numerical integration you can simply take `phi[±xo] = 0` as the boundary conditions with the bound `xo` a few times larger than `L`. (Note that the higher levels have longer range wavefunctions and you might have to move `xo` out to obtain three significant figures in the wavefunction. Take a look at the harmonic oscillator wavefunctions in Figures 6.6-1 and 6.6-2.) Use `NDSolve` and for the potential enter

```
v[x_] := -15 /; Abs[x] <  1
v[x_] :=   0 /; Abs[x] >= 1
```

(c) Use the built-in function `NIntegrate` to normalize the numerical wavefunctions. Plot the normalized eigenfunctions and check that they have the proper number of nodes. Show that the eigenfunctions are mutually orthogonal.

6.3 Energy Spectrum

We now substitute `psi[z] -> E^(-z^2/2) H[z]` into the scaled HO Schrödinger equation and obtain the Hermite differential operator. In order to have an expression we can easily work with given a generic function **H**, it's convenient to do this using total derivatives (refer to Section 2.1, Appendix V):

```
hermiteDt[H_] =
    (Dt[psi,{z,2}] + (1 + 2n - z^2) psi) E^(z^2/2) /.
        psi -> E^(-z^2/2) H //Expand

  2 H n - 2 z Dt[H, z] + Dt[H, {z, 2}]
```

When equated to zero, we obtain the *Hermite differential equation* for **H[z]**,

```
hermiteDt[H[z]] == 0

  2 n H[z] - 2 z H'[z] + H''[z] == 0
```

whose *regular solutions*, i.e. solutions finite at the origin **z = 0**, are the built-in *Hermite polynomials* **HermiteH[n,z]**. We turn now to their construction.

As we have seen in Chapter 5, we may confine ourselves to finding the even and odd parity solutions **psi[z]**. Thus, we look for series solutions **H[z]** with even or odd parity, since the factor **E^(-z^2/2)** has even parity. Since the terms in the series (6.2-r1) are independent, we can confine our attention to a general term **c[k] z^k** and *drop the sum*. The procedure will be to substitute this expression into the differential equation, collect together independent powers of **z**, and equate them to zero, because the right-hand side of the differential equation vanishes. The resulting expression gives a recursion relation for the expansion coefficients.

Thus operating on **c[k] z^k**, we obtain, after declaring **c** and **k** to be constants independent of **z**,

```
SetAttributes[{c,k}, Constant]

hermiteDt[H] /. H -> c[n,k] z^k //ExpandAll

        -2 + k                2  -2 + k                k
  -(k z        c[n, k]) + k  z        c[n, k] - 2 k z  c[n, k] +

        k
  2 n z   c[n, k]
```

which equates to zero. We now regard **k** as a dummy summation index which we can change in order to collect together *true* like powers of **z**. For example, noting the equivalence

```
(z^(-2+k) /.k -> k+2) == z^k

  True
```

we rearrange the coefficients of **z^(-2+k)** in the previous expression (labelled presently by *Mathematica* as **%%**) and add together the common terms, which are all now proportional to **z^k**:

```
Coefficient[%%,z^k] +
   (Coefficient[%%,z^(-2+k)] /.k -> k+2) //
      Collect[#,{c[n,k],c[n,k+2]}]& //Map[Factor,#]&
```

```
2 (-k + n) c[n, k] + (1 + k) (2 + k) c[n, 2 + k]
```

Alternatively, we accomplish the same thing with pattern matching and a pair of replacement rules (here we simply drop the common factor z^k):

```
%%% /.
   {k^m_. c[n,k] z^(k+q_.) -> (k-q)^m c[n,k-q],
         c[n,k] z^(k+q_.) ->          c[n,k-q]} //
   Collect[#,{c[n,k],c[n,k+2]}]& //
      Map[Factor,#]&
```

```
2 (-k + n) c[n, k] + (1 + k) (2 + k) c[n, 2 + k]
```

When equated to zero, this expression gives a two-term recursion relation which can be solved for the coefficients. At the same time, we can set up a replacement rule:

```
cRep = Solve[ % == 0, c[n,k+2] ][[1,1]] /.
   k -> k-2 //ExpandAll //MapAll[Factor,#]&
```

$$c[n, k] \rightarrow \frac{2 (2 - k + n) c[n, -2 + k]}{(1 - k) k}$$

Here the subscript [[1,1]] picks out the desired element from the solution list, effectively removing the list braces {} (see Exercise 1.3, Appendix III).

Exercise 6.3-1

Construct a series solution of the scaled Schrödinger equation,

```
(1 + 2n - z^2) psi[z] + psi''[z] == 0
```

and show that the expansion coefficients satisfy a *three-term* recursion relation.

We see that the recursion formula connects every other **k** value. Therefore, if we set **c[n,1] = 0** all higher *odd* coefficients vanish and we obtain an *even*-parity series solution. Likewise, if we set **c[n,0] = 0** we obtain an *odd*-parity solution. It's straightforward to set up a procedure using **cRep** that solves for the expansion coefficients **cEven** or **cOdd** as *dynamic-programming* functions, which call themselves. Thus, (refer to Exercise 3.1, Appendix III)

```
cEven[n_,0]  := 1
cEven[n_,1]  := 0
cEven[n_,k_] := c[n,k] /.cRep /.c->cEven //Evaluate
```

```
cOdd[n_,1]   := 1
cOdd[n_,0]   := 0
cOdd[n_,k_]  := c[n,k] /.cRep /.c->cOdd //Evaluate
```

We have simply taken the first term in each series to be one. Here **Evaluate** ensures that the replacement with **cRep** is made (for computational efficiency) even though **SetDelayed** (:=) is needed for the dynamic program. Hence, **cEven** and **cOdd** call themselves as required, which you can verify by examining **?cEven** or **?cOdd**.

Thus, we can compute the first few terms in the even-parity series

```
Sum[ cEven[n,k] z^k, {k,0,7} ]
```

$$1 - n\,z^2 + \frac{(-2 + n)\,n\,z^4}{6} - \frac{(-4 + n)\,(-2 + n)\,n\,z^6}{90}$$

or in the odd-parity series:

```
Sum[ cOdd[n,k] z^k, {k,0,7} ]
```

$$z - \frac{(-1 + n)\,z^3}{3} + \frac{(-3 + n)\,(-1 + n)\,z^5}{30} -$$

$$\frac{(-5 + n)\,(-3 + n)\,(-1 + n)\,z^7}{630}$$

It is not difficult to show (see Exercise 6.3-2) that these two series diverge *asymptotically* for large z as $E\verb|^|(+z\verb|^|2)$. Therefore, as it stands, the wavefunction **psi[z]** = $E\verb|^|(-z\verb|^|2/2)$ **H[z]** will diverge as $E\verb|^|(+z\verb|^|2/2)$, which we cannot allow since **psi[z]** must represent a probability amplitude. Hence, some modification of the series is necessary.

Exercise 6.3-2

Show that both series diverge for large z as $E\verb|^|(+z\verb|^|2)$. Note for large z the terms in the series with large k dominate. Hence, examine the ratio of the series coefficients of $z\verb|^|k$ and $z\verb|^|(k-2)$ for large k and compare with those of $E\verb|^|(+z\verb|^|2)$ for large k. (See also Problem 6.5-2.)

Up to this point, **n** is an arbitrary number. It's easy to see, however, from the series and from **cRep** that if **n** *is a positive even integer or zero* the coefficient **cEven[k]** vanishes for all $k > n$. Likewise, if **n** *is a positive odd integer* the coefficient **cOdd[k]** vanishes for all $k > n$. For example, for **n** = **4** and summing over **k** with $k > 4$, we obtain

```
Sum[ cEven[4,k] z^k, {k,0,7} ]
```

$$1 - 4\ z^2 + \frac{4\ z^4}{3}$$

Hence, both the even and odd series *truncate* to polynomials of order **n**, so that **psi[z]** is now well-behaved. In effect, we have implemented boundary conditions on the wavefunction. Moreover, we see that the evenness or oddness of **n** determines the parity of the wavefunction. These solutions of the Hermite equation are proportional to the Hermite polynomials **H[n,z]**.

We conclude that the energy can take on only discrete positive values **Energy -> eHO[n] = h w (n+1/2)** i.e. **h w/2, 3h w/2**, and so forth. We thus say that the system and its energy are *quantized*, and refer to a *spectrum of discrete energy levels* for the harmonic oscillator. Note that the energy-level separation is constant and equals **h w**, a feature peculiar to the harmonic oscillator. In particular, there are no continuum states, since the system is bounded for all energies.

It usually comes as a surprise and is one of the remarkable subtleties of Nature that the energy spectrum of the hydrogen atom can be obtained from that of the harmonic oscillator, in particular a two-dimensional isotropic oscillator, by transformation of the Schrödinger equation. We shall thus investigate the two-dimensional oscillator in Problems 18.7-2 and 20.5-3 and its connection with the hydrogen atom in Problem 20.5-4.

Problem 6.3-1

Consider again the problem of a particle in an infinitely-deep box or rectangular well from Chapter 2. As in Problem 5.0-1, shift the origin of coordinates to the middle of the box so that the potential energy has even parity. Construct the even and odd parity series solutions and show that they are equivalent to **Cos[k x]** and **Sin[kp x]** from Problem 5.0-1. (Use the built-in function **Series** to check the expansions of **Cos** and **Sin**.)

6.4 Hermite Polynomials

By convention, Hermite polynomials are normalized such that the coefficient of their highest power of **z** is **2^n**. Thus, we can renormalize the above sums by dividing by the coefficient **c[n]** of **z^n** and multiplying by **2^n**, and define a function which computes the Hermite polynomials:

```
H[n_,z_] :=
    If[EvenQ[n],
        2^n/cEven[n,n] Sum[cEven[n,k] z^k,{k,0,n}] //Expand,
        2^n/cOdd[n,n]  Sum[ cOdd[n,k] z^k,{k,0,n}] //Expand
    ]
```

For example, the first four polynomials are given by

```
{H[0,z], H[1,z], H[2,z], H[3,z]}
                  2              3
    {1, 2 z, -2 + 4 z , -12 z + 8 z}
```

which we can check are identical with the built-in functions

```
% ==
    {HermiteH[0,z],HermiteH[1,z],HermiteH[2,z],HermiteH[3,z]}

    True
```

The syntax of the above **If** statement can be clarified by entering **?If**.

Exercise 6.4-1

Plot a few Hermite polynomials and convince yourself that **n** determines the number of nodes in these functions. Observe how their parity is related to the symmetry or antisymmetry of the plots about **x = 0**.

Hermite polynomials belong to a class of functions known as *orthogonal functions*, which arise naturally in the solution of the differential equations of mathematical physics. Their properties can be obtained from a study of eigenvalue problems based on Hermitian second-order differential operators known as *Sturm-Liouville theory* (see Problem 6.4-1). One derives two especially important results, which we have already illustrated with the box eigenfunctions in Chapter 2. However, because the Hermite polynomials better represent the general case, it is appropriate here to take another look at these two properties.

Orthogonality. Two distinct polynomials are *orthogonal* with respect to their *weight* function, **E^(-z^2)**, if for **m ≠ n**

```
Integrate[
    E^(-z^2) H[m,z] H[n,z],
    {z,-Infinity,Infinity}
] == 0                                              (6.4-r1)
```

Completeness. Hermite polynomials form a complete set such that any well-behaved function **F[z]** can be approximated by the partial sum

```
F[z] == Sum[a[n] H[n,z],{n,0,nmax}]                 (6.4-r2)
```

to any desired accuracy as **nmax** is increased. Such an expansion in the limit **nmax -> Infinity** is referred to as a *generalized-Fourier series*. The expansion coefficients **a[n]** can be formally evaluated using the orthogonality condition.

The weight function **E^(-z^2)** for Hermite polynomials is seen to be the square of the asymptotic wavefunction **psi[z]** which we introduced to transform the scaled wave equation to Hermite's equation. Proving the orthogonality condition directly from the differential equation is straightforward. (See Exercise 2.3-1 and also Park [50].) Also, orthogonality of two functions with opposite parity follows from the fact that their product has odd parity (see Chapter 5).

Completeness is more difficult to prove, but can be made plausible by the following argument if **F[z]** is *analytic*. By definition, an analytic function is one that can be expanded in a power series in **z^n**. And evidently we can always write any power **z^n** as a finite linear combination of Hermite polynomials **H[k,z]** summed over **{k,0,n}** (see Problem 6.4-1). Thus, it is reasonable that an analytic function **F[z]** can be written as an infinite power series in **H[n,z]**. (The interested reader will enjoy the completeness proof given by R. Peierls [53] in his book, *More Surprises In Theoretical Physics.*) Eigenfunctions thus belonging to a complete set are often referred to as *basis states*, in analogy with basis vectors that span a finite-dimensional vector space.

Related to the orthogonality condition is the normalization integral

$$Integrate[E^{\wedge}(-z^2)\ H[n,z]^2, \{z, -Infinity, Infinity\}]\ == \\ 2^{\wedge}n\ n!\ Sqrt[Pi] \tag{6.4-r3}$$

that we derive for arbitrary **n** in Exercise 6.4-2 below. Nevertheless, *given explicit values for* **n**, such integrals are easily evaluated by the system. For this task, it is convenient to load again our own rule set **integGauss** specialized for performing integrals involving gaussians and powers, introduced in Section 4.0. Thus, for example, we check the integral (6.4-r3) for **n = 3** by entering

```
Needs["Quantum`integGauss`"]

integGauss[E^(-z^2) H[3,z]^2, {z,-Infinity,Infinity}] ==
    2^3 3! Sqrt[Pi]
```

    True

Likewise, we can verify orthogonality in particular cases.

If we multiply the partial sum (6.4-r2) through by **E^(-z^2)H[m,z]** and integrate over all **z**, we obtain an expression for the expansion coefficients **a[n]**. Owing to orthonormality, the right-hand side of (6.4-r2) reduces to **a[m] 2^m m! Sqrt[Pi]**. We therefore have that

$$a[n]\ == \\ Integrate[\ E^{\wedge}(-z^2)\ H[n,z]\ F[z],\ \{z,-Infinity,Infinity\}\ ]/ \\ (2^{\wedge}n\ n!\ Sqrt[Pi]) \tag{6.4-r4}$$

As we shall see, these formulas allow us to express any wavefunction as a linear combination of HO eigenfunctions. This means, for example, that an arbitrary wavepacket and therefore in principle the outcome of any experiment with the oscillator can be analyzed in terms of the HO eigenfunctions. Refer to Section 2.6 and also to Chapter 8.

Exercise 6.4-2

There are several formulas involving the Hermite polynomials which are useful in a variety of applications. Three such results are (i) the *Rodrigues relation*

```
H[n,z]  ==  (-1)^n E^(z^2) D[E^(-z^2),{z,n}]
```

(ii) the *recurrence relation*

```
H[n+1,z]  -  2z H[n,z]  +  2n H[n-1,z]  ==  0
```

and (iii) the *generating function*

```
E^(-t^2 + 2t z)  ==  Sum[ H[n,z] t^n/n!, {n,0,Infinity} ]
```

Analogous relations hold for other orthogonal polynomials.

(a) Verify the Rodrigues and recurrence relations for several explicit values of **n**. Verify the generating function *numerically* for several values of **t** and **z** by summing **{n,0,nmax}** and increasing **nmax** until convergence is achieved and then comparing with the built-in function **NSum** for **{n,0,Infinity}**. This will be relatively easy for values of **t, z < 1** but more difficult for **t, z > 1**.

(b) Use the Rodrigues relation and integration by parts **n** times to derive the normalization (6.4-r3).

(c) Use the Taylor expansion of **E^(-t^2 + 2t z)** in powers of **t** and the Rodrigues relation to verify the generating function.

We'll derive the Rodrigues formula in Problem 6.7-2 and use the generating function in Problem 8.4-1 to construct a coherent-state gaussian wavepacket. You might also note that recurrence relations provide the most efficient and stable means of computing orthogonal polynomials. Refer to *Numerical Recipes* [54], Chapter 5.

Problem 6.4-1

The problem of solving the Schrödinger wave equation in one dimension can be viewed as belonging to a class of boundary value problems known as *Sturm-Liouville theory*. This theory is characterized by a linear, second-order differential equation of the form

```
D[p[z] u'[z], z] + q[z] u[z]  ==  la w[z] u[z]
```

and (real) boundary conditions on the solutions **u[z]** in an interval $a \leq z \leq b$. This equation defines a second-order *Hermitian* differential operator with real eigenvalue **la**. The function **w[z]** is known as the *density* or *weight function*, and **p, q** and **w** are arbitrary real functions of **z**, except that **p > 0** for $a < z < b$. (However, both **p[a]** and **p[b]** may vanish.) For certain values of **la**, the resulting eigenfunctions **u[la,z]** are well behaved and form an orthogonal and complete set.

Thus, for any two eigenfunctions belonging to two distinct eigenvalues **la[m]** and **la[n]**

```
Integrate[ w[z] u[m,z] u[n,z], {z,a,b} ]  ==  0
```

and for any well-behaved function **F[z]** the partial sum

```
F[z] == Sum[ a[n] u[n,z], {n,0,nmax} ]
```

approximates `F[z]` to any desired accuracy as **nmax** is increased.

Using the Hermite polynomials as examples, consider the following:

(a) Transform the Hermite equation into Sturm-Liouville form and show that the weight function `w[z] = E^(-z^2)`. What forms do `p[z]`, `q[z]` and `la` take?

(b) Verify explicitly for $0 \leq$ **m, n** ≤ 4 that the Hermite polynomials are orthogonal and check the normalization integral (6.4-r3).

(c) Expand `z^4` in Hermite polynomials.

Sturm-Liouville theory and its relation to Hermitian operators are discussed in Chapter 9 of Arfken [4] and in Chapter 6 of Morse and Feshbach [49]. Orthogonal polynomials are nicely summarized in Chapter 22 of Abramowitz and Stegun [1].

6.5 Hypergeometric Functions

Our derivation of the Hermite polynomials can be well summarized by connecting with a more general point of view. We will show in a moment that the Hermite differential equation is a special case of the *confluent hypergeometric* or *Kummer's differential equation*:

```
kummerDt[f_] = z Dt[f,{z,2}] + (b - z) Dt[f,z] - a f;
```

```
kummerDt[f[z]] == 0
```

```
-(a f[z]) + (b - z) f'[z] + z f''[z] == 0
```

whose *regular* solutions are the built-in *confluent hypergeometric or Kummer's function* **Hypergeometric1F1[a,b,z]**. (In the text, we will refer to Kummer's function simply as **1F1[a, b, z]**). These functions and the closely related *hypergeometric* functions are important to mathematical physics, and will appear in various applications throughout this book. Most of the common solutions which arise from separation of variables of Laplace's equation, the diffusion equation, the classical wave equation, and the Schrödinger equation can be related to these functions and therefore studied in a unified way.

Kummer's function is defined by its series expansion about the origin, i.e. the confluent hypergeometric series

```
Hypergeometric1F1[a,b,z] + O[z]^4
```

$$1 + \frac{a\,z}{b} + \frac{a\,(1+a)\,z^2}{2\,b\,(1+b)} + \frac{a\,(1+a)\,(2+a)\,z^3}{6\,b\,(1+b)\,(2+b)} + O[z]^4$$

which is convergent for all values of **a** and **b** and for all finite values of **z** (see Abramowitz and Stegun [1], Chapter 13). By convention, these functions are "normalized" such that they *equal* unity when **z = 0**. (Note that entering `f[z] + O[z]^4` is equivalent to **Series[f[z],{z,0,3}]**.)

Problem 6.5-1

(a) Verify the confluent hypergeometric series solution of Kummer's differential equation. Hence, show that the expansion coefficients satisfy

```
                 (-1 + a + k)  c[-1 + k]
    c[k]  ->  ─────────────────────────
                   k (-1 + b + k)
```

(b) For a = b, evaluate `c[k]` explicitly in terms of `c[0]=1` and relate `1F1[a,a,z]` to a known simple function.

It's evident that when **a = -n**, a negative integer or zero, the series coefficients **c[k]** vanish for all **k** $\geq$ **n + 1**, and **1F1[-n,b,z]** reduces to a polynomial of order **n**. Hence the connection with many of the familiar functions of mathematical physics, such as Hermite polynomials.

We transform the Hermite equation to Kummer's equation by introducing a function of the form **f[z^2]**. We immediately see the connection depends on the parity of the Hermite polynomials. Thus, inserting **n** even, i.e. **n ->
2n**, into the Hermite equation, we have with **y = z^2**

```
hermiteDt[f[z^2]]/4 == 0 /.
    n -> 2n /. z -> Sqrt[y] //
        ExpandAll //Map[ Collect[#,{f[y],f'[y]}]&, #]&

              1
    n f[y] + (- - y) f'[y] + y f''[y] == 0
              2
```

which is again Kummer's equation if we make the identification **a = -n** and **b = 1/2**. (Refer to Exercise 2.1, Appendix III, for applying **Map** together with **Collect** as pure functions.) Hence, we conclude that **HermiteH[2n,z]** is proportional to **1F1[-n,1/2,z^2]**.

Exercise 6.5-1

(a) Determine the proportionality constant connecting **HermiteH[2n,z]** and **1F1[-n,1/2,z^2]**.

(b) Deduce the connection between **HermiteH** and **1F1** for **n** odd.

Finally, it is easy to set up a second linearly independent solution, with which we can construct a general solution of Kummer's equation. (Recall that every linear *second-order* differential equation has two linearly independent solutions.) Consider a function of the form **z^nu f[z]**, where **nu** is a constant to be determined.

```
SetAttributes[{nu}, Constant]

kummerDt[ z^nu f[z] ]/z^nu == 0 //
    ExpandAll //Map[Collect[#,{f[z],f'[z]}]&, #]&
```

$$(-a - nu - \frac{nu}{z} + \frac{b\ nu}{z} + \frac{nu^2}{z})\ f[z]\ +\ (b\ +\ 2\ nu\ -\ z)\ f'[z]\ +$$

$$z\ f''[z]\ ==\ 0$$

This gives again Kummer's equation if we set **nu -> 1 - b**, so that the coefficient of **f[z]** is independent of **z**:

```
% /.nu -> 1-b //ExpandAll //
   Map[ Collect[#,{f[z],f'[z]}]&, #]&
```

$$(-1\ -\ a\ +\ b)\ f[z]\ +\ (2\ -\ b\ -\ z)\ f'[z]\ +\ z\ f''[z]\ ==\ 0$$

Therefore, **z^(1-b) 1F1[1+a-b,2-b,z]** is a second linearly independent solution, which, however, is irregular (infinite) at **z = 0** if **b > 1**. In fact, any linear combination of this function and **1F1[a,b,z]** is an independent solution, and in this way the built-in function **HypergeometricU[a,b,z]** is defined. The particular choice used by *Mathematica* is conventional.

Exercise 6.5-2

(a) Deduce the relation between the built-in functions **U[a,b,z]** and **1F1[a,b,z]** by examining the series expansion of **U[a,b,z]** about **z = 0**.

(b) Find linear combinations of the form **c1 1F1[a,b,z] + c2 U[a,b,z]**, where **c1** and **c2** are constants, which reduce to the Hermite polynomials.

Exercise 6.5-3

Generate both independent solutions of Kummer's equation directly by assuming a *generalized power series solution* of the form

```
Sum[c[k]  z^(nu+k),{k,0,Infinity}]
```

an approach also known as the *method of Frobenius* (see for example Boas [9], Chapter 12).

Exercise 6.5-4

Transform Kummer's equation to the Sturm-Liouville form

```
D[z^b E^-z 1F1'[a,b,z], z]  == a z^(b-1) E^-z 1F1[a,b,z]
```

According to Problem 6.4-1, if boundary conditions are enforced then **a** becomes the eigenvalue and **z^(b-1) E^-z** the weight function, which defines orthogonality of the eigenfunctions **1F1[a,b,z]**.

Exercise 6.5-5

(a) There are many useful relations among the confluent hypergeometric functions. *Mathematica* for example automatically transforms the *n*th derivative of **1F1[a,b,z]** to another (single) **1F1**. Deduce the explicit connection.

(b) Use Kummer's equation to derive the recurrence relation

```
b(1-b+z) 1F1[a,b,z] + b(b-1) 1F1[a-1,b-1,z] -
       a z 1F1[a+1,b+1,z] == 0
```

Refer to Abramowitz and Stegun [1], eq. 13.4.7.

Problem 6.5-2

(a) By comparing coefficients of **z^k** for **k** large in the series expansions of **1F1[a,b,z]** and **z^(a-b)E^z**, demonstrate that

```
1F1[a,b,z]  ~  Gamma[b]/Gamma[a] z^(a-b) E^z
```

for **z -> +Infinity**. (The **Gamma** function is discussed in Problem 6.5-4.)

(b) Show that **1F1[a,b,z]=E^z 1F1[b-a,b,-z]**, and verify this identity by series expansion about **z = 0** and by series multiplication. Thus, indicate that **1F1[a,b,z]** ~ **Gamma[b]/Gamma[b-a](-z)^-a** for **z -> -Infinity**.

Compare results in **(a)** and **(b)** with **Series[1F1[a,b,z],{z,±Infinity,1}]**.

Important functions closely related to the confluent hypergeometric series are the *hypergeometric functions* given by the built-in quantities **Hypergeometric2F1[a,b,c,z]** (see Arfken [4], Chapter 13). They are also defined by their (hypergeometric) series expansion:

```
Hypergeometric2F1[a,b,c,z] + O[z]^3
```

$$1 + \frac{a\ b\ z}{c} + \frac{a\ (1 + a)\ b\ (1 + b)\ z^2}{2\ c\ (1 + c)} + O[z]^3$$

The differential equation that they satisfy has singular points at **z -> 1** and **z -> Infinity** which exhibit a confluence (i.e. they merge) in the appropriate limit resulting in the *confluent* hypergeometric equation. (See Problem 6.5-5 below, and refer to Morse and Feshbach [49], p. 542, and to Abramowitz and Stegun [1], Chapter 15.) Evidently, these functions also truncate to polynomials of order **n** if **a** or **b = -n** a negative integer or zero.

Hypergeometric series can be generalized to functions with any number of parameters, which when needed are given by the built-in quantity **HypergeometricPFQ**:

```
?HypergeometricPFQ
```

```
HypergeometricPFQ[numlist, denlist, z] gives the generalized
   hypergeometric function pFq where numlist is a list of the p
   parameters in the numerator  and denlist is a list of the q
   parameters in the denominator.
```

Problem 6.5-3

The built-in functions **Hypergeometric2F1[a,b,c,z]** satisfy the *hypergeometric differential equation*

$$\text{a b f[z]} + ((a + b + 1) z - c)\ f'[z] + z(z-1)\ f''[z] == 0$$

(a) Derive a series expansion for **2F1[a,b,c,z]** about **z = 0** and check your result with **Series**. By convention, the first term in the series **c[0] = 1**. Because of the singularity in the differential equation as **z -> 1**, the series converges only for **Abs[z] < 1**. Verify with some examples that these functions also reduce to polynomials of order **n** if **a** or **b = -n** a negative integer or zero.

(b) Find a second independent solution.

Problem 6.5-4

(a) Hypergeometric series are often expressed in terms of *Pochhammer symbols* as

```
1F1[a,b,z] ==
    Sum[Pochhammer[a,n]/Pochhammer[b,n] z^n/n!,{n,0,Infinity}]

2F1[a,b,c,z] ==
    Sum[Pochhammer[a,n] Pochhammer[b,n]/Pochhammer[c,n] z^n/n!,
        {n,0,Infinity}
    ]
```

Check these expressions by comparing with **Series** and with a few examples. Note that **Pochhammer[z,n]** is a built-in function defined in terms of the built-in function **Gamma[z]** by **Pochhammer[z,n] == Gamma[z+n]/Gamma[z]**.

(b) The gamma function arises in a wide variety of applications in theoretical physics. Although it is a rich function with many properties (see Abramowitz and Stegun [1], Chapter 6), only a few are generally needed. For example, it satisfies the recurrence relation **Gamma[z+1] == z Gamma[z]**, so that for nonnegative integers **Gamma[n+1] == n!**. The gamma function is analytic for all complex **z** except for simple poles at **z = -n** for **n = 0, 1, 2, ...** That is, **1/Gamma[-n] = 0**. Verify these properties with some examples. Note the special values of **Gamma[1/2]** and **Gamma[3/2]**, which are related for example to integrals of gaussian functions. Finally, express **1F1** and **2F1** in terms of gamma functions.

It is also useful to note that the built-in **Factorial** function will evaluate non-integer arguments, in which case **n! == Gamma[n+1]**. Verify this relation with some examples.

(c) The reason for introducing the Pochhammer symbols is that they are defined and finite when **z** becomes a negative integer or zero, even though it might appear that **Gamma[z+n]/Gamma[z]** becomes undefined or infinite. Prove this by showing that

```
Pochhammer[z,n] == Product[z+q,{q,0,n-1}] ==
    z(z+1)(z+2)...(z+n-1)
```

It follows that **Pochhammer[-n,n] == n! (-1)^n**, which you should verify with a few examples. Also, try evaluating **Gamma[a+n]/Gamma[a]** as replacements with

`/. a -> -n /. n -> m` and with `/.{a -> -m, n -> m}` for various (positive) integers **m**.

The values `1/Gamma[-n] = 0` for (positive) integer **n** means that **Pochhammer[** `-n,m]` vanishes for **m > n**, which you should verify. Hence, we see that `1F1[-n,b,z]` and `2F1[-n,b,c,z]` truncate to polynomials of order **n**, as required.

Problem 6.5-5

(a) Show in the limit **b -> Infinity** that `2F1[a,b,c,z/b] -> 1F1[a,c,z]`. Refer to Problem 6.5-4 and examine the limit numerically with plots.

(b) There exist a number of useful formulas for evaluation, viz. *analytic continuation*, of hypergeometric functions outside their circle of convergence **Abs[z] = 1**. Two such relations are the following:

```
2F1[a,b,c,z] ==
    Gamma[c] Gamma[b-a]/(Gamma[b] Gamma[c-a]) *
        (-z)^-a 2F1[a,1-c+a,1-b+a,1/z] +
    Gamma[c] Gamma[a-b]/(Gamma[a] Gamma[c-b]) *
        (-z)^-b 2F1[b,1-c+b,1-a+b,1/z]

2F1[a,b,c,z] ==
    Gamma[c] Gamma[c-a-b]/(Gamma[c-a] Gamma[c-b]) *
        2F1[a,b,a+b-c+1,1-z] +
    Gamma[c] Gamma[a+b-c]/(Gamma[a] Gamma[b]) *
        (1-z)^(c-a-b) 2F1[c-a,c-b,c-a-b+1,1-z]
```

The first formula connects **z** and **1/z** and is useful for example in applying boundary conditions as **z -> ±Infinity**. The second formula connects **z** and **1 - z** and is useful when **z -> ±1**. Verify these formulas with some examples. Refer to Landau and Lifshitz [41], Mathematical Appendices, and to Abramowitz and Stegun [1], Section 15.3.

Problem 6.5-6

Given an ellipse described by `(x/a)^2 + (y/b)^2 == 1`, its circumference can be shown to be

```
Pi(a+b) Hypergeometric2F1[-1/2,-1/2,1,((a-b)/(a+b))^2]
```

Compute the distance the Earth travels around the Sun given **a = 1 AU** and **b = Sqrt[** `1-e^2] a`, where **e = 0.0167** is the *eccentricity* of the Earth's elliptical orbit. Here an **AU** is an astronomical unit, which equals approximately 93,000,000 miles. (Once around the world is about 25,000 miles.) Compare this with the length of the orbit of Halley's comet for which **a = 18 AU** and **e = 0.967**.

Normalized HO Wavefunctions   **6.6**

Returning to the harmonic oscillator and collecting results, we enter the energy eigenfunctions as functions of the scaled coordinate **z** as

```
psiHOz[n_,z_] := psiHOz[n,z] =
    znorm[n] E^(-z^2/2) HermiteH[n,z]
```

where **znorm[n]** is the normalization constant, which we presently evaluate. Here we have introduced the *dynamic-programming* construct **f[x_] := f[x] =** *expr* to save **psiHOz[n,z]** as it's calculated (cf. Exercise 3.1, Appendix III). This allows us simply to retrieve results from computer memory when needed rather than have to recompute them.

By convention, the phase is chosen such that the eigenfunctions are real. We thus normalize them by requiring that

```
psiHOz[n,z]^2

               2           2
  HermiteH[n, z]   znorm[n]
  --------------------------
               2
              z
             E
```

integrated over all **z** is unity. This ensures that the probability of finding the particle somewhere is unity. However, rather than integrate this expression explicitly, we can simply introduce the normalization integral (6.4-r3) as a replacement rule to obtain the appropriate equation for the normalization constant. That is, the explicit request **Integrate[** *integrand*, **{z,-Infinity,Infinity}]** isn't necessary since we already know the answer. We just need a rule of the form **/.** *integrand* **->** *integral*. (See also Exercise 2.7, Appendix III.) We thus obtain

```
% == 1 /. f_. E^(-z^2) HermiteH[n,z]^2 -> f 2^n n! Sqrt[Pi]

   n                      2
  2  Sqrt[Pi] n! znorm[n]   == 1
```

as if we had integrated explicitly. This result determines the normalization constant for arbitrary **n**:

```
znorm[n_] = znorm[n] /. Solve[ %, znorm[n] ][[1]]

            1
  -----------------------
   n/2   1/4
  2    Pi    Sqrt[n!]
```

Recalling that **z = Sqrt[m w/h] x**, we also obtain the normalized eigenfunctions as a function of the particle's actual displacement **x** according to

```
Needs["Quantum`PowerTools`"]

psiHO[n_,x_] := psiHO[n,x] =
    (m w/h)^(1/4) psiHOz[n,z] /.z -> Sqrt[m w/h] x //
        PowerContract
```

(The function **PowerContract** was introduced in Section 4.1 as an inverse to the built-in function **PowerExpand**.) For example, we easily calculate the normalized ground- and first-excited-state wavefunctions, as functions of either the scaled coordinate **z** or of the actual displacement **x**:

```
{psiHOz[0,z],psiHOz[1,z],psiHO[0,x],psiHO[1,x]}
```

$$\left\{\frac{1}{E^{z^2/2}\,Pi^{1/4}}, \frac{Sqrt[2]\,z}{E^{z^2/2}\,Pi^{1/4}}, \frac{\left(\frac{m\,w}{h\,Pi}\right)^{1/4}}{E^{(m\,w\,x^2)/(2\,h)}}, \frac{Sqrt[2]\,\left(\frac{m\,w}{h}\right)^{3/4}\,x}{E^{(m\,w\,x^2)/(2\,h)}\,Pi^{1/4}}\right\}$$

We can also verify directly that these are solutions of the original HO Schrödinger equation. Thus, for **n = 2**

```
hamiltonian[VHO] @ psiHO[2,x] == eHO[2] psiHO[2,x] //
    ExpandAll
```

```
True
```

(The **hamiltonian** operator was defined in Section 1.0.) Since these functions involve gaussians, we can quickly check their normalization using **integGauss**. We find with **PowerExpand** that

```
Needs["Quantum`integGauss`"]
```

```
{integGauss[psiHO[0,x]^2,{x,-Infinity,Infinity}],
 integGauss[psiHO[1,x]^2,{x,-Infinity,Infinity}]} //
    PowerExpand
```

```
{1, 1}
```

as required. Moreover, different eigenfunctions are *orthogonal* in the sense that their *overlap integrals* vanish. Consider, for example,

```
integGauss[psiHO[0,x] psiHO[2,x],{x,-Infinity,Infinity}]
```

```
0
```

a case which doesn't vanish already by parity. Completeness of the HO eigenfunctions follows from the completeness of the Hermite polynomials. We thus refer to the *HO basis set* or simply the *HO basis*.

Exercise 6.6-1

Explain the normalization of **psiHO[n,x]**, in particular, the extra factor **(m w/h)^(1/4)**.

Problem 6.6-1

Use the shooting method to compute the first five eigenfunctions of the harmonic oscillator. For convenience, integrate the scaled HO Schrödinger equation from Section 6.1. For accuracy, use the asymptotic values **psi[-zo] = E^(-zo^2/2)**, and therefore **psi'[-zo] = +zo psi[-zo]**, for **zo** sufficiently large. Refer to Section 2.2 and to Problem 6.2-1.

Problem 6.6-2

Expand the rectangular wave in Figure 2.4-1 in a generalized Fourier series of **psiHOz[n, z]**. Thus, compute several partial sums and use **NIntegrate** to evaluate the expansion coefficients. For convenience, shift the coordinate origin so that the wave has even parity. This wave form is very sharp compared with the HO functions and convergence is slow. For better convergence, take the full width of the rectangular wave to be **2**, which is closer to the full widths of the **psiHOz** (see Figures 6.6-1 and 6.6-2 below). Make plots analogous to Figures 2.4-2 and 2.4-3 to investigate convergence.

The dimensionless functions **psiHOz** are particularly convenient for plotting. In Figure 6.6-1, we show the first six eigenfunctions together in a **GraphicsArray**, as in Figures 2.4-3 and 2.5-1. The wavefunctions are shifted by the appropriate eigenenergy and plotted along with the potential. In a similar fashion, but all on one graph, we plot in Figure 6.6-2 the first four probability densities. The constant level separation **h w**, peculiar to the HO, is evident.

These figures make clear several general features. The quantum number **n** specifies the number of nodes in the wavefunction beginning with the ground state, which has no node. Wavefunctions with **n** even are symmetric about **z = 0** and therefore have even parity. Odd-**n** wavefunctions are antisymmetric and have odd parity, and in general the parity of the *nth* level is **(-1)^n**. This is a consequence of the definite parity of the potential. The wavefunctions penetrate into the classically forbidden region outside the potential well. That is, the wavefunctions extend beyond the classical turning points. Quantum mechanically, this means there is a finite probability of finding the particle beyond the walls of the well. We have seen similar effects in the case of rectangular potential wells, although wavefunctions of the infinite box vanish outside the box (see Problems 5.0-1 and 6.2-1).

At the classical turning points the kinetic energy vanishes, and beyond the turning points in the classically forbidden region it becomes negative. This happens when the second derivative in the wavefunction vanishes and therefore at the two *inflection points* in the wavefunction which coincide with the turning points. These are the points where the curvature of the wavefunction goes from up to down. As **n** increases, the probability density is maximum just inside the classical turning points in correspondence with the classical probability: the likelihood of finding the particle is highest in regions through which it moves most slowly. This effect is illustrated in Problem 6.6-4 and Figure 6.6-3.

Exercise 6.6-2

Generate Figure 6.6-2.

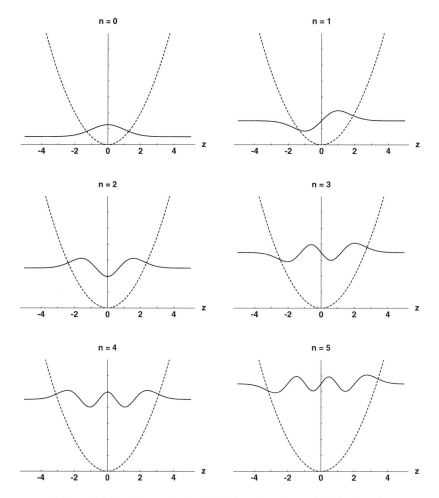

Figure 6.6-1. Dimensionless HO eigenfunctions `phiHOz[n,z]`.

Exercise 6.6-3

Because HO eigenfunctions are useful in a variety of applications, it is often convenient to retain an arbitrary scale parameter `x -> a x` and work with normalized functions `psiHO[n,a,x]`. Verify that a correct definition is

```
psiHO[n_,a_,x_] := Sqrt[a] psiHOz[n,z] /.z -> a x
```

Hence, check that `psiHO[n,a,x]` is still an eigenfunction of `hamiltonian[k x^2/2]` with eigenvalue `h w (n + 1/2)` if the spring constant `k` and therefore the frequency `w` relate to `a` according to

$$\{k \to \frac{a^4\,h^2}{m}, \quad w \to \frac{a^2\,h}{m}\}$$

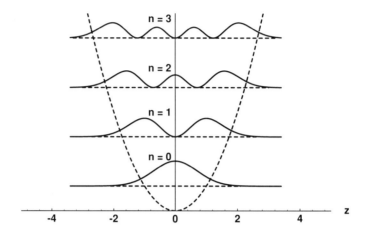

Figure 6.6-2. HO energy levels and probability densities.

Problem 6.6-3

Introduce an HO approximation to estimate the ground-state energy of a particle of mass **m** moving in the model *6-12* potential

$$\text{Vo} \left(\frac{-2 \ xo^6}{x^6} + \frac{xo^{12}}{x^{12}} \right)$$

Use the results of Exercise 6.6-3 to write down an approximate wavefunction. Make a plot of **V** along with the appropriate HO potential. Show that **h/2 Sqrt[72Vo/ (m xo^2)]** approximates the ground-state energy. Refer to Section 7.4.

Problem 6.6-4

Obtain an expression for the *classical* HO probability distribution by calculating the fraction of a *half* period a classical oscillator spends within a small interval **dx** between **-A** $\le$ **x** $\le$ **A**, where **A** is the amplitude. Compare a plot of your result with one of the quantum distribution for **n = 15** and generate Figure 6.6-3. For convenience, take **Sqrt[m w/h] = 1** to determine the magnitude of **A** from the energy.

Problem 6.6-5

Determine the probability of finding the particle in the classically forbidden regions for the first few HO eigenfunctions. Use **NIntegrate** and take **Sqrt[m w/h] = 1**, for convenience.

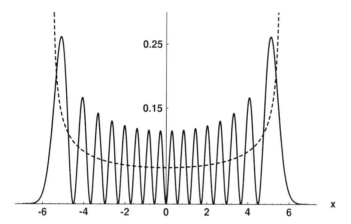

Figure 6.6-3. HO probability distribution for **n = 15**. Dashed curve is the classical distribution.

Raising and Lowering Operators 6.7

Finally, we mention for later reference that the HO eigenvalue problem can be solved in an elegant and purely algebraic fashion by introducing the so-called *raising* and *lowering operators* defined by the position **x** and the momentum (operator) **p** according to

```
p @ psi_ := -I h D[psi,x]
aR @ psi_ := Sqrt[m w/(2h)] (x psi - I p @ psi/(m w))
aL @ psi_ := Sqrt[m w/(2h)] (x psi + I p @ psi/(m w))
```

We shall examine this solution in detail later on in Section 15.4 and so just take a quick look here at a few of the properties of these operators.

For example, we can apply the raising operator **aR** to the ground-state to generate the first-excited state:

```
aR @ psiHO[0,x] == psiHO[1,x] //
   PowerExpand //Map[Together,#]&
```

```
True
```

And we can go back to the ground state with the lowering operator:

```
aL @ psiHO[1,x] == psiHO[0,x] //
   PowerExpand //Map[Together,#]&
```

```
True
```

By applying these operators repeatedly to **psiHO[n,x]**, we can generate in principle any other eigenstate. This process, like climbing up and down the rungs of a ladder, is referred to as the *ladder method*. The ground state is the bottom rung and operating with **aL** gives zero:

```
aL @ psiHO[0,x]
```

  0

This property can also be turned around and used as a defining equation for the ground state of the HO. In effect, it reduces the solution of the Schrödinger equation to a first-order differential equation, since **aL** is a first-order differential operator. We can thus think of the HO hamiltonian operator as being factored by **aR** and **aL** in the sense that

```
hamiltonian[VHO] @ psi[x] ==
    h w aL @ aR @ psi[x] - h w/2 psi[x] ==
    h w aR @ aL @ psi[x] + h w/2 psi[x] //
        ExpandAll //PowerExpand //PowerContract
```

  True

The raising and lowering operators are not hermitian and therefore do not have real eigenvalues. We shall see later on (Section 8.4), however, that the lowering operator **aL** has normalizable eigenstates with complex eigenvalues, known as *coherent states*. On the other hand, the raising operator has no normalizable eigenstates.

Because of their role in quantum electrodynamics and the theory of the emission and absorption of photons, the raising and lowering operators are also called the *creation* (emission) and *annihilation* (absorption) operators, respectively.

Problem 6.7-1

(a) Determine for the HO ground-state wavefunction by solving **aL @ psi[x]== 0** as a first-order differential equation. Try using the built-in function **DSolve**.

(b) Generate the first few excited states of the oscillator by applying the raising operator repeatedly to your normalized result from Part (a). Deduce the **n**-dependent constant which must be introduced if the *nth* state is to be normalized.

Problem 6.7-2

Use the result of the previous problem to derive the Rodrigues relation introduced in Exercise 6.4-2 for Hermite polynomials.

7
Variational Method and Perturbation Ideas

Suppose we were not familiar with the analysis in the previous chapter or we simply wanted to guess a ground-state wavefunction for the harmonic oscillator. Given a reasonable "trial" function defined by one or a few adjustable parameters, how might we go about determining the best set of parameters? The *variational principle* gives us a clear-cut answer to this question if we are just looking for a trial energy closest to the true ground-state energy. This principle directs us to compute simply the expectation value of the hamiltonian with the trial function and minimize it with respect to all the adjustable parameters.

Showing that the resulting trial energy provides a least upper bound on the true ground-state energy is straightforward (see Park [50], Section 7.8, and Merzbacher [45], Section 4.6). And given a reasonable guess at the true eigenfunction, we are ensured of an equally reasonable estimate of the ground-state energy. In fact, if the trial function differs from the eigenfunction by at most a small amount δ, then the error in the expectation value will be only of order δ^2. Also, the difference in the trial function and the eigenfunction has to be small only in the coordinate range important to the energy integral.

Finally, if the trial function is orthogonal to the true ground-state wavefunction, by symmetry for example, then the trial energy will give a least upper bound on the first excited state.

7.1 HO Ground State Variationally

Let's try out the method with a simple example. If we make a slightly educated guess and introduce just the "bare bones" asymptotic analysis from Section 6.2, we might take for a trial ground-state HO wavefunction a normalized gaussian with one adjustable parameter. Thus, consider

```
psi[0,a_,x_] = Sqrt[a] Pi^(-1/4) E^(-a^2 x^2/2)
```

$$\frac{\text{Sqrt}[a]}{E^{(a^2 x^2)/2} Pi^{1/4}}$$

where the parameter **a** is related to the reciprocal of the function's width. Consider first the expectation values of position and momentum squared, **x^2** and **p^2**. Calling again on the package **Quantum`integGauss`** to integrate the gaussians, we find

```
Needs["Quantum`integGauss`"]

xsq[0] =
    integGauss[x^2 psi[0,a,x]^2,{x,-Infinity,Infinity}] //
        PowerExpand
```

$$\frac{1}{2 a^2}$$

and

```
p @ psi_ := -I h D[psi,x]

psq[0] =
    integGauss[
        psi[0,a,x] p @ p @ psi[0,a,x],
        {x,-Infinity,Infinity}
    ] //PowerExpand
```

$$\frac{a^2 h^2}{2}$$

With these quantities, we can form the expectation value of the hamiltonian simply as kinetic plus potential energy:

```
eTrial[0] = psq[0]/(2m) + m w^2/2 xsq[0]
```

$$\frac{a^2 h^2}{4 m} + \frac{m w^2}{4 a^2}$$

Minimizing, we calculate the optimum value for **a** and use it to evaluate the trial energy.

```
Needs["Quantum`PowerTools`"]

Solve[ D[eTrial[0],a] == 0, a][[1,1]] //PowerContract
             m w
   a -> Sqrt[---]
              h


eTrial[0] /.%
   h w
   ---
    2
```

Here the subscript **[[1,1]]** picks out the desired subelement from the solution list and effectively removes the list braces **{}** (see Exercise 1.3, Appendix III). Not surprisingly, we have obtained the exact ground-state wavefunction and energy (see Exercise 6.6-3). This will always happen if we guess the true form of the wavefunction for our trial function. The variational method is of greatest utility, however, when we do not know the true wavefunction, the situation usually encountered.

At the same time, we should stress that a good variational energy doesn't necessarily mean that the trial wavefunction is everywhere close to the true wavefunction. Large discrepancies in the two functions can occur over a coordinate range not particularly important to the energy integral. This means we risk obtaining rather approximate results during certain computations, such as expectation values of physical quantities which emphasize the coordinate ranges poorly represented. Since we generally do not know the true wavefunction, we are often forced to rely on physical intuition to reduce the risks.

Problem 7.1-1

Estimate the HO ground-state energy using the lorentzian function **nrm/(a^2 + x^2)** as the trial wavefunction, where **nrm** is a normalization constant. Evaluate the necessary integrals with the system function **Integrate**.

HO Excited State Variationally   7.2

Let us continue for a moment with gaussian trial functions and estimate the first-excited HO state variationally. In order to obtain this energy, our trial function has to be orthogonal to the ground state. From general principles, we require a function with odd parity and one node. The odd-parity requirement automatically ensures orthogonality, since **psi[0,a,x]** has even parity.

The simplest gaussian function which satisfies these requirements is one multiplied by **x**. After normalizing, we thus take (see again Exercise 6.6-3)

```
psi[1,a_,x_] = Sqrt[2a] Pi^(-1/4) a x E^(-a^2 x^2/2)
```

```
                3/2
     Sqrt[2] a     x
     ─────────────────
        2  2
      (a  x )/2   1/4
     E           Pi
```

Calculating the trial energy now in one step

```
eTrial[1] =
    integGauss[
         psi[1,a,x] (p @ p @ psi[1,a,x]/(2m) +
              m (w x)^2/2 psi[1,a,x]),
         {x,-Infinity,Infinity}
    ] //PowerExpand
```

```
     2  2            2
   3 a  h        3 m w
   ────────  +  ───────
     4 m             2
                  4 a
```

we minimize as before:

```
Solve[ D[eTrial[1],a] == 0, a][[1,1]] //PowerContract
```

```
                m w
   a -> Sqrt[───]
                 h
```

```
eTrial[1] /.%
```

```
   3 h w
   ──────
     2
```

As now expected, this is the exact energy of the first excited state.

Problem 7.2-1

Estimate the HO first excited-state energy by modifying the lorentzian trial function **nrm/(a^2 + x^2)** from Problem 7.1-1. Calculate the uncertainty product for this state and the ground state from Problem 7.1-1. Hint: Some care is required here. You not only need to make sure that this state is orthogonal to the ground state, but also that all the required integrals are convergent.

```
Plot[
    {VHO, V},
    {z,-10,10},
    PlotRange -> {-2,2}, AxesLabel -> {" z","V[z]"},
    PlotStyle ->
        {{Dashing[{.01,.01}]},{GrayLevel[0.0]}}
];
```

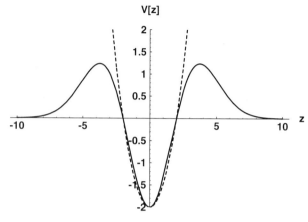

Figure 7.3-1. Model potential well with finite barriers. Dashed curve is an HO potential.

Model Hamiltonian   **7.3**

As an introduction to a useful model, let's estimate variationally the ground state of the hamiltonian defined by

```
hamiltonian[V_] @ psi_ := -1/2 D[psi,{z,2}] + V psi

V   = VHO E^(-b z^2);    b  =  0.1;
VHO = Vo + z^2/2;        Vo = -2.0;
```

where **VHO** is a harmonic oscillator potential with its minimum lowered by **Vo**. For convenience, we retain the scaled coordinate **z** and express energies in units of **h w**. (In effect, we set **h = m = w = 1**.) These potentials are plotted together in Figure 7.3-1.

The finite barriers at the top of the well, which can be adjusted by changing the well minimum **Vo** and the barrier-width parameter **b**, make this model potential relevant to many physical systems. Classically, for energies below the barriers, the potential supports bound orbits, although quantum mechanically only those orbits with **Energy < 0** correspond to true bound states. Orbits below the barriers but with **Energy $\geq$ 0** tunnel out of the well quantum mechanically and move without bound. Hence, the negative-energy spectrum is discrete while the positive-energy spectrum is continuous. Also, the number of bound states is finite, unlike the HO (see Problem 7.3-1 below).

Clearly, the model potential is very close to the HO potential for **Energy < 0**, so that the bound states of the model hamiltonian must be close to those of the HO. In particular, the ground-state energy must be close to **Vo + 1/2 = -3/2** in units of **h w**. These ideas form the basis of perturbation theory, which we will examine briefly in the next section. Since the model potential **V** is essentially gaussian, expectation values of the model hamiltonian with gaussian trial functions are still easy to calculate with **integGauss**.

With the trial wavefunction **psi[0,a,z]** from our previous example but now as a function of **z** (see also Exercise 6.6-3), we can calculate the trial energy in one step according to

```
eTrial[0] =
    integGauss[
        psi[0,a,z] hamiltonian[V] @ psi[0,a,z],
        {z,-Infinity,Infinity}
    ] //PowerExpand
```

$$\frac{a^2}{4} + \frac{a}{4\ (0.1 + a^2)^{3/2}} - \frac{2.\ a}{\mathrm{Sqrt}[0.1 + a^2]}$$

Here **eTrial[0]** and therefore **D[eTrial[0],a]** are complicated enough that it is appropriate to search for a minimum as a function of **a** numerically. A good first step in that direction is to plot **eTrial[0]** as a function of **a**, as in Figure 7.3-2.

```
Plot[eTrial[0],{a,0,5}, AxesLabel -> {" a","eTrial[0]"}];
```

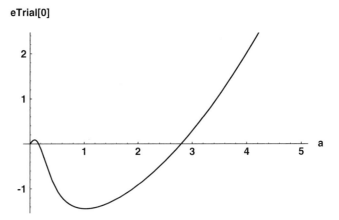

Figure 7.3-2. Dependence of the ground-state trial energy of the model hamiltonian on the variational parameter **a**.

There is clearly a minimum around **a = 1**, which we can locate precisely with the built-in function **FindMinimum** according to

```
FindMinimum[ eTrial[0], {a,1} ]

   {-1.44084, {a -> 1.02603}}
```

Here the first list element gives the minimum trial energy (in units of **h w**) which is close to the value **Vo + 1/2 = -3/2** we expected.

Exercise 7.3-1

How do we know that **eTrial[0]** in Figure 7.3-2 doesn't turn over for large **a** and dip into another lower minimum?

Exercise 7.3-2

Determine the minimum trial energy by locating the *appropriate* root of **D[eTrial[0], a] == 0** using the built-in function **FindRoot**. (See **?FindRoot**.) Try also using the built-in function **NSolve**.

Problem 7.3-1

Estimate the energy of the first excited state of the model hamiltonian using the gaussian trial function **psi[1,a,z]** from the HO example in the text (and Exercise 6.6-3). Show that the model potential probably supports no other bound states by constructing a gaussian trial function **psi[2,a,z]** which is orthogonal to both **psi[0,a,z]** and **psi[1,a,z]** and showing that it gives a *positive* variational energy for all values of **a**. Compare with Exercise 7.4-2 below and see Chapter 10.

Problem 7.3-2

(a) Using the gaussian trial functions **psi[n,a,z]** from the text, estimate the ground-state and first excited-state energies of an *anharmonic oscillator* with potential function **V[x] = b x^4**. Adjust the width in each case and show that (see Park [50], Sec. 7.8)

```
eTrial[0] = 1.082 (h^2 Sqrt[b]/(2m))^(2/3)
eTrial[1] = 3.847 (h^2 Sqrt[b]/(2m))^(2/3)
```

(b) Use the shooting method to show that the exact values of the coefficients are **1.060** and **3.800**. Hence, derive and work with a scaled Schrödinger equation independent of parameters to obtain results for arbitrary **b**.

(c) An anharmonic potential could arise as a perturbation in the following way. Imagine three particles in a line in equilibrium with forces which depend only on the distance between them. The particle in the middle is given a small, sudden displacement **x** perpendicular to the line. Show that the potential energy generated by the displacement is proportional to **x^4**. (Compare with Marion and Thornton [44], Example 3.5.)

7.4 First-Order Perturbation Energy

Finally, we can use the variational principle to connect with quantum perturbation theory. Suppose we have difficulty solving the Schrödinger equation for a potential **V** but can solve it for a similar potential **Vo** which differs from **V** by a small amount, **V -> Vo + (V – Vo)**. One refers to the difference **VP = V – Vo** as a *perturbation* of the *unperturbed* potential **Vo**. Thus, we assume we can solve **hamiltonian[Vo]@psio[n,x]==eo[n]psio[n,x]** for the unperturbed or *zeroth-order* eigenenergies and eigenfunctions **eo[n]** and **psio[n,x]**.

If we then introduce the zeroth-order functions **psio[n,x]** as trial functions and let **V -> Vo + VP**, the variational theorem gives us an upper-bound on the eigenenergies **e[n]** of the perturbed hamiltonian **hamiltonian[V] == hamiltonian[Vo] + VP**. Evidently, **e[n]** $\leq$ **eo[n] + <VP>**, where the expectation value is calculated with **psio[n,x]**. This estimate agrees with the most basic prediction of perturbation theory and finds wide application. The difference Δ**e = <VP>** is referred to as the *first-order correction* to the unperturbed energies **eo[n]**. (Refer to Park [50], Chapter 7, and to Merzbacher [45], Section 4.6.)

As an example, let's take another look at our model potential, shown in Figure 7.3-1. Clearly, the HO potential **VHO** defines the appropriate unperturbed hamiltonian with unperturbed eigenfunctions **psiHOz[n,z]** and eigenenergies **-3/2 + n** (in units of **h w**). Relative to **VHO**, the model hamiltonian thus introduces the perturbation

```
VP = V - VHO;
```

In Exercise 7.4-1 below, we show that **VP** vanishes as the barrier-width parameter **b -> 0**, and therefore we can also refer to **b** as a *perturbation parameter*. Recall that we set **b = 0.1**. We find that **VP** has the expectation value

```
VPExp[0] =
    integGauss[
        VP psiHOz[0,z]^2,
        {z,-Infinity,Infinity}
    ] //PowerExpand //N
```

```
0.0597709
```

in the unperturbed ground state. We obtain the first-order perturbation estimate of the ground-state energy by adding this expectation value to the unperturbed ground-state energy:

```
-3/2 + %
```

```
-1.44023
```

As expected, this estimate is very close to but somewhat above our previous variational estimate **FindMinimum[eTrial[0]]**.

We point out in passing that, although **VP** diverges as **-VHO** for large **z** and any fixed value of **b**, we only require **VP** to be small over the range of the wavefunction whose energy is being estimated, since this is what contributes to the expectation value **VPExp[0]**.

We shall return to the model hamiltonian in Chapters 10, 12 and 14 and examine its eigenvalue spectrum and scattering properties in some detail.

Exercise 7.4-1

Make a plot of **VP** along with **VHO**. Use **Series** to show that **VP -> 2b z^2 (1 - z^2/4)** in the limit **b -> 0**. Compute the first-order correction to the ground-state energy with this expression setting **b = 0.1** and compare it with the expectation value **VPExp[0]** calculated in the text.

Exercise 7.4-2

Estimate the excited bound-state energy spectrum of the model hamiltonian using first-order perturbation theory. Compare with Problem 7.3-1 above.

Problem 7.4-1

Consider an harmonic oscillator of mass **m** and spring constant **k** in its ground state. If a second spring with spring constant **kp** is attached alongside the original one, by how much does the ground-state energy increase?

(a) Assuming **kp << k**, estimate the energy increase using first-order perturbation theory.

(b) Determine the exact energy increase by changing **k -> k + kp** in the HO ground-state energy. Expand your result in powers of **kp** and check your estimate in Part **(a)**.

8
Squeezed States

As an illustration of the completeness of the HO basis set, we will now examine the motion of a wavepacket dropped into the harmonic oscillator potential well. The procedure is to expand the wavepacket in HO eigenstates as a generalized Fourier series, analogous to the expansion of a free-particle wavepacket in planewaves we constructed in Chapter 4. Since each eigenfunction is a solution of the HO Schrödinger equation, the superposition describes a particle undergoing simple harmonic motion. If we assume the wavepacket to be initially a gaussian, we obtain a remarkably simple result for the time-dependent probability density, namely, another gaussian. We can initially displace the peak of the wavepacket from the origin and give it a velocity in analogy with the excitation of a classical oscillator. We can also start the wavepacket with a width different from that of the HO ground-state wavefunction.

Interest in these states goes back to the very early days of quantum mechanics and a paper by Schrödinger [60] in 1926 who sought to construct wavefunctions which follow the classical motion of a particle as closely as possible. In recent times, these wavepackets have proven to be of great interest in the theory of measurement of weak signals, especially in quantum optics but even in gravitational-wave detection. In view of the insight into the Heisenberg uncertainty principle these states provide, it is interesting to note that Schrödinger's paper appeared before Heisenberg published his ideas in 1927.

Although our final results are simple, their derivation is protracted, and it is helpful to suppress constants by an appropriate scaling similar to what we did in defining the dimensionless eigenfunctions `psiHO[n,z]`. We shall simply set Planck's constant `h = 1` but retain the mass `m` explicitly to facilitate comparison with the classical motion. In these units, the wavenumber

k equals the momentum **p** so that **k/m** is a velocity. We thus define HO eigenfunctions with **h = 1** by entering

```
psiHOw[n_,x_] = psiHO[n,x] /.h -> 1
```

$$\frac{\left(\dfrac{m\,w}{Pi}\right)^{1/4}\ \text{HermiteH}[n,\ \text{Sqrt}[m\,w]\,x]}{2^{n/2}\ E^{(m\,w\,x^2)/2}\ \text{Sqrt}[n!]}$$

As in Section 4.2, we again set up an initial gaussian wavepacket **psi[x,0]** with momentum **kPo** but we displace it from the origin to **x = xPo**. It's convenient to do this simply by displacing the HO ground-state wavefunction **psiHOw[0,x]** and giving it a momentum boost. At the same time, we can also choose a width different from that of the ground state by replacing the oscillator's fundamental frequency **w** by an effective frequency **wPo**. We therefore take

```
psi[x_,0] = E^(I kPo x) psiHOw[0,x-xPo] /.w -> wPo
```

$$E^{I\ kPo\ x\ -\ (m\ wPo\ (x\ -\ xPo)^2)/2}\ \left(\frac{m\ wPo}{Pi}\right)^{1/4}$$

If we take the parameter **wPo** to be greater than **w**, the initial wavepacket will be narrower than the oscillator's ground-state wavefunction, and such wavepackets are commonly referred to as *squeezed states*. Physically, the wavepacket can be thought of as belonging to a potential well which is narrower than the oscillator potential **m w^2 x^2/2**. Likewise, if **wPo** is less than **w**, the initial wavepacket will be wider than the ground-state wavefunction. Although it is really anti-squeezed, it is also called a squeezed state. As we shall see shortly, the state **wPo = w** is very special in that the width and therefore the shape of the wavepacket remains constant in time. Schrödinger was most interested in these wavepackets, commonly known as *coherent states*.

Since squeezed states are initially gaussians, they are initially minimum-uncertainty wavepackets. This essentially follows from our analysis in Chapter 4 of a gaussian free-particle wavepacket. Coherent states have the additional distinction that they remain minimum-uncertainty wavepackets in time, since their width does not change.

Squeezed states and many of their properties were elucidated by Saxon [58] in his introductory quantum mechanics book long before the term squeezed states became fashionable. Squeezed states and their connection with quantum optics are discussed in the book by Cohen-Tannoudji, Dupont-Roc and Grynberg [15]. Their potential usefulness for gravitational-wave detection was pointed out by Caves [13] (see also Caves et al. [14]). Recent applications to atomic physics are discussed in the book by Friedrich [24].

Exercise 8.0-1

Solve the classical equations of motion for a particle of mass **m** moving in the HO potential m w^2 x^2/2 subject to the initial conditions

```
{x[0] == xPo, x'[0] == kPo/m}
```

Try the built-in function **DSolve** and show that the particle's displacement as a function of time is given by

```
x[t_] := xPo Cos[t w] + kPo Sin[t w]/(m w)
```

and, hence, that the classical velocity is given by

```
v[t_] := kPo/m Cos[t w] - w xPo Sin[t w]
```

Finally, show that energy is conserved so that m v[t]^2/2 + m w^2 x[t]^2/2 ==

$$\frac{kPo^2}{2\,m} + \frac{m\,w^2\,xPo^2}{2}$$

We shall relate these quantities to the motion of the centroid of the squeezed states in Section 8.3.

Exercise 8.0-2

Calculate the energy expectation value in the initial state **psi[x,0]** and show that

$$\frac{kPo^2}{2\,m} + \frac{w}{4\,wPo} + \frac{wPo}{4} + \frac{m\,w^2\,xPo^2}{2}$$

This result is of course greater than the HO ground-state energy, as required by the variational principle. (We worked through a **xPo = kPo = 0** version of **psi[x,0]** in Section 7.1 to demonstrate the variational theorem.) Show that the coherent-state limit **wPo -> w** is obtained by minimizing the energy with respect to **wPo**, and that in this limit the energy becomes the quantum ground-state energy of the oscillator plus the classical energy from Exercise 8.0-1. (Keep in mind that **h = 1**.)

Eigenfunction Expansion **8.1**

We proceed to expand the initial wavepacket **psi[x,0]** in the HO eigenstates **psiHOw[n,x]**. (If you have not already done so, now is a good time to acquaint yourself with the algebraic methods of Section 2.0, Appendix III.) It follows from the orthonormality of the basis functions that the expansion coefficients **c[n]** are the usual overlap integrals of the basis functions with the initial wavepacket. That is, **c[n]** is the integral of

```
Needs["Quantum`PowerTools`"]

cintegrand =
    psi[x,0] psiHOw[n,x] /.
        E^q_ :> E^Map[Factor,Collect[q,x]] //
        PowerContract
                       2                    2
    -(m (w + wPo) x )/2 - (m wPo xPo )/2 + x (I kPo + m wPo xPo)
   (E


        2
      m  w wPo 1/4                                      n/2
     (—————————)     HermiteH[n, Sqrt[m w] x]) / (2     Sqrt[n!])
          2
        Pi
```

over all **{x,-Infinity,Infinity}**. Recall that **RuleDelayed :>** is required here, rather than **Rule ->**. With **Rule**, the system would evaluate **Map[Factor,Collect[q,x]]** before the exponent is substituted for **q** and nothing would happen. Compare with Section 4.2 and refer to Exercise 2.2, Appendix III.

Although we can evaluate these integrals for specific values of **n** with **integGauss** (from the package **Quantum`integGauss`**), we can also derive a closed-form expression valid for arbitrary **n**. For brevity, we shall simply adopt a formula from a table of integrals. As in determining the HO wavefunction normalization in Section 6.6, all we need is a replacement rule of the form **cintegrand /. integrateRule -> *integral***.

For example, consider Eq. 7.374.8 on page 837 of Gradshteyn and Ryzhik [28] which we can enter almost directly as

```
integrateRule =
    E^(c_ + b_(x-y_)^2) HermiteH[n,a_ x] :>
        E^c Sqrt[Pi/(-b)] Sqrt[1+a^2/b]^n *
        HermiteH[n,a y/Sqrt[Together[1+a^2/b]]]

                        2
    (c_) + (b_) (x - (y_))
   E                              HermiteH[n, x (a_)] :>

                        2
        c       Pi     a   n                          a y
       E  Sqrt[——] Sqrt[1 + —] HermiteH[n, ——————————————————————]
               -b           b                                2
                                                            a
                                            Sqrt[Together[1 + —]]
                                                              b
```

although we have included a change of variables to handle an arbitrary exponential width **b_** and the system function **Together** to help simplify our result for **c[n]** below.

However, we still need to complete the square in the exponent of **cintegrand** before we can apply **integrateRule**. To that end, and to help simplify various expressions later on, it is convenient to define a function which completes the square in an expression **expr** with respect to the variable **x** (see Exercise 2.2, Appendix III). Let's thus define

```
completeSquare[expr_,x_] := expr /.
   a_. x^2 + b_. x + c_. :>
   a(x + b/(2a))^2 + Together[b^2/(-4a) + c]
```

Then, for example,

```
E^(b x - a x^2) //completeSquare[#,x]&
    2                        2
   b /(4 a) - a (-b/(2 a) + x)
  E
```

which we easily check by expanding out again:

```
% //ExpandAll
          2
   b x - a x
  E
```

Hence, we transform **cintegrand** into the desired form

```
cintegrand //completeSquare[#,x]&
             2                  2        2
      -(kPo  - 2 I kPo m wPo xPo + m  w wPo xPo )
 (Power[E, ─────────────────────────────────────── -
                    2 m (w + wPo)

                 I kPo + m wPo xPo 2
   m (w + wPo) (x - ────────────────)          2
                     m (w + wPo)            m  w wPo 1/4
  ──────────────────────────────────] (──────────)
                 2                            2
                                           Pi

                       n/2
   HermiteH[n, Sqrt[m w] x]) / (2    Sqrt[n!])
```

to which we can apply **integrateRule** and define the *nth* expansion co-efficient **c[n]** and simplify:

```
c[n_] = % /.integrateRule //
             PowerExpand //PowerContract //
             MapAll[Together,#]& //
             MapAll[Collect[#,n!]&,#]&
                             2
    1/2 - n/2              -kPo        I kPo wPo xPo
  (2          Power[E, ───────────── + ──────────── -
                       2 m (w + wPo)     w + wPo

              2
    m w wPo xPo           1/4   -w + wPo n/2
   ────────────] (w wPo)     (──────────)
    2 (w + wPo)                w + wPo

                               w
   HermiteH[n, Sqrt[-(─────────────────────)]
                      m (w - wPo) (w + wPo)

   (I kPo + m wPo xPo)]) / Sqrt[(w + wPo) n!]
```

As desired, this result is valid for arbitrary **n** and **w** and initial conditions **xPo**, **kPo** and **wPo**.

Exercise 8.1-1

Verify `c[n]` by direct integration with `integGauss` for a few of the lowest values of n. Hint: Be on the lookout that you do not inadvertently transform patterns of the form `Sqrt[(-a)(-b)]` to `I^2 Sqrt[a b]` with `PowerExpand`.

8.2   Time Evolution

We now set up the *nth* term `psi[n,x,t]` in the time-dependent superposition `psi[x,t]` by including the time-development factors `E^(-I eHO[n] t/h)` in the usual way. Problem 8.2-2 summarizes the results we obtain in this section.

Recalling that `eHO[n]=hw(n+1/2)`, we obtain after some simplification

```
c[n_,t_] := c[n] E^(-I w (n+1/2) t)

psi[n_,x_,t_] =
    c[n,t] psiHOw[n,x] /.
        1/(Sqrt[n_] Sqrt[f_ n_]) -> 1/(n Sqrt[f]) /.
        Sqrt[f_/m] :> Sqrt[ExpandAll[f]/m] /.
        E^q_ :> E^Expand[q] //PowerContract
```

$$
(2^{1/2-n} \; \text{Power}[E, \; \frac{-I}{2} t w - I n t w - \frac{kPo^2}{2 m (w + wPo)} - \frac{m w x^2}{2} +
$$

$$
\frac{I \; kPo \; wPo \; xPo}{w + wPo} - \frac{m w wPo xPo^2}{2 (w + wPo)}] \; (\frac{m w \; wPo^2}{Pi})^{1/4} \; (\frac{-w + wPo \; n/2}{w + wPo})
$$

$$
\text{HermiteH}[n, \; \text{Sqrt}[m w] \; x]
$$

$$
\text{HermiteH}[n, \; \text{Sqrt}[-(\frac{w}{m (w^2 - wPo^2)})]] \; (I \; kPo + m \; wPo \; xPo)]) \; /
$$

```
(Sqrt[w + wPo] n!)
```

We require the sum of this quantity over all `{n,0,Infinity}`. In the coherent-state limit `wPo -> w`, the sum can be evaluated in closed form using the generating function for Hermite polynomials (see Problem 8.4-1 below). For squeezed states with `wPo ≠ w`, the sum can also be evaluated in closed form but a generalization of the generating function is necessary. In any case, all that is needed is a replacement rule of the form `psi[n,x,t] /. sumRule -> ` *sum* analogous to `integrateRule`.

Problem 8.2-1

(a) Show in the coherent-state limit `wPo -> w` that `Abs[c[n,t]]^2` becomes the *Poisson distribution*

$$\frac{E^{-kPo^2/(2\,m\,w)\ -\ (m\,w\,xPo^2)/2}\ \left(\frac{kPo^2}{2\,m\,w}\ +\ \frac{m\,w\,xPo^2}{2}\right)^n}{n!}$$

Hint: Use the fact that Hermite polynomials are normalized in the sense that the coefficient of their highest power of **z** is **2^n**. See also the hint in Exercise 8.1-1.

(b) Show that the coherent-state expectation value of the energy is given by the sum

```
Sum[Abs[c[n,t]]^2 w(n+1/2),{n,0,Infinity}]
```

(cf. the discussion in Section 2.5) and that the sum equals

$$\frac{kPo^2}{2\,m}\ +\ \frac{w}{2}\ +\ \frac{m\,w^2\,xPo^2}{2}$$

in agreement with the initial-state energy from Exercise 8.0-2 above. This of course explicitly demonstrates conservation of energy. Hint: Load the package **Algebra`Symbolic Sum`** to evaluate the sum symbolically.

(c) Finally, calculate the uncertainty **deln** in the excitation quantum number **n** and show that it equals

$$\frac{Sqrt\left[\frac{kPo^2}{2\,m}\ +\ \frac{m\,w^2\,xPo^2}{2}\right]}{Sqrt[w]}$$

This result tells us that the number of HO levels excited in the coherent state is large if the classical energy of the state is large compared with **w** (i.e. with **h w** in normal units). Make some plots of the Poisson distribution as a function of **n**.

Consider a formula from Morse and Feshbach [49], p. 786, which we can enter directly as

```
sumRule =
    z^n 2^-n/n! HermiteH[n,x_] HermiteH[n,y_] ->
    E^((-x^2 - y^2 + 2 x y z)/(1-z^2))/Sqrt[1-z^2] E^(x^2 + y^2)
```

$$\frac{z^n\ HermiteH[n,\ x\_]\ HermiteH[n,\ y\_]}{2^n\ n!}\ ->$$

$$\frac{E^{x^2\ +\ y^2\ +\ (-x^2\ -\ y^2\ +\ 2\ x\ y\ z)/(1\ -\ z^2)}}{Sqrt[1\ -\ z^2]}$$

in order to replace the sum of **z^n 2^-n/n! HermiteH[n,x] HermiteH[n,y]** over all **n** by a gaussian in **x**, **y** and **z**. (This expression contains, however, a correction to Morse and Feshbach: **Sqrt[1+z^2]** in their formula has been changed here to **Sqrt[1-z^2]**.)

It is evident, however, (and easily checked) that `psi[n,x,t]` has some minor pattern differences with `sumRule` such that as it stands `sumRule` will not work for us. We need to accommodate the **n**-dependent exponent and the root `(...)^(n/2)` in `psi[n,x,t]`, which we can easily do by replacing **z** by `z -> E^a Sqrt[zp]` in `sumRule`. Allowing for some other extra factors, we thus reenter

```
sumRule =
    (f_ E^c_ Sqrt[2] z^n 2^-n/n! *
      HermiteH[n,x_] HermiteH[n,y_] /.
       z -> E^a_ Sqrt[zp_] ) ->
    (f E^c Sqrt[2] E^((-x^2 - y^2 + 2 x y z)/(1-z^2))/
      Sqrt[1-z^2] E^(x^2 + y^2) /.
       z -> E^a Sqrt[zp] ) //
      PowerExpand
```

$$
\left(2^{1/2-n}\ E^{n(a\_)+(c\_)}\ \text{HermiteH}[n,\ x\_]\right.
$$
$$
\left.\text{HermiteH}[n,\ y\_]\ (f\_)\ (zp\_)^{n/2}\ \right)\ /\ n!\ \to
$$
$$
\left(\text{Sqrt}[2]\ \text{Power}\!\left[E,\ c + x^2 + y^2 + \right.\right.
$$
$$
\left.\left.\frac{-x^2 - y^2 + 2\,E^a\,x\,y\,\text{Sqrt}[zp]}{1 - E^{2a}\,zp}\right]\ f\right)\ /
$$
$$
\text{Sqrt}[1 - E^{2a}\,zp]
$$

in order to perform the sum over `psi[n,x,t]`. We therefore obtain after simplification

```
psi[x_,t_] = psi[n,x,t] /.sumRule /.
                Sqrt[a_] :> PowerExpand[Sqrt[Factor[a]]] /.
                1/Sqrt[a_] :> 1/Sqrt[Together[a]] //
                PowerContract
```

$$
\left(\text{Sqrt}[2]\ \text{Power}\left[E,\ \frac{-I}{2}\,t\,w - \frac{kPo^2}{2\,m\,(w+wPo)} + \frac{m\,w\,x^2}{2} + \frac{I\,kPo\,wPo\,xPo}{w+wPo} - \right.\right.
$$
$$
\frac{m\,w\,wPo\,xPo^2}{2\,(w+wPo)} - \frac{w\,(I\,kPo + m\,wPo\,xPo)^2}{m\,(w^2 - wPo^2)} +
$$
$$
\left.\frac{-I\,t\,w}{(-(m\,w\,x)^2 + \frac{2\,E^{-I\,t\,w}\,w\,x\,(I\,kPo + m\,wPo\,xPo)}{w+wPo} +}\right.
$$

$$\frac{w \ (I \ kPo + m \ wPo \ xPo)}{m \ (w^2 - wPo^2)}) \ / \ (1 - \frac{E^{-2 \ I \ t \ w} \ (-w + wPo)}{w + wPo})]$$

$$(\frac{m \ w^2 \ wPo}{Pi})^{1/4}) \ / \ Sqrt[E^{-2 \ I \ t \ w} \ (w + E^{2 \ I \ t \ w} \ w - wPo +$$

$$E^{2 \ I \ t \ w} \ wPo)]$$

for the wavefunction of the moving wavepacket, where we have taken extra care to avoid splitting up the complex factors in the **Sqrt** in the denominator. We can partially check this result by making sure we get back **psi[x,0]** when we set **t -> 0**:

```
psi[x,0] == psi[x,t] /.t -> 0 /.
        E^q_ :> E^Expand[Together[q]]
```

    True

Note that we have to introduce a replacement here to extract the limit **t -> 0** from **psi[x,t]**. Otherwise, *Mathematica* would use our original definition of **psi[x,0]** (from Section 8.0) on both the right- and left-hand sides of this comparison. (Take a look at **?psi**.)

In Problem 8.2-2, we show that **psi[x,t]** can be considerably simplified to make clear the moving gaussian form of the squeezed state.

Problem 8.2-2

Simplify **psi[x,t]** and show that it can be expressed as

```
psi[x_,t_] :=
    E^(I xphase - (m wP[t] (x-xP[t])^2)/2) (m wP[t]/Pi)^(1/4)
```

where **wP[t]**, **xP[t]** and **xphase** are conveniently defined by the replacement rules

```
wPRule := wP[t] -> w^2 wPo/(w^2 Cos[t w]^2 + wPo^2 Sin[t w]^2)

xPRule := xP[t] -> xPo Cos[t w] + kPo Sin[t w]/(m w)

xphaseRule := xphase ->
    -w t/2 -
    ArcTan[(wPo-w)Cos[t w] Sin[t w]/
    (w Cos[t w]^2 + wPo Sin[t w]^2)]/2 + ((-kPo^2/(2m) +
    (m (-w^2 + wPo^2) x^2)/2 + m wPo^2 xPo^2/2) *
    Cos[t w] Sin[t w] wP[t])/(w wPo) +
    (kPo wPo xPo Sin[t w]^2 wP[t])/w^2 +
    (x (kPo w Cos[t w] - m wPo^2 xPo Sin[t w]) wP[t])/(w wPo)
```

Hint: This problem entails a lot of algebra. Keep your intermediate results as compact as possible. Set up replacements of the form

```
w + E^(2 I t w) w - wPo + E^(2 I t w) wPo ->
   2 E^(I t w) (w Cos[t w] + I wPo Sin[t w])
```

8.3 Newton's Laws

From the results of Problem 8.2-2, we see that the absolute square of the wavefunction has as advertised a remarkably simple form, namely,

```
Needs["Quantum`QuickReIm`"]

psisq[x_,t_] = Conjugate[psi[x,t]] psi[x,t]
```

$$\frac{Sqrt\left[\dfrac{m\ wP[t]}{Pi}\right]}{E^{m\ wP[t]\ (x\ -\ xP[t])^2}}$$

which is of the same form as the initial probability density except that the peak position **xPo** and the width parameter **wPo** have been replaced by the time-dependent quantities **xP[t]** and **wP[t]**, defined in Problem 8.2-2. (Here the package **Quantum`QuickReIm`** loads our symbolic **Conjugate** rule from Exercise 2.4, Appendix III.)

It follows from the similarity of **psisq[x,t]** with the initial wavepacket (and with the free-particle wavepacket in Chapter 4 as well) that **xP[t]** determines the expectation value of the position:

```
xExp[t_] = xP[t] /.xPRule

   xPo Cos[t w] +  kPo Sin[t w]
                   ───────────
                       m w
```

We show in Problem 8.3-1 below that the expectation value of the momentum is given by

```
pExp[t_] := kPo Cos[t w] - m w xPo Sin[t w]
```

(Keep in mind we have set **h = 1**.) The peak of the wavepacket thus follows exactly the trajectory **x[t]** of a classical oscillator subject to the initial conditions **x[0] = xPo** and **x'[0] = kPo/m**. Comparing with the classical solution from Exercise 8.0-1 above, we therefore find that

```
{xExp[t] == x[t], pExp[t] == m v[t]} //ExpandAll

   {True, True}
```

Note that the peak of the wavepacket thus accelerates according to the classical force such that

```
{m D[xExp[t],t] == pExp[t], D[pExp[t],t] == -m w^2 xExp[t]} //
   ExpandAll
```

{True, True}

which again (cf. Section 4.2) are just examples of Ehrenfest's theorem (see Problem 18.2-2).

It also follows from the similarity of **psisq[x,t]** with the initial wave-packet that **wP[t]** determines the uncertainty in **x** according to

```
delx[t_] = 1/Sqrt[2m wP[t]] /.wPRule //
               PowerExpand //PowerContract
```

$$\text{Sqrt}\left[\cfrac{\text{Sqrt}\left[\dfrac{w^2\ \text{Cos}[t\ w]^2\ +\ wPo^2\ \text{Sin}[t\ w]^2}{m\ wPo}\right]}{\text{Sqrt}[2]\ w}\right]$$

Exercise 8.3-1

Verify **xExp[t]** and **delx[t]** directly by integrating **x** and **x^2** with **psisq[x,t]**.

Thus, the width of the wavepacket oscillates in time. Moreover, as we shall show in Problem 8.3-1 below, the momentum uncertainty also oscillates but not in a way reciprocal to the position uncertainty. Hence, squeezed states do not remain minimum-uncertainty wavepackets at all times. Rather, they oscillate in and out of minimum uncertainty at twice the rate at which they shuttle back and forth inside the well. Their uncertainty product is plotted in Figure 8.3-1 as a function of time.

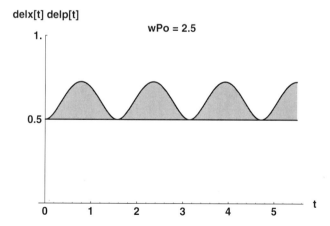

Figure 8.3-1. Squeezed-state uncertainty product **delx[t] delp[t]** as a function of time for **w = 1** and therefore with period of oscillation **Pi**. Plotted using the function **FilledPlot** in the same way we plotted the uncertainty product of the two-component box wavefunction in Figure 3.0-1.

This time development is depicted in Figure 8.3-2 in a **GraphicsArray** by showing time sequences of the probability density **psisq[x,t]** for three different cases: **wPo < w**, **wPo = w**, and **wPo > w**. Each sequence is generated by passing a table of plots of the probability density in equal time steps to the function **StackGraphics**, as we did in generating Figure 4.2-1. Since the motion is periodic with frequency **w** and the squeezing with frequency **2w**, it is only necessary to plot for one-half period **Pi/w**. The sinusoidal trajectories are nevertheless evident.

Exercise 8.3-2

Examine the time development of squeezed states for various sets of initial conditions. For example, try a stationary wavepacket with **xPo = kPo = 0**.

Problem 8.3-1

Starting with **psi[x,t]** from Problem 8.2-2, show that the momentum expectation value and uncertainty are given by

```
pExp[t_] := kPo Cos[t w] - m w xPo Sin[t w]
delp[t_] := Sqrt[m/(2wPo) (wPo^2 Cos[t w]^2 + w^2 Sin[t w]^2)]
```

We shall verify these quantities later on in Section 11.11 (Problem 11.11-1) with considerably less effort when we derive the momentum distribution **phi[k,t]** reciprocal to the wavefunction **psi[x,t]**.

Show that the uncertainty product **delx[t] delp[t]** for a squeezed state can be written as

$$
\text{Sqrt}\left[-\frac{1}{4} + \frac{(w^2 - wPo^2)^2 \; Sin[2 \; t \; w]^2}{16 \; w^2 \; wPo^2}\right]
$$

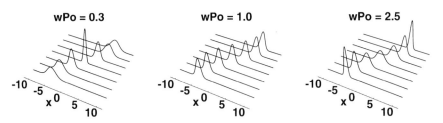

Figure 8.3-2. Time development of squeezed wavepackets for three different initial-width parameters **wPo** but otherwise with the same initial displacements **xPo = -5** and initial momenta **kPo = 0**. Here we've set **m = w = 1** for computational convenience. Time increases from front to back starting with **t = 0** in equal time steps for one-half period **Pi/w**. The middle packet with **wPo = 1** is a coherent state. (**StackGraphics** is a computationally demanding function. This **GraphicsArray** was generated in just over 3.5 minutes on a 33 MHz, 16MB (NeXT Cube) computer.)

which makes clear that coherent states with `wPo = w` are minimum-uncertainty states for all time and that `2w` is the frequency at which squeezed states achieve minimum uncertainty.

Finally, evaluate the energy expectation value in the state `psi[x,t]` and show that it equals the initial-state value from Exercise 8.0-2 above and hence that energy is conserved. Compare with Problem 8.2-1.

Quasi-Classical States 8.4

Figure 8.3-2 makes clear that coherent states oscillate back and forth within the well without change of shape. Thus, the wavepacket width in the limit `wPo -> w`

```
delx[t] /.wPo -> w /.
    Cos[z_]^2 -> 1 - Sin[z]^2 //
    ExpandAll //PowerExpand //PowerContract
          1
    ───────────────
    Sqrt[2] Sqrt[m w]
```

is independent of time and equal to the width of the HO ground-state wavefunction, as advertised. We also see from Problem 8.3-1 that the coherent-state momentum uncertainty

```
delp[t] /.wPo -> w /.
    Cos[z_]^2 -> 1 - Sin[z]^2 //
    ExpandAll //PowerExpand //PowerContract
    Sqrt[m w]
    ─────────
     Sqrt[2]
```

is independent of time. Hence, coherent states remain at all times minimum-uncertainty states, and the coherent-state probability distribution is just the HO ground-state distribution shifted by `xP[t]`:

```
psisq[x,t] //.{wP[t] -> wPo, wPo -> w}
             m w
        Sqrt[───]
             Pi
    ─────────────────────
                         2
      m w (x - xP[t])
    E
```

Coherent states thus display an absence of certain quantum correlations and hence dispersion which leads ordinarily to incoherence and spreading of the wavepacket. Moreover, coherent-state expectation values of important observable quantities coincide with the corresponding classical variables. Coherent states are thus often given the special designation *quasi-classical*.

It's interesting to note in this connection that the portion of the coherent-state phase from Problem 8.2-2 which is linear in **x**

```
xphase /.xphaseRule //.{wP[t] -> wPo, wPo -> w} //
    Select[#,!FreeQ[#,x]&]& //Expand //Factor

x (kPo Cos[t w] - m w xPo Sin[t w])
```

is determined by the classical momentum **m v[t]** from Exercise 8.0-1 above. (Here the built-in function **Select** with the test **!FreeQ[#,x]&** has been applied to extract the desired terms. See Exercise 1.9, Appendix III.) Hence, this portion of the phase is simply

```
E^(I %) == E^(I m v[t] x) //ExpandAll

True
```

a momentum boost.

Finally, we should point out that coherent states are usually defined (see for example Merzbacher [45], Chapter 15) as eigenstates of the lowering operator we introduced in Section 6.7. Here we need the raising and lowering operators defined for **h = 1** and thus enter

```
p @ psi_ := -I D[psi,x]
aL @ psi_ := Sqrt[m w/2] (x psi + I p @ psi/(m w))
aR @ psi_ := Sqrt[m w/2] (x psi - I p @ psi/(m w))
```

As in the solution of the HO eigenvalue problem, these operators can be used to derive algebraically in a compact and direct way many of the properties of coherent and squeezed states we have examined here. (See for example Cohen-Tannoudji, Dupont-Roc and Grynberg [15] and also Friedrich [24].)

Since the lowering operator is not hermitian, its eigenvalues are not real. Operating on the initial state and dividing by the initial state in the coherent-state limit, we thus obtain the complex eigenvalue

```
alpha = aL @ psi[x,0]/psi[x,0] /.wPo -> w //
            Expand //PowerExpand //PowerContract

      I kPo              Sqrt[m w] xPo
 ──────────────────  +  ──────────────
 Sqrt[2] Sqrt[m w]         Sqrt[2]
```

This quantity can take on any value since it determines the classical energy of the oscillator according to

```
w Conjugate[alpha] alpha //Expand

     2          2    2
  kPo        m w  xPo
  ─────  +  ──────────
  2 m          2
```

Non-hermiticity also means that two coherent states belonging to two different eigenvalues are not orthogonal. Rather, the distance between the two eigenvalues in the complex plane determines the degree to which the two coherent eigenstates are approximately orthogonal. See Exercise 8.4-2.

We shall return to the squeezed states of an oscillator in Section 11.11 and examine briefly their momentum representation.

We might note in passing that much effort has gone into constructing minimum-uncertainty wavepackets for other potentials in addition to the harmonic oscillator. For example, recent studies have indicated that it should be possible to prepare a *Kepler wavepacket* representing a localized state of an electron moving along a Kepler ellipse with minimum-uncertainty fluctuations. For a brief review, see Alber and Zoller [3], and refer also to Friedrich [24].

Exercise 8.4-1

Show that the time-dependent coherent state `psi[x,t]/.wPo->w` is also an eigenfunction of the lowering operator `aL`. Show that it is not, however, an eigenfunction of the raising operator `aR`.

Exercise 8.4-2

Consider two coherent initial states `psiC[x,0]` and `psiCp[x,0]` corresponding to two sets of initial conditions `{xPo,kPo}` and `{xPop,kPop}` and therefore two complex eigenvalues `alpha` and `alphap`, respectively. Show that the absolute square of the overlap integral of these two states equals

```
E^(-Conjugate[alpha-alphap] (alpha-alphap))
```

Problem 8.4-1

Derive the coherent-state wavefunction by applying the limit `wPo->w` to the initial state `psi[x,0]` and retracing our derivation of `psi[x,t]` but this time performing the sum over `n` using the generating function for Hermite polynomials from Exercise 6.4-2.

9
Basic Matrix Mechanics

It is possible and often very effective to represent a physical quantity **Q** in terms of its integrals between a complete set of wavefunctions such as the HO basis states:

```
Integrate[psiHO[n,x] Q @ psiHO[k,x],{x,-Infinity,Infinity}]
```
$$(9.0\text{-}r1)$$

for all **n** and **k**. These integrals are conveniently arranged into a square array or *matrix* with rows labelled by **n** and columns by **k**, and each entry in the array is referred to as the *n-kth matrix element* of **Q**. The dynamical description which this representation affords is known as *matrix mechanics* and was introduced by Heisenberg just before Schrödinger's discovery of the wave equation. As we shall see, it is completely equivalent to wave mechanics and when the eigenvalue spectrum is discrete can provide not only a convenient but also a powerful representation.

We enter and work with matrices as lists of lists. Denoting the matrix elements of **Q** as **Qme[n,k]**, we form the matrix as a **Table** over **n** and **k**. **MatrixForm** formats the result for display.

```
nmax = 4;
Table[Qme[n,k],{n,0,nmax},{k,0,nmax}] //MatrixForm
```

Qme[0, 0]	Qme[0, 1]	Qme[0, 2]	Qme[0, 3]	Qme[0, 4]
Qme[1, 0]	Qme[1, 1]	Qme[1, 2]	Qme[1, 3]	Qme[1, 4]
Qme[2, 0]	Qme[2, 1]	Qme[2, 2]	Qme[2, 3]	Qme[2, 4]
Qme[3, 0]	Qme[3, 1]	Qme[3, 2]	Qme[3, 3]	Qme[3, 4]
Qme[4, 0]	Qme[4, 1]	Qme[4, 2]	Qme[4, 3]	Qme[4, 4]

Of course, we enter and display only a *truncated* version of the matrix in the case of the HO basis, since it has infinite dimension. Here we work with **nmax = 4** and thus with five dimensions. The diagonal elements (from left top to bottom right) are just the expectation values of **Q** in the HO basis. Evidently, we can set up a matrix representation in any basis. The HO basis simply affords a good example, which is easily pursued. In general, the appropriate choice of basis states will depend on the system and the physical quantities under consideration.

9.1 HO Coordinate and Momentum Matrix Elements

Consider the matrix elements of the position **x** and the momentum **p** of the oscillator. Although there are several ways to evaluate these integrals, for example from relations among the Hermite polynomials, it is easiest to deduce them algebraically with the raising and lowering operators we introduced in Section 6.7. We shall examine this approach in Section 15.4. However, here for the purpose of demonstrating the matrix representation, we simply introduce the results from Problem 15.4-1:

```
xme[n_,k_] = Sqrt[h/(2m w)] *
        (Sqrt[k+1] KD[n,k+1] + Sqrt[k] KD[n,k-1]);

pme[n_,k_] = I Sqrt[h m w/2] *
        (Sqrt[k+1] KD[n,k+1] - Sqrt[k] KD[n,k-1]);
```

Here **KD[n,k]** is the *Kronecker delta function*, which equals one if **n = k** and zero if **n ≠ k**. It arises here from the orthonormality of the eigenfunctions. We thus enter the rules

```
KD[n_, n_] := 1
KD[n_?NumberQ, k_?NumberQ] := 0
```

The tests with **?NumberQ** attached to the second rule ensure that **KD[n,k]** evaluates only if **n** and **k** are numbers. This will allow us to combine and simplify expressions involving **KD[n,k]** symbolically (see also Exercise 2.6, Appendix III). The delta function gives the matrix elements of the identity matrix, which has ones along the diagonal and zeros off. The identity matrix is also given by the built-in function **IdentityMatrix[d]** for a **d**-dimensional matrix. Thus, for a 5 × 5 matrix, we verify that

```
IdentityMatrix[5] == Table[ KD[n,k],{n,0,4},{k,0,4} ]
```

```
True
```

Exercise 9.1-1

Check our definition of the Kronecker delta function with a few examples.

The ordering (see **?KD**) of the rules in our definition of **KD** is important. Explain why. Hint: More specialized rules generally have to be tested first. Use **Clear[KD]** and try reversing the order of the rules.

HO Coordinate and Momentum Matrices   9.2

Let's set up truncated **x** and **p** matrices. In order to obtain compact results for display, we divide out the scale factors before formatting with **TableForm**. Sticking to five dimensions with **nmax = 4**, we have

```
xMatrix = Table[xme[n,k],{n,0,nmax},{k,0,nmax}];

xMatrix/Sqrt[h/(2m w)] //
   TableForm[#,TableAlignments -> Center]&
```

0	1	0	0	0
1	0	Sqrt[2]	0	0
0	Sqrt[2]	0	Sqrt[3]	0
0	0	Sqrt[3]	0	2
0	0	0	2	0

```
pMatrix = Table[pme[n,k],{n,0,nmax},{k,0,nmax}];

pMatrix/Sqrt[h m w/2] //
   TableForm[#,TableAlignments -> Center]&
```

0	-I	0	0	0
I	0	-I Sqrt[2]	0	0
0	I Sqrt[2]	0	-I Sqrt[3]	0
0	0	I Sqrt[3]	0	-2 I
0	0	0	2 I	0

Exercise 9.2-1

Verify these truncated matrices by integrating **x** and **p** explicitly with **integGauss**. Which matrix elements vanish on the basis of parity alone?

Exercise 9.2-2

Determine the matrix elements of **x** and **p** between the scaled HO wavefunctions **psiHOz[n,z]** and between the functions **psiHO[n,a,x]** from Exercise 6.6-3.

We see that each matrix equals the complex conjugate of its *transpose*, i.e. its complex conjugate with rows and columns interchanged. Such matrices are called *hermitian*. However, since **xMatrix** is a real matrix, it equals simply its transpose and is therefore *symmetric*. Using the built-in function **Transpose** and our symbolic function **Conjugate** from the package **Quantum`QuickReIm`**, we thus readily verify that

```
Needs["Quantum`QuickReIm`"]

{pMatrix == Conjugate[Transpose[pMatrix]],
 xMatrix == Transpose[xMatrix]}
```

    {True, True}

We therefore have, for example, that **pme[n,k] == Conjugate[pme[k,n]]**, which we recognize as the definition of hermiticity of the momentum operator (see Section 1.1 and Problem 1.1-1). In fact, *we require all physical quantities to be represented by hermitian matrices in matrix mechanics.* Their diagonal matrix elements are then the (real) expectation values of the physical quantity being represented in the chosen basis.

9.3 HO Hamiltonian Matrix

We can essentially perform all calculations by familiar matrix algebra. For example, the matrix product **pMatrix.pMatrix** squares the momentum matrix, and we can calculate the hamiltonian matrix directly as

```
pMatrix.pMatrix/(2m) + m w^2/2 xMatrix.xMatrix //
    TableForm[#,TableAlignments -> Center]&
```

$$
\begin{array}{ccccc}
\dfrac{h\,w}{2} & 0 & 0 & 0 & 0 \\[2mm]
0 & \dfrac{3\,h\,w}{2} & 0 & 0 & 0 \\[2mm]
0 & 0 & \dfrac{5\,h\,w}{2} & 0 & 0 \\[2mm]
0 & 0 & 0 & \dfrac{7\,h\,w}{2} & 0 \\[2mm]
0 & 0 & 0 & 0 & 2\,h\,w
\end{array}
$$

which we see is a diagonal matrix with the eigenenergies along the diagonal, except for the last diagonal element. We will show in a moment that this is an error caused by working with truncated matrices.

Exercise 9.3-1

Verify this result by integrating **hamiltonian[VHO]** explicitly with **integGauss**.

We can easily understand why the hamiltonian matrix is diagonal. Since the HO basis states are eigenfunctions of the hamiltonian, matrix elements of the hamiltonian are just the eigenenergies **h w (k + 1/2)** times the overlap integrals of the basis functions with each other. Since the basis functions are orthonormal, the overlap integrals are simply Kronecker delta functions. That is,

```
hHOme[n_,k_] := h w (k + 1/2) KD[n,k]
```

When this equation is expanded out into a 5D matrix, the previous result is obtained, except that the last diagonal element is now the correct eigenenergy **9/2 h w**.

Now let's go back and calculate matrix elements of the hamiltonian exactly by performing matrix multiplication as an infinite summation over all component matrix elements. Recall that the *n-kth* element of the product **A.B** of two matrices is a contraction or scalar product of the *nth* row of **A** with the *kth* column of **B**. Therefore, **hHOme[n,k]** is given by

```
Needs["Quantum`PowerTools`"]

pme[n,l] pme[l,k]/(2m) + m w^2/2 xme[n,l] xme[l,k] //
    Expand //PowerContract
```

$$\frac{h \; \text{Sqrt}[(1 + k) \; l] \; w \; \text{KD}[l, \; 1 + k] \; \text{KD}[n, \; -1 + l]}{2} \; +$$

$$\frac{h \; \text{Sqrt}[k \; (1 + l)] \; w \; \text{KD}[l, \; -1 + k] \; \text{KD}[n, \; 1 + l]}{2}$$

when summed over all **{l,0,Infinity}**. In this case, however, the two sums are trivial to perform since the delta functions **KD[l,±1 + k]** reduce each sum to a single term with **l -> ±1 + k**. Hence, we can perform the sums simply as pattern matches which introduce replacement rules for **l**. Thus, we obtain

```
% /. f_ KD[l,k_] :> (f /.l->k) //
        PowerExpand //Expand //Collect[#,{h,w,KD[n,k]}]&
```

$$h \; (\frac{1}{2} + k) \; w \; \text{KD}[n, \; k]$$

as desired.

Although matrix and wave mechanics lead to the same conclusions, for example equivalent eigenvalue spectrums, the great utility of matrix formalism is the efficiency of machines to compute the eigenvalues and eigenvectors of hermitian matrices. We shall demonstrate this approach in the next chapter.

Exercise 9.3-2

Set up the commutator of **x** and **p** as a matrix relation and verify that [x,p] == I h. Refer to Exercise 1.1-2. Do this first with the truncated matrices **xMatrix** and **pMatrix** and then exactly as sums over the component matrix elements.

Problem 9.3-1

Calculate the uncertainty matrices Δx and Δp and show that the diagonal elements of their product Δx Δp are given by

$$\frac{h \ (1 + 2 \ n)}{2}$$

Do this first with truncated matrices and then exactly as sums over matrix elements. Refer to Chapter 3. Hint: Add the following two rules to the Kronecker delta function from Exercise 2.6, Appendix III:

```
KD[n_,n_ + m_] := 0
KD[n_ + m_,n_] := 0
```

Problem 9.3-2

With the analytic results **xme[n,k]** and **pme[n,k]**, we can obtain closed-form expressions for the matrix elements of many different functions of **x** and **p** in the HO basis. Show generally that the expectation values of the anharmonic-oscillator potential energy **V[x] = b x^4** are given by

$$\frac{3 \ b \ h^2 \ (1 + 2 \ n + 2 \ n^2)}{4 \ m^2 \ w^2}$$

Use the Kronecker-delta rules from the previous problem. Check this result by direct integration for a few values of **n**. Obtain the expectation value of **V[x]** between the functions **psiHO[n,a,x]** from Exercise 6.6-3. Compare with your variational calculations from Problem 7.3-2.

Problem 9.3-3

Set up the **x** and **p** matrices for the box eigenfunctions from Chapter 2 and check their properties (cf. Problem 2.6-1).

10
Partial Exact Diagonalization

Suppose we retain HO eigenfunctions as a basis set and calculate the matrix elements of another hamiltonian, which describes a system other than the HO. Clearly, the resulting matrix would not be diagonal, and we now examine how we might go about extracting the dynamics of the system defined by the new hamiltonian.

The idea we need is to treat the Schrödinger equation as a matrix equation in order to compute matrix eigenvalues and eigenvectors. The resulting matrix formalism provides a general and often powerful technique for estimating the energy level spectrum of the new system. Like the variational method, matrix mechanics is particularly useful if a solution in terms of familiar functions is not known, which is usually the case.

Since modern "canned routines" take care of the computing (refer to the built-in functions **Eigenvalues** and **Eigenvectors** and to *Numerical Recipes* [54], Chapter 11), the effort goes into calculating the matrix elements and hence choosing an appropriate basis set, one with enough of the physics built-in but one still convenient to work with. In effect, one constructs linear combinations of the basis functions which diagonalize the new hamiltonian. The method is generally approximate, however, since in practice the basis set and hence the hamiltonian matrix have to be truncated, a constraint usually imposed by the computer. The approach is thus called *partial exact diagonalization*. It is closely related to the variational technique and can in fact be derived from a *linear* variational method (see for example Morrison, Estle and Lane [47], Chapter 4, or Pauling and Wilson [52], Section 26). Also, both approaches can be combined by introducing additional variational parameters into the basis functions (see for example Bethe and Salpeter [6], Sections 32 to 34).

As an introduction to the method, consider again the model hamiltonian we introduced in Section 7.3 defined by

```
hamiltonian[V_] @ psi_ := -1/2 D[psi,{z,2}] + V psi

V   = VHO E^(-b z^2);    b  =  0.1;
VHO = Vo + z^2/2;        Vo = -2.0;
```

As in Section 7.3, we use for convenience the scaled coordinate **z** and express energies in units of **h w**. In effect, we set **h = m = w = 1**. Recall that **V** is the model potential and **VHO** a harmonic oscillator potential with its minimum shifted by **Vo**. These potentials were plotted together in Figure 7.3-1. As before, it is appropriate to retain HO wavefunctions as basis functions.

We shall estimate here both the bound and continuum energy spectrum and the corresponding eigenfunctions. As we showed in Sections 7.3 and 7.4 (see Problem 7.3-1 and Exercise 7.4-2), only those orbits with **Energy < 0** correspond to true bound states. Orbits below the barriers but with **Energy ≥ 0** are unbound, as are of course orbits above the barriers. Hence, the negative-energy spectrum is discrete while the positive-energy spectrum is continuous. Nevertheless, the barriers give rise to special continuum states having much in common with bound states. Such states are referred to as *metastable or resonance states*, and we will estimate their energies here. We will examine them more systematically in Chapter 14. Discrete-energy basis functions used to approximate a continuum, such as the HO wavefunctions retained here, are sometimes called *pseudo-continuum states*.

10.1 Model-Hamiltonian Matrix

Our task is to obtain solutions of the Schrödinger (differential) equation **hamiltonian[V]@ psi[n,z]== e[n]psi[n,z]**. To begin, we expand the state **psi[n,z]** in a truncated (generalized-Fourier) series in HO basis states **psiHOz[n,z]** (from Section 6.6) by forming the partial sum

```
psi[n_,z_] := Sum[ psiHOz[k,z] c[k,n], {k,0,nmax}]
```

where **nmax** determines our working dimension. If we substitute this expansion into the Schrödinger equation and multiply through by **psiHOz[n,z]** and integrate over all **z**, we convert the Schrödinger equation to a matrix equation of the form **hMatrix.cvec[n] == e[n] cvec[n]** of dimension **nmax**. Here **hMatrix** is the (nondiagonal) hamiltonian matrix in the HO basis and **cvec[n]** its eigenvector for a given state **n**. These single-column matrices determine the expansion coefficients according to

```
c[k_,n_] := cvec[n][[k+1]]
```

One can show by a variational argument that the eigenvalues `e[n]` of the truncated matrix provide upper bounds on the corresponding exact eigenenergies. If we include, however, the complete set by letting `nmax -> Infinity`, the matrix eigenvalues `e[n]` become identical with the eigenenergies of the Schrödinger equation. (Refer to Morrison, Estle and Lane [47], Section 4.4, and to Pauling and Wilson [52], Section 26.)

Exercise 10.1-1

Fill in the details of the transformation of the Schrödinger equation to a matrix equation and show that `c[k_,n_] := cvec[n][[k+1]]`. This is again one of those exercises which is perhaps better done with paper and pencil. Nevertheless, it is important to get the *Mathematica* syntax right.

In order to facilitate calculation of matrix elements, we add and subtract `VHO` and rewrite the potential energy as `V -> VHO + (V - VHO)`. This allows us to express the hamiltonian in terms of the HO hamiltonian as `hamiltonian[V] -> hamiltonian[VHO] + (V - VHO)`. Then, when operating on HO basis states, we can simply substitute `hamiltonian[V] -> Vo + n + 1/2 + (V - VHO)`. (Recall that energies are expressed in units of `h w` and shifted by `Vo`.) We bypass in this way calculation of the kinetic energy matrix elements.

We shall refer to the difference `VP = V - VHO` as a perturbation, as in Section 7.4, since it is small compared with the energies of the unperturbed system at least over the range of the wavefunctions being estimated. We can thus anticipate poorer results for the positive-energy levels.

We can cut our efforts roughly in half by invoking hermiticity of the hamiltonian matrix. In fact, the `VP` matrix is real and therefore symmetric, since `VP` and the HO basis functions are real. Hence, symmetric elements above and below the diagonal are equal, i.e. `VPme[k,n] == VPme[n,k]`. We can cut our work in half again by recognizing that matrix elements of `VP` between states with the same parity vanish, since `VP` has definite parity. That is, every other element along a row or column is zero, viz. `VPme[n,k] = 0` if `n + k = odd`. We thus build these properties into a procedure for calculating matrix elements of `VP` with `integGauss`:

```
Needs["Quantum`integGauss`"]

VPme[n_,k_] := VPme[n,k] =
    If[ OddQ[n + k], 0,
        integGauss[
            psiHOz[n,z] (V - VHO) psiHOz[k,z],
            {z,-Infinity,Infinity}
        ] //N
    ]

VPme[k_,n_] := VPme[n,k] /; k > n
```

Now we can calculate the hamiltonian matrix by adding these perturbation matrix elements to the matrix elements of the (unperturbed) HO hamiltonian from Section 9.3. In the spirit of perturbation theory but mostly for simplicity, we work in *5D* with **nmax = 4**, which gives us two pseudo states above the potential barriers. (This might take a couple of minutes to compute.)

```
nmax = 4;
hMatrix =
    Table[
        (Vo + n + 1/2) KD[n,k] + VPme[n,k],
        {n,0,nmax}, {k,0,nmax}
    ];

hMatrix //MatrixForm
```

-1.44023	0	0.0336936	0	-0.0524215
0	-0.392579	0	-0.0464838	0
0.0336936	0	0.526665	0	-0.19011
0	-0.0464838	0	1.3451	0
-0.0524215	0	-0.19011	0	2.08463

Since **b** and therefore **VP** are relatively small, this matrix is nearly diagonal and its diagonal elements approximately equal the unperturbed HO energies **Vo + n + 1/2**. If we were to ignore the off-diagonal elements altogether, the eigenenergies we are looking for would be just these diagonal elements. In fact, we recognize the diagonal elements as estimates of the energies from first-order perturbation theory (see Section 7.4 and Exercise 7.4-2). For the first two bound states with **Energy < 0**, this looks to be just fine since the levels are discrete. But for **Energy ≥ 0** the true energy spectrum is continuous, and a discrete set of eigenenergies wouldn't appear to be a very good approximation. However, it turns out that for positive eigenenergies we actually obtain position estimates of the resonance states mentioned at the beginning of the chapter.

10.2 Matrix Eigenvalues and Eigenvectors

We now solve **hMatrix.cvec[n] == e[n] cvec[n]** for the eigenvalues **e[n]** and eigenvectors **cvec[n]**. In order for the corresponding set of homogeneous equations to have a nontrivial solution, we require in the usual way that the determinant of the matrix of coefficients vanishes (see Arfken [4], Chapter 4). The roots of the resulting secular or characteristic equation determine the eigenvalues **e[n]**:

```
Solve[ Det[ hMatrix - e IdentityMatrix[5] ] == 0, e ]
    {{e -> -1.4415}, {e -> -0.393822}, {e -> 0.504182},
     {e -> 1.34634}, {e -> 2.10838}}
```

Alternatively, we can use the built-in function **Eigenvalues** to set up and solve the secular equation. Here the built-in function **Sort** arranges the results in ascending order:

```
Eigenvalues[ hMatrix ] //Sort

  {-1.4415, -0.393822, 0.504182, 1.34634, 2.10838}
```

In general, we have to do this numerically since we can't always find the roots of the secular equation algebraically above dimension four. As expected, the eigenvalues are very close to the diagonal elements of **hMatrix**. There appears to be only one resonance at **e[2] = 0.504**, since the next eigenvalue **e[3] = 1.35** is above the barriers in the potential energy (see Figure 7.3-1). As we find in Exercise 10.2-2 below, however, this is an artifact of not having included enough basis states to begin with, i.e. **nmax** is too small.

Exercise 10.2-1

Is there any way of predicting beforehand how many basis states should be included? The answer depends of course on how accurate we want our results to be, although some rules of thumb can be formulated. To that end, consider a two-level system represented by the *2D* hamiltonian matrix

```
h2D = {{h11, h12}, {h12, h22}}
```

whose eigenvalues determine the system's two eigenenergies. Assume for simplicity that all elements are real and show that **Eigenvalues[h2D]** can be written as

$$\left\{\frac{h11 + h22}{2} + \frac{(-h11 + h22)\ \text{Sqrt}\left[1 + \dfrac{4\ h12^2}{(-h11 + h22)^2}\right]}{2},\right.$$

$$\left.\frac{h11 + h22}{2} - \frac{(-h11 + h22)\ \text{Sqrt}\left[1 + \dfrac{4\ h12^2}{(-h11 + h22)^2}\right]}{2}\right\}$$

Apply **Series** to this result in the case that the interaction matrix element **h12** between the two levels is small compared to their energy separation **-h11 + h22**. Thus, show that if the matrix is nearly diagonal its eigenvalues are approximately

$$\left\{h22 + \frac{h12^2}{-h11 + h22} + O[h12]^3,\ h11 - \frac{h12^2}{-h11 + h22} + O[h12]^3\right\}$$

One refers in this case to *weakly interacting energy levels*.

If we now consider several interacting levels, it should be evident that the interaction between the *ith* and *jth* levels is determined by **hij/(hjj - hii)**. We thus conclude: When setting up a partial-exact diagonalization, we can expect better results if the levels included are well separated in energy from those excluded, or if the interaction matrix elements between the two sets of levels are small, or, optimally, if both of these conditions are satisfied.

In practice, one usually sets an accuracy goal and simply includes more and more states until the desired accuracy is achieved, if machine time and memory allow. This is the procedure you will be asked to follow in the next exercise.

Exercise 10.2-2

Check convergence of the eigenvalues of our model hamiltonian by increasing **nmax** in steps of **2**, for example, up to at least **nmax = 14**. Refer to the previous exercise and discuss convergence. Show that a second resonance exists below the barriers. Can you rule out a third resonance?

The eigenvalues and eigenvectors are conveniently calculated with the built-in function **Eigensystem**, which returns a list of lists of the eigenvalues and the corresponding eigenvectors. (The eigenvectors alone can also be computed with the built-in function **Eigenvectors**.) Here **Chop** replaces list elements less than **10^-10** in absolute value by zero.

```
{es,cvecs} = Eigensystem[hMatrix] //
              Chop //Transpose //Sort //Transpose
  {{-1.4415, -0.393822, 0.504182, 1.34634, 2.10838},
    {{-0.999778, 0, 0.015762, 0, -0.0140135},
     {0, 0.999643, 0, 0.0267219, 0},
     {0.01397, 0, 0.992691, 0, 0.119872},
     {0, -0.0267219, 0, 0.999643, 0},
     {0.0158005, 0, 0.11965, 0, -0.99269}}}
```

The first inner list **es** contains the eigenvalues, while the second inner list of lists **cvecs** contains the corresponding eigenvectors. Since purely numerical output from **Eigensystem** is not ordered (unlike symbolic output), we have sorted the results into ascending order of eigenvalues by applying a combination of **Transpose** and **Sort**. (You can see how this works by entering {es,cvecs} //Transpose.) Thus, we can label individual eigenvalues and eigenvectors according to

```
e[n_]    := es[[n+1]]
cvec[n_] := cvecs[[n+1]]
```

We'll verify in a moment that these numerical eigenvectors were normalized by **Eigensystem**.

We thus see that each vector is dominated by a single element. This is just another consequence of the approximately diagonal nature of **hMatrix**, i.e. of the smallness of the perturbation matrix in this example. In particular, the elements of an eigenvector and therefore the expansion coefficients **c[k,n]** of a perturbed state `psi[n,z]` approximately equal `KD[k,n]`. From the defining superposition in Section 10.1, we therefore see that the perturbed wavefunction `psi[n,z]` is dominated by the unperturbed basis state `psiHOz[k=n,z]`.

Hermiticity of the hamiltonian matrix means, in addition to having real eigenvalues, the eigenvectors are orthogonal. That is, the complex dot products of the vectors with each other `Conjugate[cvec[n]].cvec[k]` vanish. This ensures that the perturbed states `psi[n,z]` are orthogonal to each other in the sense of an integration over `z` (see Exercise 10.3-1). Since here `hMatrix` is real and therefore symmetric, the eigenvectors are real as well, and we thus find that

```
Table[ cvec[n].cvec[k], {n,0,nmax},{k,0,nmax}] //
    Chop //MatrixForm
```

```
 1.    0    0    0    0

 0    1.    0    0    0

 0    0    1.    0    0

 0    0    0    1.    0

 0    0    0    0    1.
```

The ones along the diagonal show that the eigenvectors are normalized.

We can check matrix eigenvalues and eigenvectors and their ordering by substituting them back into the Schrödinger matrix equation. Thus,

```
Table[ hMatrix.cvec[n] == e[n] cvec[n], {n,0,nmax}]
```

```
{True, True, True, True, True}
```

Perturbed Eigenfunctions   **10.3**

We have thus determined the normalized perturbed states. For example, for the ground state we have that

```
psi[0,z]
```

$$\frac{0.999778}{E^{z^2/2} \, Pi^{1/4}} - \frac{0.015762 \, (-2 + 4 \, z^2)}{2^{3/2} \, E^{z^2/2} \, Pi^{1/4}} +$$

$$\frac{0.0140135 \, (12 - 48 \, z^2 + 16 \, z^4)}{Sqrt[384] \, E^{z^2/2} \, Pi^{1/4}}$$

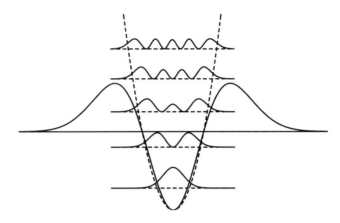

Figure 10.3-1. Perturbed energy levels and probability densities. The potentials **V** and **VHO** from Figure 7.3-1 are shown in the background, and the solid horizontal line marks the *continuum threshold* **e = 0**.

As discussed above, we see that this perturbed state is dominated by the unperturbed HO ground-state wavefunction **psiHOz[0,z]**. We can also verify its (approximate) normalization directly:

```
integGauss[psi[0,z]^2,{z,-Infinity,Infinity}] ==
    KD[0,0] //Chop

    True
```

Exercise 10.3-1

Show by direct integration that the overlap matrix of the perturbed states **psi[n,z]** with each other is the identity matrix, and therefore that the perturbed states are orthonormal, since the HO basis is.

This means of course we have completed our main task, at least approximately. Namely, we have obtained approximate solutions of the Schrödinger *differential* equation **hamiltonian[V] @ psi[n,z] == e[n] psi[n,z]**. We can get a fair idea of the accuracy of our results by plotting the perturbed probability densities, as in Figure 10.3-1.

The bound states look to be o.k., since their inflection points appear to co-incide with the classical turning points in the model potential (cf. Section 6.6). The first pseudo-continuum level below the barrier, however, misses its classical turning point. This is a consequence of not having included enough basis states in the diagonalization, i.e. of not having converged numerically. Moreover, as we found in Exercise 10.2-2 above, if **nmax** is increased the second pseudo-continuum level will drop below the barriers, and a second resonance exists below the barriers. The pseudo states above the barriers clearly follow the unperturbed potential **VHO**.

<h1 style="text-align:right">Local Energy   10.4</h1>

We can make another useful check of our results by plotting a quantity called the *local energy*. By definition, the local energy of a state **psi** equals **hamiltonian[V] @ psi/psi** so that its expectation value in the state **psi** determines the energy, i.e. the expectation value of the hamiltonian. If a perturbed wavefunction **psi[n,z]** is accurate, its local energy should be nearly constant and equal to the corresponding perturbed eigenvalue **e[n]**. In Figure 10.4-1, we plot the local energy of the ground state and see that convergence is satisfactory, at least where the probability density **psi[0,z]^2** is nonvanishing. (That is, it's acceptable that the local energy diverges (here exponentially) from **e[0]** in the tails of the wavefunction, since the error won't contribute significantly to the energy expectation value.) When computed for the excited states in Figure 10.3-1, the local energy is found to be less satisfactory (see Exercise 10.4-1 below).

Exercise 10.4-1

Plot the local energy **hamiltonian[V] @ psi[n,z]/psi[n,z]** of the perturbed excited states as a function of **z** as in Figure 10.4-1. Increase **nmax** as in Exercise

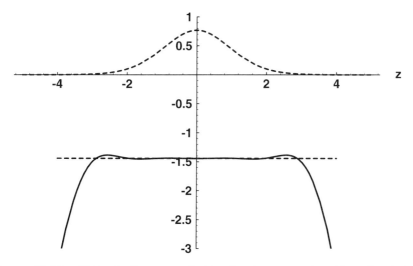

Figure 10.4-1. Plot of the ground-state local energy **hamiltonian[V] @ psi[0,z]/psi[0,z]** (lower solid curve). Convergence to **e[0]** (dotted straight line) is satisfactory over the range of **psi[0,z]** (upper dotted curve). The vertical axis is energy.

10.2-2 to improve convergence. Evaluate the local energy for a few values `z = zo` with
`Series[hamiltonian[V]@ psi[n,z]/psi[n,z]//N,{z,zo,1}]` and compare
with `e[n]`.

10.5   Pseudo States and Resonances

In any case, the perturbed levels shown in Figure 10.3-1 have **n** nodes inside
the well beginning with **n = 0** for the ground state. Note especially that
the resonance levels follow the bound-state sequence and hence that the first
resonance has **n = 2** nodes. Thus, the resonance levels below the barriers
have much in common with bound states, including exponential tails under
the barriers.

There is, however, one important aspect of resonance states which our
pseudo states do not represent. True continuum states describe a free particle
far away from the center of force **x = 0** and must therefore have infinite
range with nonvanishing amplitude. So although a resonance level below
the barriers may behave like a bound state inside the well, the wavefunction
must pass through the barriers (tunnel) and behave like a free-particle state
far from the center of force. Resonances above (and even below) the barriers
arise from multiple reflections of the wavefunction off the barriers (i.e. off the
impedance mismatch), which are in phase and hence constructively interfere.
Nevertheless, the pseudo states approximate this interference between the
barriers.

Later on in Chapter 14, we will determine these resonance states more
accurately and examine the scattering of a particle off the model potential.
See also Section 12.6 and the discussion there in connection with Figure 12.6-3.

Problem 10.5-1

Use the shooting method to estimate the bound states and the first two resonance
states of the model hamiltonian. For the resonance levels, begin the integration
under a barrier between turning points with **psi = 0**. Refer to Exercise 10.2-2 for
improved estimates of the energies. Normalize the wavefunctions with **NIntegrate**
and compare plots of the resulting probability densities with those in Figure 10.3-1.
Finally, plot the local energies and compare with your results from Exercise 10.4-1.

10.6   Diagonalization

Orthogonality of the eigenvectors `cvec[n]` also means that the matrix formed
with columns given by the eigenvectors has an *inverse* given by its transpose.
(By definition, a matrix multiplied by its inverse equals the identity matrix.)
Thus, if we form the matrix

```
cMatrix =
    Transpose[{cvec[0],cvec[1],cvec[2],cvec[3],cvec[4]}];
```

we verify that **Transpose[cMatrix]** is the inverse by evaluating the matrix product

```
Transpose[cMatrix].cMatrix //Chop //MatrixForm
```

1.	0	0	0	0
0	1.	0	0	0
0	0	1.	0	0
0	0	0	1.	0
0	0	0	0	1.

cMatrix is said to be an *orthogonal* matrix. In general, however, it is *unitary* since the eigenvectors are complex so that its inverse is given by the complex conjugate of its transpose. (Refer to Arfken [4], Chapter 4, and to *Numerical Recipes* [54], Chapter 11.)

Because the columns of **cMatrix** are the eigenvectors, we *diagonalize* **hMatrix** by multiplying by **cMatrix** and its inverse in the following way:

```
Transpose[cMatrix].hMatrix.cMatrix //Chop //MatrixForm
```

-1.4415	0	0	0	0
0	-0.393822	0	0	0
0	0	0.504182	0	0
0	0	0	1.34634	0
0	0	0	0	2.10838

Here the diagonal elements are just the new eigenenergies. Such a matrix product is referred to as a *similarity transformation*.

However, the desired diagonalization of the model hamiltonian **hamiltonian[V]** is perhaps most convincing if we simply calculate its matrix elements directly by integrating over the new basis functions **psi[n,z]**. (This may take 30 minutes.)

```
Table[
    integGauss[
        psi[n,z] hamiltonian[V] @ psi[k,z],
        {z,-Infinity,Infinity}
    ] //N,
    {n,0,nmax}, {k,0,nmax}
] //Chop //MatrixForm
```

-1.4415	0	0	0	0
0	-0.393822	0	0	0

0	0	0.504182	0	0
0	0	0	1.34634	0
0	0	0	0	2.10838

One also speaks of the *orthogonal transformation*, which is in general a *unitary transformation*, from the unperturbed basis to the new perturbed basis which preserves normalization and diagonalizes the hamiltonian matrix. Thus, if we form a vector of unperturbed basis functions (suppressing the coordinate **z** for compactness)

```
psiHOvec = Table[ psiHOz[n], {n,0,nmax} ]

   {psiHOz[0], psiHOz[1], psiHOz[2], psiHOz[3], psiHOz[4]}
```

we obtain the perturbed basis states **psi[n]** by the matrix product

```
cMatrix.psiHOvec //TableForm

   0.999778 psiHOz[0] + 0.01397 psiHOz[2] - 0.0158005 psiHOz[4]

   -0.999643 psiHOz[1] - 0.0267219 psiHOz[3]

   -0.015762 psiHOz[0] + 0.992691 psiHOz[2] - 0.11965 psiHOz[4]

   -0.0267219 psiHOz[1] + 0.999643 psiHOz[3]

   0.0140135 psiHOz[0] + 0.119872 psiHOz[2] + 0.99269 psiHOz[4]
```

In summary, partial exact diagonalization provides a powerful technique for estimating the energy levels of a system, as long as we can calculate the necessary matrix elements of the hamiltonian. The method generally proves superior for systems with several degrees of freedom.

Problem 10.6-1

Introduce a variational parameter **a** and diagonalize our model hamiltonian with the HO basis states **psiHO[n,a,z]** with h = m = 1 from Exercise 6.6-3. Use the fact that these are eigenfunctions of **hamiltonian[a^4 z^2/2]** with eigenvalue a^2 (n + 1/2) to set up the perturbation matrix. Minimize numerically each eigenvalue as a function of **a**.

Problem 10.6-2

Redo Problem 10.6-1 but with the model potential energy

```
V = z^2/2               /; z <  0
V = z^2/2 E^(-b z^2)    /; z >= 0
```

with **b = 0.15**. Hint: Extract the even and odd parity parts of **V** in order to calculate matrix elements. (Refer to Chapter 5.)

Problem 10.6-3

Diagonalize the model potential energy $V = x^2/2 - 0.1\ x^3$ with HO pseudo-continuum states and estimate its resonance energies. Note this potential supports no bound states. Derive and simplify closed-form expressions for the matrix elements of the potential using the HO matrix elements **xme[n,k]** of **x** from Section 9.1. Start with four pseudo states but study convergence by pushing **nmax** to large values, such as **nmax** $\sim$ **100**. Plot the perturbed resonance functions and their local energies as in Figures 10.3-1 and 10.4-1.

Problem 10.6-4

Estimate the eigenvalues of the anharmonic potential energy $V = b\ x^4$ from Problem 7.3-2 by partial exact diagonalization. Derive and work with a scaled Schrödinger equation independent of parameters to obtain results for arbitrary **b**. Calculate matrix elements of the potential by extending the results of Problem 9.3-2. Start with four pseudo states but study convergence by pushing **nmax** to larger values. Compare with the estimates from Problem 7.3-2.

Problem 10.6-5

Diagonalize the model potential energy defined by the infinite rectangular box from Section 2.0 but with a finite rectangular barrier inside. Shift the origin of coordinates to the middle of the box as in Problem 5.0-1. Thus, inside the box for $-L/2 < x < L/2$ take

```
V = 0 /;          x < -L/4
V = b /; -L/4 <= x <= L/4
V = 0 /;          x <  L/4
```

and take **b = 0.2 e[0]**, where **e[0]** is the ground-state energy of the unperturbed box. Compare plots of the perturbed eigenfunctions with Figure 2.1-1. Can you obtain results that make sense for **b = e[0]**?

11
Momentum Representation

We gain considerable insight into the dynamics of a particle's motion with a Fourier analysis of the wavefunction. We generally have to do this, however, as a Fourier integral rather than as a Fourier series in order to be able to describe a particle that can be found anywhere in space and unconfined by impenetrable walls. For example, in Section 2.6 we constructed wavepackets in a box by summing over eigenstates of the box. In Section 4.0, however, we integrated over a continuous range of planewaves as a function of wavenumber **k** to construct a wavepacket free to move over all space. We shall see shortly that the coefficients of the planewave components are determined by the Fourier transform of the wavepacket as a function of **k**.

We conveniently investigate many of the properties of Fourier transforms using harmonic oscillator wavefunctions, in particular, gaussian functions. As in Chapters 4 and 8, it's helpful if we load our gaussian integration rules **integGauss** as well as our function **PowerContract**, as an inverse to **PowerExpand**. Since Fourier transforms generally involve complex functions, it's also helpful if we load our symbolic **Conjugate** rule. We thus enter

```
Needs["Quantum`integGauss`"];
Needs["Quantum`PowerTools`"];
Needs["Quantum`QuickReIm`"]
```

Tools 11.1

We define the *Fourier transform* **FT** of a function **psi[x]** and the *inverse Fourier transform* **InvFT** of a function **phi[k]** by entering the following rules:

```
FT[psi_,x_][k_] := 1/Sqrt[2Pi] *
             integGauss[psi E^(-I k x),{x,-Infinity,Infinity}]
InvFT[phi_,k_][x_] := 1/Sqrt[2Pi] *
             integGauss[phi E^( I k x),{k,-Infinity,Infinity}]
```

(Refer to Park [50], Appendix II and to Boas [9], Chapter 15.) Here **FT** is an integral over **x** and therefore a function of **k** (and of course any other parameters which **psi[x]** depends upon), and the syntax **FT[psi,x][k]** is intended to make this clear (cf. Exercise 1.4, Appendix III). Thus, for example, we calculate the Fourier transform of a simple gaussian function of **x** as

FT[E^(-w x^2),x][k] //PowerExpand

$$\frac{1}{\text{Sqrt}[2]\ E^{k^2/(4\ w)}\ \text{Sqrt}[w]}$$

which returns a function of **k** (and **w**). (That the Fourier transform is also a gaussian is a property somewhat peculiar to gaussian functions.) Likewise, **InvFT** is an integral over **k** and therefore a function of **x**. Hence,

InvFT[%,k][x] //PowerExpand

$$E^{-(w\ x^2)}$$

The integrals **FT** and **InvFT** are referred to as *Fourier transform pairs*. This example is illustrated in Figure 11.1-1.

In principle, we can calculate the Fourier transform with respect to any parameter, although **integGauss** restricts us to an overall gaussian dependence (see again **?integGauss**). In general, we would have to replace **integGauss** by the built-in function **Integrate**, or develop other sets of rules. The *Mathematica* package **Calculus `FourierTransform`** is an example designed for wide application.

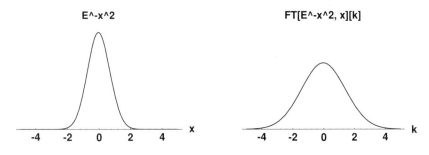

Figure 11.1-1. A gaussian function and its Fourier transform.

Exercise 11.1-1

Calculate the Fourier transform of the lorentzian function `1/(a^2+x^2)` using the built-in function **Integrate**. Check that the inverse Fourier transform of your result gives back the original lorentzian.

Exercise 11.1-2

Our rule for the *inverse* Fourier transform is somewhat redundant in the sense that the following identity is easily established:

```
InvFT[phi[k],k][x] == FT[phi[k],k][-x]
```

Verify this expression with some examples. One thus often refers to Fourier transform pairs as simply Fourier transforms of each other.

Exercise 11.1-3

Fourier transform pairs satisfy two important integral relations.

(a) One is *Parseval's Theorem*:

```
Integrate[Conjugate[f] g,{x,-Infinity,Infinity}] ==
Integrate[Conjugate[FT[f,x][k]] FT[g,x][k],{k,-Infinity,Infinity}]
```

(See Park [50], Appendix II, and also Morse and Feshbach [49], Chapter 4.) This relation ensures for example that the Fourier transform `phi[k]` of a wavefunction `psi[x]` is normalized if `psi[x]` is, and vice versa. The Fourier integral theorem discussed in the next section is an almost immediate consequence of Parseval's theorem.

(b) The other is the *Convolution Theorem*: Given the *convolution* of two functions `f[x]` and `g[x]` defined by

```
1/Sqrt[2Pi] Integrate[f[y] g[x-y],{y,-Infinity,Infinity}]
```

one can show that

```
FT[convolution][k] == FT[f,x][k] FT[g,x][k]
```

In words, the Fourier transform of the convolution is just the product of the individual Fourier transforms. This relation can be very helpful in the solution of integral equations and in data analysis.

Verify these relations with some examples involving gaussian functions and also with the lorentzian function from Exercise 11.1-1.

Problem 11.1-1

Certain symmetries in a function `psi` as a function of `x` lead to relationships in its Fourier transform as a function of `k` and `-k`. Using the definition of **FT**, prove the following with paper and pencil:

(a) If `psi[x]` is real then

```
FT[psi,x][-k] == Conjugate[ FT[psi,x][k] ]
```

What if **psi[x]** is imaginary?

(b) If **psi[x]** is even (odd) then **FT[psi,x][k]** is even (odd).

(c) If **psi[x]** is real and even, then **FT[psi,x][k]** is real and even. If **psi[x]** is real and odd, then **FT[psi,x][k]** is pure imaginary and odd.

Verify these relations with some examples involving gaussian functions.

11.2 Momentum Wavefunctions

Evidently, the inverse Fourier transform of the Fourier transform (and vice versa) is an identity transformation which returns the original function. This fundamental result is known as the *Fourier integral theorem* and is of course contained in the definitions of **FT** and **InvFT**, albeit in a factored form (see Park [50], Appendix II). It justifies through the definition of **InvFT** the useful notion that any reasonable wavefunction **psi[x,t]** can be expanded in free-particle planewaves **E^(I k x)/Sqrt[2Pi]** and instructs that the expansion coefficients **phi[k,t]** are simply the Fourier transform of the wavefunction as a function of **k**. This idea in fact completes our analysis of the free-particle wavepacket in Chapter 4. For example, starting with the *stationary* free-particle wavepacket from the beginning of Section 4.1

```
psi[x_] = (w/Pi)^(1/4) E^(-w x^2/2)

      w  1/4
    (---)
      Pi
  -----------
        2
    (w x )/2
  E
```

we calculate its Fourier transform according to

```
phi[k_] = FT[psi[x],x][k] //PowerExpand //PowerContract

              1
  -------------------------
      2
    k /(2 w)        1/4
  E           (Pi w)
```

which we recognize as the spectral or momentum distribution we began Section 4.1 with. Writing out the definition of **InvFT**, we connect the two functions

```
psi[x] ==
    integGauss[
        phi[k] E^(I k x)/Sqrt[2Pi],
        {k,-Infinity,Infinity}
    ] //PowerExpand //PowerContract

  True
```

as in Section 4.1. We generalize this result in the next problem to the *moving* (time-dependent) free-particle wavepacket **psi[x,t]** from Section 4.2 and show that its Fourier transform gives back the time-dependent momentum distribution **phi[k-kP] E^(-I h k^2/(2m) t)** we used to construct the wavepacket with. The class of "reasonable" functions for which the Fourier integral theorem holds is quite large. (Refer to Morse and Feshbach [49], Chapter 4, for details.)

The Fourier transform of a wavefunction thus determines the wavefunction's spectral or momentum distribution. It follows from Parseval's theorem (see Problem 11.1-3 above) that the momentum distribution **Abs[phi[k,t]]^2** is normalized if the coordinate distribution **Abs[psi[x,t]]^2** is. As we have already seen in Section 4.1 in connection with the free-particle wavepacket, this property leads us to interpret the momentum distribution as the probability density of finding the particle with wavenumber **k** and therefore with momentum **p = h k**.

We thus call the Fourier transform of the coordinate wavefunction the *momentum wavefunction*. The momentum wavefunction completely specifies the state of the system just as the coordinate wavefunction does, since if we're given the momentum wavefunction we can determine the coordinate wavefunction, and vice versa. The two functions thus provide two completely equivalent descriptions of the state of the system known as the *coordinate and momentum representations*.

Problem 11.2-1

Calculate the Fourier transform of the *moving* free-particle wavepacket **psi[x,t]=**

$$
\mathrm{Power}\left[E, \frac{-(h\ kP^2\ t)}{2\ (-I\ m + h\ t\ w)} + \frac{kP\ m\ x}{-I\ m + h\ t\ w} + \frac{-\dfrac{I}{2}\ m\ w\ x^2}{-I\ m + h\ t\ w}\right]\ \left(\frac{w}{Pi}\right)^{1/4}
$$

$$
\mathrm{Sqrt}\left[\frac{m}{m + I\ h\ t\ w}\right]
$$

we constructed in Section 4.2 and check that it gives back the time-dependent momentum distribution

```
phi[k_,t_] := phi[k-kP] E^(-I h k^2/(2m) t)
```

that we used to construct the wavepacket with.

Conventions 11.3

The utility of an expansion in planewaves in quantum mechanics accounts for the choice of signs of the exponents **E^(±I k x)** in the definitions of **InvFT** and **FT**, respectively. In engineering applications, for example, these signs are usually reversed. The factor **1/Sqrt[2Pi]** included in both **FT** and **InvFT** is also appropriate in quantum mechanics, although we could simply include a single factor **1/(2Pi)** in the definition of either **FT** or **InvFT** (see for example Boas [9], p. 649). In signal analysis, one usually encounters Fourier transform pairs *in time and in frequency* (see e.g. *Numerical Recipes* [54], Chapter 12). When comparing our expressions here with such references we can assume **x** plays the role of time and **k** the role of frequency, although the reverse point of view is permissible from the overall symmetry of Fourier transform pairs (see Exercise 11.1-2 and Problem 11.1-1 above). Otherwise, of course, we shall take **x** to be a length and **k** a wavenumber measured in inverse units of length.

11.4 HO Momentum Wavefunctions

We will now construct the Fourier transform of the harmonic oscillator eigenfunctions **psiHO[n,x]** from Section 6.6. At the same time, it's convenient to set up the Fourier transform of the dimensionless eigenfunctions **psiHOz[n,z]**. We thus define HO *momentum* eigenfunctions according to

```
 phiHO[n_,k_]   :=   phiHO[n,k]   =  FT[psiHO[n,x],x][k]
phiHOkz[n_,kz_] := phiHOkz[n,kz] = FT[psiHOz[n,z],z][kz]
```

and obtain for the ground- and first-excited states

```
{phiHO[0,k], phiHO[1,k]} //PowerExpand //PowerContract
```

$$\left\{ \frac{\left(\dfrac{h}{m\,Pi\,w}\right)^{1/4}}{E^{(h\,k^2)/(2\,m\,w)}}, \quad \frac{-I\,Sqrt[2]\,k\,\left(\dfrac{h}{m\,w}\right)^{3/4}}{E^{(h\,k^2)/(2\,m\,w)}\,Pi^{1/4}} \right\}$$

The momentum wavefunctions **phiHOkz[n,kz]** are dimensionless just like their counterparts **psiHOz[n,z]** in the coordinate representation and are therefore convenient for many tasks including plotting. We can also obtain them directly from the **phiHO[n,k]** by an appropriate change of variables (see Problem 11.4-2 below). We thus have for the ground- and first-excited states that

```
{phiHOkz[0,kz], phiHOkz[1,kz]} //PowerExpand //PowerContract
```

$$\left\{ \frac{1}{E^{kz^2/2}\,Pi^{1/4}}, \quad \frac{-I\,Sqrt[2]\,kz}{E^{kz^2/2}\,Pi^{1/4}} \right\}$$

These two wavefunctions are plotted in Figure 11.4-1. You'll notice that the odd-**n** momentum functions are proportional to **I** and need to be wrapped

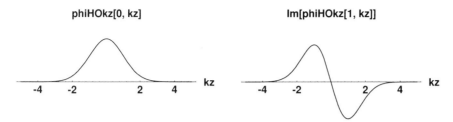

Figure 11.4-1.   Ground- and first-excited state HO momentum wavefunctions.

with **Im** for plotting. This phase follows from the corresponding reality and oddness of the coordinate wavefunctions **psiHO[n,x]** for **n** odd (see Problem 11.1-1 above and also Problems 11.4-1 and 11.9-1 below).

As with the coordinate eigenfunctions, the momentum eigenfunctions also constitute an orthonormal set, a property ensured by Parseval's theorem (Exercise 11.1-3 above). We find for example with the first two HO momentum wavefunctions that

```
Table[
    integGauss[
        Conjugate[phiHO[np,k]] phiHO[n,k],
        {k,-Infinity,Infinity}
    ],
    {np,0,1},{n,0,1}
] //PowerExpand //MatrixForm

  1   0

  0   1
```

Problem 11.4-1

(a) Generate several HO momentum wavefunctions and show that they're polynomials of order **n** and orthonormal. Plot your results using **phiHOkz[n,kz]** and check that **n** gives the number of nodes as well as the "parity" of the wavefunction defined by **kz -> -kz**. Show that an inverse Fourier transform of **phiHO[n,k]** gets back **psiHO[n,x]**.

(b) Show that the symmetry of the momentum-space wavefunctions under inversion **k -> -k** follows from the parity of the coordinate-space wavefunctions and the definition of the Fourier transform. Explain why **phiHO[n,k]** is purely imaginary when **n** is odd. Refer to Problem 11.1-1 above.

Problem 11.4-2

Show that the coordinate-space scaling **z == Sqrt[m w/h] x** translates into the momentum-space scaling **kz == Sqrt[h/(m w)] k**. Then show by a change of variables in the Fourier transform that **phiHO[n,k]== (h/(m w))^(1/4)phiHOz[n,kz]** and verify with some examples.

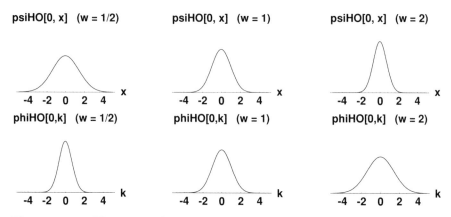

Figure 11.4-2. Three ground-state HO wavefunctions (top row) and their Fourier transforms (bottom row) defined for different frequencies **w**. Computed with `psiHO[0,x]/. {h->1, m->1}`.

Before going on, we should note that a relatively narrow coordinate wavefunction corresponds to a relatively wide momentum wavefunction, and vice versa. Figure 11.4-2 illustrates this *reciprocity* with three ground-state HO wavefunctions. This of course is just a symptom of the uncertainty principle, for as we showed in Chapter 3 if the distribution of **k** values specified by the momentum wavefunction is large, then the distribution of **x** values given by the coordinate wavefunction is small, and vice versa. (Refer to the discussion around Figure 3.0-2.) We inferred this behavior in squeezed states in Section 8.3 by calculating their position and momentum uncertainties, and we shall examine it explicitly in Section 11.11 using the momentum representation (see Figure 11.11-1).

11.5 Dirac Delta Function

The Fourier integral theorem leads to another useful notion with wide formal application. Let's look at this theorem again with the HO ground- and first-excited states:

```
{psiHO[0,xo] == InvFT[ FT[psiHO[0,x],x][k], k][xo],
 psiHO[1,xo] == InvFT[ FT[psiHO[1,x],x][k], k][xo]} //
     ExpandAll //PowerExpand
```

    {True, True}

The useful idea derives from the fact that we can formally rearrange these relations into what is generally regarded as the definition of the *Dirac delta function* **DiracDelta[x-xo]**:

```
psi[xo] ==
    Integrate[
        psi[x] DiracDelta[x-xo],
        {x,-Infinity,Infinity}
    ]                                          (11.5-r1)
```

At the same time, we obtain one of many representations of the delta function known as the *Fourier representation*:

```
DiracDelta[x-xo] == 1/(2Pi) *
    Integrate[E^(I k (x-xo)),{k,-Infinity,Infinity}]    (11.5-r2)
```

Here we have replaced **integGauss** by the built-in function **Integrate** to indicate the generality of these relations.

Exercise 11.5-1

With paper and pencil, derive relations (11.5-r1) and (11.5-r2) by substituting the definitions of **FT** and **InvFT** into the Fourier integral theorem and formally interchanging the order of integration.

The utility of these relations may at first be difficult to accept in light of the fact that the integral (11.5-r2) does not exist. However, the integral (11.5-r1) does (as we just demonstrated with **psiHO[n,x]** as long as **psi[x]** vanishes sufficiently fast as **x -> ±Infinity**. Hence, a representation like (11.5-r2) is meaningful as long as we stipulate that it be used only in combination with a well-behaved wavefunction under an integral like in (11.5-r1).

Evidently, the delta function is not a function in the ordinary sense. For example, since **psi[x]** appears on both sides of (11.5-r1), **DiracDelta[x-xo]** behaves as though it vanishes everywhere except when **x** is very close to **xo**. Moreover, inserting **psi[x] = 1** into (11.5-r1), we see that **DiracDelta[x-xo]** must be infinite for **x = xo** in such a way that it has unit area under its curve. We can think in this way of the delta function as a generalization of the Kronecker delta **KD[n,m]** of the discrete variables **n** and **m** (see Section 9.1) to the case of continuous variables. Thus, the delta function is really a property, or even just a particular collection of symbols, which turns out to be useful enough to deserve its own special notation. Perhaps a more secure point of view is to think of the Dirac delta function as a limit of a sequence of regular functions.

Exercise 11.5-2

(a) Evaluate the integral (11.5-r2) by replacing **k -> ±Infinity** by **k -> ±kmax** and thus show that **DiracDelta[x-xo]** is equivalent to

```
Sin[kmax (x - xo)]
──────────────────
   Pi (x - xo)
```

in the limit **kmax -> Infinity**.

(b) By introducing a suitable limiting process in (11.5-r2), show that the expression

$$\frac{1}{E^{(-x \, + \, xo)^2 /w} \quad Sqrt[Pi \ w]}$$

in the limit **w -> 0** provides an alternative representation of the delta function. Hint: Use **integGauss**.

(c) Show in a similar fashion that

$$\frac{w}{Pi \ (w^2 + (x \, - \, xo)^2 \,)}$$

is also a suitable alternative representation of the delta function in the limit **w -> 0**.

Make a sequence of plots of the representation in **(a)** as a function of increasing **kmax** and of those in **(b)** and **(c)** as functions of decreasing **w**.

Although we shall have only occasional need for the delta function in this book, there is enough general interest in it to consider briefly how one might go about implementing it on the computer. Consider for example the following rule:

```
integGauss[
    f_ . DiracDelta[x_ + xo_.],
    {x_,-Infinity,Infinity}
] := f /. x -> -xo
```

This then gives

```
integGauss[f[x] DiracDelta[x-xo],{x,-Infinity,Infinity}]

    f[xo]
```

as desired. The blanks _. in the rule also cover cases, however, for which **xo = 0** and **psi = 1** (see also Exercise 2.2, Appendix III) so that, for example,

```
integGauss[DiracDelta[x],{x,-Infinity,Infinity}]

    1
```

Alternatively, we could attach this rule to the built-in function **Integrate** using the **TagSet** syntax **DiracDelta /: Integrate[...]**. Refer to Exercise 1.13, Appendix III, and see also the *Mathematica* package **Calculus `DiracDelta`**.

Exercise 11.5-3

Check that `DiracDelta[x-xo]` has the desired effect on `psiHO[n,x]` for various values of **n**. Evaluate integrals of `psiHO[n,x]` and `D[psiHO[n,x],x]` with `DiracDelta[x]` for various values of **n** and explain.

We can also do some physics. If, for example, the coordinate wavefunction of a particle were simply a delta function at the origin `DiracDelta[x]`, its momentum wavefunction would be the Fourier transform

```
FT[ DiracDelta[x], x ][k]
        1
   ─────────
   Sqrt[2 Pi]
```

and therefore a constant for all momenta. Physically, this means that maximum certainty in locating a particle at the coordinate origin corresponds to maximum uncertainty in the particle's momentum, as required by the uncertainty principle. The particular value of the (normalization) constant ensures that the inverse Fourier transform gives back the delta function.

If on the other hand the momentum wavefunction of a particle were a delta function `DiracDelta[k-ko]`, its coordinate wavefunction would be given by

```
InvFT[ DiracDelta[k-ko], k ][x]
     I ko x
   E
   ─────────
   Sqrt[2 Pi]
```

and therefore a free-particle planewave with momentum **ko**, which is also in compliance with the uncertainty principle.

Momentum Operator 11.6

A slight generalization of the Fourier integral theorem introduced in Section 11.2 gives us still another very useful identity. Suppose we take the Fourier transform of a function, multiply the result by the wavenumber **k**, and take the inverse transform of the product. Working again with **E^(-w x^2)**, we thus find that

```
InvFT[ k FT[E^(-w x^2),x][k], k][x] //PowerExpand
    2 I w x
   ─────────
       2
     w x
   E
```

which we recognize is proportional to the *derivative* of **E^(-w x^2)**. In particular,

```
-I D[E^(-w x^2),x] == %
```

    True

In fact, if we multiply this identity by Planck's constant **h**, we obtain the fundamental connection between the momentum as a derivative operator in the coordinate representation and **h k** in the momentum representation:

```
p @ psi_ := -I h D[psi,x]

p @ (E^(-w x^2)) ==
    InvFT[(h k) FT[E^(-w x^2),x][k], k][x] //
        PowerExpand
```

    True

That is, operating on the coordinate wavefunction with the derivative operator **p** is equivalent to multiplying the momentum wavefunction by **h k** (cf. Park [50], Section 2.5). For example, operating on the HO ground state,

```
p @ psiHO[0,x] == InvFT[(h k) phiHO[0,k], k][x] //
                    PowerExpand
```

    True

We can also do this the other way around using **FT**, so that alternatively

```
(h k) phiHO[0,k] == FT[p @ psiHO[0,x], x][k] //
                    PowerExpand
```

    True

This equivalence of course means that we can evaluate momentum expectation values and generally momentum matrix elements simply by integrating **h k** with the corresponding momentum wavefunctions, as we have seen for example with both free-particle and squeezed-state wavepackets. Thus, with the ground-state and first excited-state HO momentum eigenfunctions, we find that

```
Table[
    integGauss[
        Conjugate[phiHO[np,k]] (h k) phiHO[n,k],
        {k,-Infinity,Infinity}
    ],
    {np,0,1},{n,0,1}
] //PowerExpand //PowerContract //MatrixForm
```

$$
\begin{pmatrix}
0 & \dfrac{-I\ \text{Sqrt}[h\ m\ w]}{\text{Sqrt}[2]} \\[2ex]
\dfrac{I\ \text{Sqrt}[h\ m\ w]}{\text{Sqrt}[2]} & 0
\end{pmatrix}
$$

in agreement with the first 2×2 element of the matrix **pMatrix** from Section 9.2 (cf. especially Exercise 9.2-1). Note that here the diagonal elements vanish by parity, as in the coordinate representation, since **(-1)^n** is also the parity of the *nth* momentum eigenfunction as a function of **k** (see Problem 11.4-1 above and Exercise 9.2-1).

Local Energy 11.7

We can evaluate higher-order derivatives of the wavefunction in a similar fashion. For example, we apply the kinetic energy operator to the HO ground state according to

```
InvFT[ (h k)^2/(2m) FT[psiHO[0,x],x][k], k ][x] //
    PowerExpand //Expand //PowerContract
   3/4   m  1/4  5/4                   5/4  9/4  2
  h     (--)    w                5/4  w    x
         Pi                           m    w    x
 ------------------------   -  ------------------------
          2                             2
    (m w x )/(2 h)              (m w x )/(2 h)       1/4
  2 E                        2 E              (h Pi)
```

Dividing by the ground-state wavefunction, we thus obtain the *local kinetic energy* (cf. Section 10.4):

```
%/psiHO[0,x] //PowerExpand //Expand
               2 2
   h w    m w  x
   --- - -------
    2       2
```

This little theorem will prove to be very handy in evaluating the kinetic energy numerically in the lattice representation (in Section 12.4). Here, of course, we can check this result by applying the kinetic energy operator directly:

```
1/(2m) p @ p @ psiHO[0,x]/psiHO[0,x] //Expand
               2 2
   h w    m w  x
   --- - -------
    2       2
```

We can further check this result by adding in the potential energy and thus obtain the *total* local energy

```
% + m w^2/2 x^2
   h w
   ---
    2
```

which is just the HO ground-state eigenenergy and therefore independent of **x**.

Exercise 11.7-1

Redo this calculation for some of the HO excited states. Check the total local energy in the momentum representation by starting with **phiHO[n,k]**.

11.8 Coordinate Operator

The analysis of the momentum in the previous section can be turned around somewhat to answer an obvious question: How do we calculate the expectation value of the position **x** in the momentum representation?

We actually get a good hint from the reciprocity of the coordinate and momentum wavefunctions, and easily find that multiplication of the coordinate wavefunction by **x** is equivalent to *differentiation* of the momentum wavefunction by **k**. In particular, in the momentum representation the position **x** is defined by

```
x @ phi_  := I D[phi,k]

x @ phiHO[0,k] == FT[ x psiHO[0,x], x][k] //PowerExpand

    True
```

Note that here on the left **x** is the derivative operator which operates on **phiHO[0,k]**, as indicated by the notation **x @**, whereas on the right **x** is the coordinate which simply multiplies **psiHO[0,x]**. Alternatively, we have

```
x psiHO[0,x] == InvFT[ x @ phiHO[0,k], k][x] //PowerExpand

    True
```

Of course, these definitions hold for any wavefunction of **x** and not just for the HO ground state. Again using the ground-state and first excited-state HO momentum eigenfunctions, we thus obtain for the position matrix elements in the momentum representation

```
Table[
    integGauss[
        Conjugate[phiHO[np,k]] x @ phiHO[n,k],
        {k,-Infinity,Infinity}
    ],
    {np,0,1},{n,0,1}
] //PowerExpand //PowerContract //
    Together //MatrixForm
```

```
                         h
                  Sqrt [ ─── ]
                         m w
                  ────────────
       0             Sqrt [2]

            h
     Sqrt [ ─── ]
            m w
     ────────────
      Sqrt [2]        0
```

in agreement with the first 2×2 element of the matrix **xMatrix** from Section 9.2.

Exercise 11.8-1

Check the fundamental commutator **[x,p]== I h** in the momentum representation. Refer to Exercise 1.1-2 and to Section 18.2.

Momentum-Space Hamiltonian   11.9

The position operator defined in the previous section allows us to transform the HO hamiltonian to the momentum representation. We replace **p** by **h k** and **x phi** by **I D[phi,k]** and thus obtain

```
hamiltonian @ phi_ :=
    (h k)^2/(2m) phi + m w^2/2 x @ x @ phi
```

so that

```
hamiltonian @ phi[k]
```

$$
\frac{h^2 \ k^2 \ phi[k]}{2 \ m} - \frac{m \ w^2 \ phi''[k]}{2}
$$

The overall similarity of this result with its coordinate-space counterpart is a consequence of the special symmetry of the harmonic oscillator. In any case, we easily check the momentum eigenfunctions **phiHO[n,k]** and the HO eigenvalue spectrum **h w (n + 1/2)**. For example, operating on the ground state

```
hamiltonian @ phiHO[0,k] == h w/2 phiHO[0,k] //ExpandAll
```
```
    True
```

We use this hamiltonian in Problem 11.9-1 to set up and solve the Schrödinger equation as a differential equation in the momentum representation and therefore to rederive the momentum eigenfunctions. We also deduce in this way a direct (scaling) connection between the **phiHO[n,k]** and the **psiHO[n,x]**.

Exercise 11.9-1

Compare the HO kinetic and potential energies in the coordinate and momentum representations.

Exercise 11.9-2

Redefine the raising and lowering operators introduced in Section 6.7 in the momentum representation and verify their properties using the HO momentum eigenfunctions. Redo Problem 6.7-1 in the momentum representation.

Problem 11.9-1

Solve the HO Schrödinger equation in the momentum representation and show that the same energy spectrum is obtained. Write down the normalized solutions by comparison with the coordinate-representation Schrödinger equation and show that

```
phiHO[n_,k_] := (-I)^n Sqrt[h] psiHO[n,x] //.
    {x -> h k, m w -> 1/(m w)} //PowerExpand //PowerContract
```

Compare results derived with this expression with the Fourier transforms defined in the text and show for a few special cases that an inverse Fourier transform gets back **psiHO[n,x]**, which justifies the choice of phase **(-I)^n**.

11.10 Exponential Operators

Evidently, identities analogous to the Fourier integral theorem (Section 11.2) can be defined for any power of an operator and therefore for any (analytic) function of an operator that has a power series expansion. We examine here three important examples of exponential operators of the form (1.4-r2).

Consider first introducing **E^(-I k xP)** between the **FT** and **InvFT** of a coordinate function. For example, with our simple gaussian we obtain

```
InvFT[ E^(-I k xP) FT[E^(-w x^2),x][k], k ][x] //
    Simplify //PowerExpand

      -(w (-x + xP) )
  E
```

which we recognize as a translation or shift of the gaussian by the fixed distance **xP**. Likewise, we can shift the HO ground state **psiHO[0,x]** to **psiHO[0,x-xP]**:

```
psiHO[0,x-xP] ==
    InvFT[ E^(-I k xP) FT[psiHO[0,x],x][k], k ][x] /.
        I x - I xP -> I (x - xP) //PowerExpand

    True
```

Note that we did essentially the same thing in Section 4.2 when we shifted the peak of the free-particle wavepacket by phase shifting each of its planewave components by the phase **E^(-I k xP)**. The equivalence is clear from the interpretation of **InvFT** as an expansion in planewaves.

The operator **E^(-I k xP)** is thus known as the *translation or displacement operator* in the momentum representation. We note that its coordinate representation **E^(-I xP p/h)**, where **p** is the momentum operator and hence a derivative, is not easily implemented on the computer. Its application via the momentum representation as we've done here can be therefore a very useful tool. (You might compare this with **?MatrixExp**.)

As a second, similar example, consider multiplying a coordinate function by **E^(I kP x)** and taking the **FT** of the product. We find the result to be a shift of the Fourier transform of the function by the fixed momentum **kP**. For example, we obtain with the simple gaussian

```
FT[ E^(I kP x) E^(-w x^2), x ][k] //
    Simplify //PowerExpand
```

$$\frac{1}{\text{Sqrt}[2]\ E^{(-k\ +\ kP)^2/(4\ w)}\ \text{Sqrt}[w]}$$

i.e. the **FT** of the gaussian *shifted by* **kP**:

```
FT[ E^(-w x^2), x ][k-kP] //
    Simplify //PowerExpand
```

$$\frac{1}{\text{Sqrt}[2]\ E^{(-k\ +\ kP)^2/(4\ w)}\ \text{Sqrt}[w]}$$

Similarly, we can shift a coordinate wavefunction and thus obtain the corresponding *momentum* wavefunction shifted by **kP**. For example, with the HO ground state

```
phiHO[0,k-kP] == FT[ E^(I kP x) psiHO[0,x], x ][k] /.
    -I k + I kP -> -I (k - kP) //PowerExpand
```

```
    True
```

We thus refer to the operator **E^(I kP x)** as a *momentum boost*. As with the displacement operator, we introduced it already in Section 4.2 in an essentially equivalent fashion in order to set up a wavepacket with initial momentum **h kP** (see also Section 8.0).

We note in passing that a momentum boost arises naturally when one performs a (nonrelativistic) *Galilean transformation* of the Schrödinger equation between two frames of reference in relative motion. See Park [50], p. 157.

As a final example, let's examine again the time-development of a *free-particle* gaussian wavepacket. In Section 4.2, we defined the initial gaussian wavepacket

```
psi[x_,0] = E^(I kP x - w x^2/2) (w/Pi)^(1/4)
```

$$
E^{I\ kP\ x\ -\ (w\ x^2)/2}\ (\frac{w}{Pi})^{1/4}
$$

with momentum **h kP**. We ensure free-particle motion if we replace the hamiltonian in the time-development operator in (1.4-r1) by just the kinetic energy operator **p @ p/(2m)**. Nevertheless, as with a displacement, the resulting time-development operator is difficult to evaluate unless we first Fourier transform the wavepacket to the momentum representation. The time-development operator then becomes simply the phase **E^(-I h k^2/(2m)t)**. An inverse Fourier transform gets back the coordinate representation

```
InvFT[ E^(-I h k^2/(2m) t) FT[psi[x,0],x][k], k ][x] //
       ExpandAll //MapAll[Together,#]& //
       PowerExpand //PowerContract
```

$$
Power[E,\ \frac{-(h\ kP^2\ t)}{2\ (-I\ m\ +\ h\ t\ w)}\ +\ \frac{kP\ m\ x}{-I\ m\ +\ h\ t\ w}\ +\ \frac{-\frac{I}{2}\ m\ w\ x^2}{-I\ m\ +\ h\ t\ w}]\ (\frac{w}{Pi})^{1/4}
$$

$$
Sqrt[\frac{m}{m\ +\ I\ h\ t\ w}]
$$

of the time-developed wavefunction, in agreement with the result of Section 4.2 and Problem 11.2-1 above. That is,

```
psi[x,t] == %
```

```
True
```

Exercise 11.10-1

Redo this calculation for the time-reversed wavepacket with initial state **Conjugate[psi[x,to > 0]]** from the future of **psi[x,t]** and show that the result is **Conjugate[psi[x,-(t-to)]]** which develops into **Conjugate[psi[x,0]]** when **t = to**. Refer to Exercise 5.0-6.

In the general case of a nonvanishing potential, this trick for evaluating the time development doesn't work. The reason is the time-development operator can't be simply split into kinetic and potential factors, i.e. **E^(-I(K+V) t/h) ≠ E^(-I K t/h) E^(-I V t/h)**. This is because the kinetic energy involves derivatives with respect to **x** while the potential energy is a function of **x** and the ordering of these operators in a series expansion like (1.4-r2) is important. The identity which allows these operators to be split thus introduces the commutator of **K** and **V** in a complicated way and is known as the *Baker-Hausdorff formula*. (See Merzbacher [45], Exercise 8.18, for a simple version of this theorem.)

We note, however, that we can evaluate the general case *approximately* if we take a short enough interval of time `dt` such that `E^(-I(K+V)dt/h)` $\sim$ `1 - I(K+V)dt/h + O[dt]^2`. Then, we can use `E^(-I(K+V)dt/h)` $\sim$ `E^(-I K dt/h)E^(-I V dt/h) + O[dt]^2` and evaluate the kinetic-energy contribution in the momentum representation as we did above with the free particle. We can integrate over time by applying this approximate *split* operator repeatedly. We shall exploit this idea shortly in Section 12.6 in order to evaluate the time-development of wavepackets on a lattice.

Problem 11.10-1

Show with paper and pencil by comparing series expansions that the approximate split

```
E^(-I K/2 dt/h) E^(-I V dt/h) E^(-I K/2 dt/h)
```

of the time-development operator `E^(-I(K+V)dt/h)` involves errors only of `O[dt]^3`. Hint: Be sure to pay attention to operator ordering.

More Squeezed States   11.11

We can now complete our discussion of squeezed states from Chapter 8 by deriving the momentum representation of the time-dependent squeezed state `psi[x,t]`. Although in principle this is just the Fourier transform of `psi[x,t]`, we obtain a compact and useful result in a familiar way simply by retracing our derivation from Chapter 8. Here, however, we shall start with the Fourier transform of the initial wavepacket `psi[x,0]` and expand in the momentum-space eigenfunctions `phiHO[n,k]` from Problem 11.9-1.

As in Section 8.0, it is convenient to set `h = 1`. Introducing the results from Problem 11.9-1, we thus define normalized momentum eigenfunctions

```
phiHOw[n_,k_] = phiHO[n,k] /.h -> 1

      n    1    1/4                     1
   (-I)  (-----)     HermiteH[n, k Sqrt[---]]
         m Pi w                         m w
   ---------------------------------------------
                 2
            n/2  k /(2 m w)
           2    E             Sqrt[n!]
```

reciprocal to the coordinate-space eigenfunctions `psiHOw[n,x]` we defined in Section 8.0. These functions are of course Fourier transform pairs.

We obtain the initial momentum distribution by taking the Fourier transform of the initial coordinate-space wavepacket `psi[x,0]` from Section 8.0:

```
phi[k_,0] = FT[psi[x,0],x][k] /.
                E^q_ :> E^Collect[q,xPo] //
                completeSquare[#,k]& //
                PowerExpand //PowerContract
```

$$\frac{E^{-(k - kPo)^2/(2\ m\ wPo)\ -\ I\ (k\ -\ kPo)\ xPo}}{(m\ Pi\ wPo)^{1/4}}$$

where we've completed the square with respect to **k** (cf. Section 8.1). The phase **E^(-I (k - kPo) xPo)** in this result derives from the displacement of the coordinate-space origin to **x = xPo** (see Section 11.10).

We now easily verify the momentum of the initial wavepacket as

```
integGauss[
    k Conjugate[phi[k,0]] phi[k,0],
    {k,-Infinity,Infinity}
]
```

    kPo

Likewise, we can calculate the initial kinetic energy according to

```
integGauss[
    k^2/(2m) Conjugate[phi[k,0]] phi[k,0],
    {k,-Infinity,Infinity}
] //Factor //Expand
```

$$\frac{kPo^2}{2\ m} + \frac{wPo}{4}$$

We check this result by adding to it the expectation value of the potential energy in the initial state **psi[x,0]**:

```
integGauss[
    m w^2 x^2/2 Conjugate[psi[x,0]] psi[x,0],
    {x,-Infinity,Infinity}
] //Factor //Expand //PowerExpand //PowerContract
```

$$\frac{w^2}{4\ wPo} + \frac{m\ w^2\ xPo^2}{2}$$

We thus obtain the total energy of the squeezed state

```
% + %%
```

$$\frac{kPo^2}{2\ m} + \frac{w^2}{4\ wPo} + \frac{wPo}{4} + \frac{m\ w^2\ xPo^2}{2}$$

in agreement with Exercise 8.0-2.

Exercise 11.11-1

Verify the energy expectation value in the momentum representation with the initial state `phi[k,0]` by introducing the coordinate operator `x @ phi_ := I D[phi,k]` in the potential energy. This problem is the momentum-representation analogue of Exercise 8.0-2.

The time-dependent momentum-space representation `phi[k,t]` of the squeezed state can now be calculated in analogy with our derivation of the coordinate-space representation `psi[x,t]` in Section 8.2. The details are relegated to Problem 11.11-1.

Problem 11.11-1

Expand `phi[k,t]` in the HO momentum basis states `phiHOw[n,k]` by following our derivation of `psi[x,t]` from Section 8.2. Show that

```
phi[k_,t_] :=
    E^(I kphase - (k-kP[t])^2/(4 delp[t]^2))/
        ((2Pi)^(1/4) Sqrt[delp[t]])
```

where `kP[t]` and `delp[t]` are conveniently defined by the replacement rules

```
kPRule := kP[t] -> kPo Cos[t w] - m w xPo Sin[t w]
delpRule := delp[t] ->
    Sqrt[m/(2 wPo) (wPo^2 Cos[t w]^2 + w^2 Sin[t w]^2)]
```

and where `kphase` is a phase which depends on `k` and `t`, analogous to `xphase` of `psi[x,t]`. Keep in mind that `h = 1` so that `p = k`. Also, if you use these results, be sure to clear `kP` and `delp` to avoid collision with any previous definitions.
Hint: Before summing the series, make the replacement `(-I)^(2n) -> E^(I n Pi)`. Refer to Problem 8.2-2.

Problem 11.11-2

Verify the results of the previous problem by taking the Fourier transform of `psi[x,t]` directly. Note: This is a messy problem which requires a good deal of algebra in order to obtain the form of `phi[k,t]` given in the previous problem.

Problem 11.11-3

Obtain the squeezed-state momentum distribution `phi[k,t]` for `kPo = 0` simply by scaling the coordinate-space wavefunction `psi[x,t]` appropriately. Determine the necessary replacements by transforming `psi[x,t]` into `phi[k,t]`. Compare with Problem 11.9-1.

With the results of Problem 11.11-1, we easily verify the momentum expectation value and uncertainty in the squeezed-state. The square of the momentum distribution is given by

```
phisq[k_,t_] = Conjugate[phi[k,t]] phi[k,t]
```

$$\frac{1}{E^{(k - kP[t])^2 /(2\ delp[t]^2)}\ Sqrt[2\ Pi]\ delp[t]}$$

Hence, the expectation value of the momentum is verified to be

```
pExp[t_] =
    integGauss[k phisq[k,t],{k,-Infinity,Infinity}] //
        PowerExpand

  kP[t]
```

while the expectation value of the square of the momentum is found to be

```
psqExp[t_] =
    integGauss[k^2 phisq[k,t],{k,-Infinity,Infinity}] //
        PowerExpand //Factor

  delp[t]^2 + kP[t]^2
```

Thus, **delp[t]** is the root-mean-square uncertainty in the momentum

```
Sqrt[psqExp[t] - pExp[t]^2] //PowerExpand

  delp[t]
```

as desired. Introducing the replacement rules from Problem 11.11-1, we have that

```
{kP[t], delp[t]} /.{kPRule, delpRule}

  {kPo Cos[t w] - m w xPo Sin[t w],
```

$$\frac{Sqrt[\dfrac{m\ (wPo^2\ Cos[t\ w]^2 + w^2\ Sin[t\ w]^2)}{wPo}]}{Sqrt[2]}\}$$

in agreement with the results of our coordinate-space evaluation from Problem 8.3-1.

Therefore, squeezed-state wavepackets in the momentum representation also oscillate in and out of minimum uncertainty just as their coordinate-space counterparts do. Because of the uncertainty principle, the momentum-space distribution is narrowest when the coordinate-space distribution is widest, and vice versa. Of course, in the coherent-state limit **wPo -> w**, the momentum distribution oscillates without change of shape, just as its coordinate-space counterpart **psisq[x,t]/.wPo -> w** does. That is,

```
phisq[k,t] /.delpRule /.wPo -> w /.
           Cos[z_]^2 -> 1 - Sin[z]^2 //
           ExpandAll //completeSquare[#,k]& //
           PowerExpand //PowerContract
```

$$\frac{1}{E^{(k - kP[t])^2 /(m w)} \quad Sqrt[m \ Pi \ w]}$$

Hence, as we showed in Section 8.4, `delp[t] -> Sqrt[m w/2]` in the coherent-state limit, and the coherent state is a minimum-uncertainty wavepacket. This momentum-space motion is displayed in Figure 11.11-1 along with the coordinate-space motion from Figure 8.3-2.

Problem 11.11-4

Verify the position expectation value `xExp[t]` and uncertainty `delx[t]` from Section 8.3 by introducing the coordinate operator `x @ phi_ := I D[phi,k]` in the momentum representation. Note that you will need the `k` dependence of `kphase` from Problem 11.11-1.

In a similar fashion, evaluate the energy expectation value in the momentum representation with the state `phi[k,t]` and show that it equals the initial-state value derived in the text. See Problem 8.3-1.

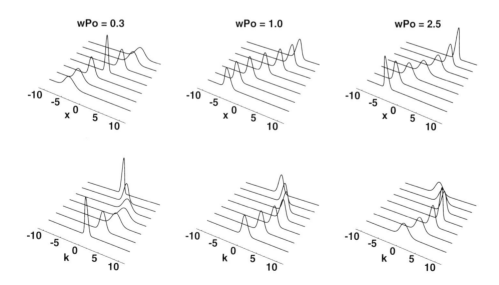

Figure 11.11-1. Time development of the squeezed wavepackets from Figure 8.3-2 but now in both the coordinate (top row) and momentum (bottom row) representations. In all cases the initial displacement is `xPo = -5` and the initial momentum `kPo = 0` with `m = w = 1`. Time increases from front to back starting with `t = 0` in equal time steps for one-half period `Pi/w`. The middle packet with `wPo = 1` is a coherent state.

12
Lattice
Representation

Some of the notions we developed in the momentum representation can be turned into a powerful numerical technique for solving the *time-dependent* wave equation if we are willing to approximate the wavefunction by a finite grid or lattice of points. In particular, we can apply the hamiltonian and even the time-development operator to the wavefunction by switching back and forth between the coordinate and momentum representations with Fourier transforms. On a discrete lattice, the transforms can be computed efficiently and relatively quickly with an algorithm known as the *fast Fourier transform* or *FFT* for short.

Of course, whenever we compute a wavefunction numerically we introduce a lattice representation, and we've already seen many examples of this including plotting and determining stationary bound states with **NDSolve**. In Chapter 14, we shall use **NDSolve** in this way to compute stationary continuum states and potential scattering in a time-independent framework. The method we now describe extends and complements this approach and is capable of computing the time development directly.

Coordinate Lattice 12.1

Let's begin by setting up a coordinate lattice and evaluating some harmonic oscillator wavefunctions. For computational convenience, we will work with the scaled HO functions **psiHOz[n,z]** on a **z** lattice. In addition, it's convenient and appropriate for fast Fourier transforms that we set up a uniform grid with fixed spacings **dz** between points.

There are only two essential lattice parameters, the length **L** and the total number of grid points **nmax**. These two quantities determine a third basic element, the lattice spacing **dz = L/nmax**. Let's thus enter

```
{nmax = 2^5, L = 12.0, dz = L/nmax}
   {32, 12., 0.375}
```

and generate the lattice according to

```
zGrid = Table[ -L/2 + (n-1) dz, {n,1,nmax} ]
   {-6., -5.625, -5.25, -4.875, -4.5, -4.125, -3.75, -3.375, -3.,
     -2.625, -2.25, -1.875, -1.5, -1.125, -0.75, -0.375, 0.,
     0.375, 0.75, 1.125, 1.5, 1.875, 2.25, 2.625, 3., 3.375,
     3.75, 4.125, 4.5, 4.875, 5.25, 5.625}
```

It is convenient for list operations to take the lattice index, or grid-point label, **n** to be a nonzero, positive integer. Here we used the shift **n-1** to facilitate comparison with the built-in *FFT* later on. Because it is desirable (optimal) for the *FFT* algorithm, we take **nmax** to be an even integer. Also, to facilitate comparison with the built-in *FFT*, our spacing **dz** actually corresponds to a lattice with an odd number of points **nmax + 1** in order to include the origin **z = 0** explicitly in the lattice. This has the slight drawback, as long as we require **nmax** to be even, that **zGrid** is not symmetric about the origin (see also Problem 12.3-1).

We can compute the HO ground state on the lattice simply by passing **zGrid** to the argument of the wavefunction **psiHOz[0,z]** from Section 6.6. The result is a list of **nmax** equally-spaced function values corresponding to each grid point:

```
psiHOzGrid[0] = psiHOz[0,zGrid] //N
                 -8              -7              -7              -6
   {1.14396 10  , 1.01167 10  , 7.77305 10  , 5.18887 10  ,
     0.0000300941,0.000151641, 0.000663865, 0.00252505,
     0.00834425, 0.023957, 0.0597592, 0.12951, 0.243855,
     0.39892, 0.566979, 0.700126, 0.751126,0.700126, 0.566979,
     0.39892, 0.243855, 0.12951, 0.0597592, 0.023957,
     0.00834425, 0.00252505, 0.000663865, 0.000151641,
                 -6              -7              -7              -7
     0.0000300941, 5.18887 10  , 7.77305 10  , 1.01167 10  }
```

These values can be plotted directly as a function of the lattice index **n** using **ListPlot**, as in Figure 12.1-1. They clearly reflect the even parity of the ground state.

```
ListPlot[
    psiHOzGrid[0],
    AxesLabel -> {" n",""}, Axes -> {Automatic,None},
    PlotRange -> {-0.2,1}
];
```

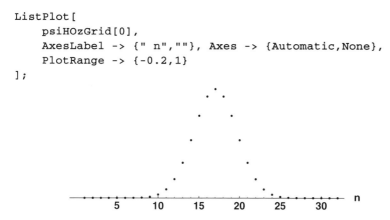

Figure 12.1-1.   Lattice representation of the HO ground-state wavefunction as a function of the lattice index **n**.

We can also plot this as a function of **z** by threading together the two lists **zGrid** and **psiHOzGrid[0]** using **Thread**. The result is a list of **nmax** pair lists, each pair list containing a lattice point and the corresponding wavefunction value. For example, recalling that the subscript **[[n]]** selects the *nth* list element, we can grab the first lattice point with

```
Thread[{zGrid,psiHOzGrid[0]] [[1]]
```

$\{-6., 1.14396\ 10^{-8}\}$

If we pass the full list to **ListPlot**, we obtain as desired a plot as a function of **z**, as in Figure 12.1-2.

```
ListPlot[
    Thread[{zGrid,psiHOzGrid[0]}],
    AxesLabel -> {" z",""}, Axes -> {Automatic,None},
    PlotRange -> {{-6,6},{-0.2,1}}
];
```

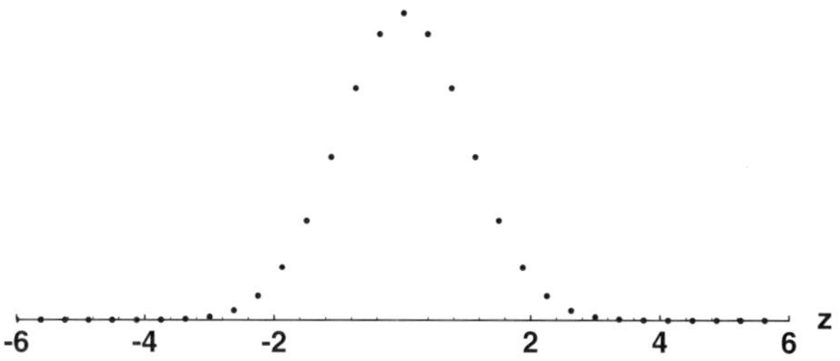

Figure 12.1-2.   Lattice representation of the HO ground-state wavefunction as a function of **z**.

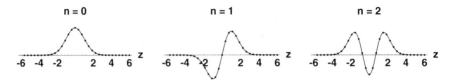

Figure 12.1-3. Lattice representations (dots) of the first three HO wavefunctions as a function of **z**. The solid curves are for comparison plots of the wavefunctions as a function of the continuous variable **z**.

Likewise, we can compute the first two excited states as

```
psiHOzGrid[1] = psiHOz[1,zGrid] //N;
psiHOzGrid[2] = psiHOz[2,zGrid] //N;
```

Here, we've suppressed the output with a semicolon. Instead, we plot these three lattice wavefunctions in Figure 12.1-3 together with **psiHOz[n,z]** as a function of the continuous variable **z** for comparison.

These plots make clear that the length **L** should be chosen large enough that it includes the tails of all the wavefunctions of interest. Of course, we can only do this approximately because of the (usual) exponential falloff, i.e. the wavefunctions only approach zero and never actually vanish asymptotically. At the same time, **nmax** should be chosen large enough that all of the "wiggles" in the wavefunctions of interest are faithfully sampled, in other words, that the lattice spacing **dz = L/nmax** is fine enough to resolve all of the structure in the wavefunction. For example, **dz** certainly has to be smaller than the separation between nodes in the wavefunction. Of course, in order to reduce the numerical effort required, we shall want to keep the number of grid points **nmax** as small as possible. In many respects, **nmax** serves the same function as the option **PlotPoints** in **Plot** or **Plot3D**.

We shall see shortly that **L** and **nmax** are further constrained by momentum-space considerations.

The lattice representation of a wavefunction by a list of points means that we can approximate overlap integrals and matrix elements simply as sums over elements of the list, in particular, as list dot products. For example, we verify the ground-state normalization according to

```
psiHOzGrid[0].psiHOzGrid[0] dz
```

```
1.
```

and orthonormality of the first three states as

```
Table[
    psiHOzGrid[np].psiHOzGrid[n] dz,
    {np,0,2},{n,0,2}
] //Chop //MatrixForm
```

```
1.   0    0
0    1.   0
0    0    1.
```

where the **Chop** has been introduced to replace off-diagonal elements smaller than 10^-10 by zero. These simple integration approximations are thus analogous to a *trapezoidal integration rule* (see Abramowitz and Stegun [1], Chapter 25) and are more than adequate for what we will do here. Note that if the wavefunctions were complex-valued, we would have to compute the dot products as **Conjugate[psi[np]].psi[n]** instead (cf. the orthonormality of the vectors **cvec[n]** in Section 10.2).

Exercise 12.1-1

Compute the dot products using a **Sum** over list elements **list[[n]]**.

Exercise 12.1-2

The finite length **L** of the lattice and the finite number **nmax** of lattice points means that the diagonal elements of the overlap matrix are only approximately unity, just as the off-diagonal elements are only approximately zero. Have *Mathematica* display their true values or compute the differences of the true values with one and then improve these values by bumping up **L** and **nmax**.

Likewise, we can compute coordinate-space matrix elements of **z** according to

```
Table[
    psiHOzGrid[np].(zGrid psiHOzGrid[n]) dz,
    {np,0,2},{n,0,2}
] //Chop //MatrixForm
```

```
0            0.707107   0
0.707107     0          1.
0            1.         0
```

which agrees with the matrix **xMatrix** from Section 9.2, since we have **h** = m = w = 1 and **1/Sqrt[2]** $\sim$ **0.707107** (see also Section 11.8). Note that the product **zGrid psiHOzGrid[n]** simply multiplies the two lists together element by element.

Exercise 12.1-3

Show that if **nmax** is even, the parity operation of changing **z -> -z** can be effected by applying the built-in function **Reverse** to the lattice wavefunction in order to reverse the order of the list. Explain why this doesn't work, at least directly, when **nmax** is odd.

12.2 Momentum Lattice

Let's consider now momentum-lattice wavefunctions reciprocal to the coordinate-lattice wavefunctions we just introduced. First, we need to construct a momentum lattice **kzGrid** reciprocal to the coordinate lattice **zGrid** in order to approximate Fourier transforms by discrete sums. In effect, we shall approximate the Fourier integrals by Fourier series, in particular, by Fourier partial sums with **nmax** terms.

The Fourier series approximation introduces implicitly the assumption that the lattice wavefunction is periodic in the lattice length, i.e. **psi[z+L] == psi[z]** (see Park [50], Appendix 2). This in turn requires the planewave components **E^(±I kz z)** in the Fourier integrals to have the same periodicity, namely, **E^(±I kz (z + L)) == E^(±I kz z)**. It follows (cf. Exercise 2.1-1) that the wavenumbers **kz** are restricted to discrete values **kz = q dkz**, where **q** is an integer momentum-lattice index and **dkz = 2Pi/L** is the momentum-lattice spacing.

In order that the coordinate and momentum representations contain the same information, we require the momentum lattice to contain exactly the same number of points **nmax** as its coordinate counterpart. Since in addition Fourier series require a balance between positive and negative wavenumbers, the value **kzmax = nmax/2 dkz** is the largest wavenumber allowed. We thus enter

```
{dkz = 2Pi/L //N, kzmax = nmax/2 dkz}

  {0.523599, 8.37758}
```

and define a *reciprocal lattice* in analogy with the coordinate lattice according to

```
kzGrid = Table[-kzmax + (q-1) dkz, {q,1,nmax}]

  {-8.37758, -7.85398, -7.33038, -6.80678, -6.28319, -5.75959,

    -5.23599,-4.71239, -4.18879, -3.66519, -3.14159, -2.61799,

    -2.0944, -1.5708,-1.0472, -0.523599, 0., 0.523599, 1.0472,

    1.5708, 2.0944, 2.61799, 3.14159, 3.66519, 4.18879, 4.71239,

    5.23599, 5.75959, 6.28319, 6.80678, 7.33038, 7.85398}
```

As with the coordinate lattice index **n**, it is convenient for list operations to take **q** to be a nonzero positive integer and to introduce the shift **q-1** to facilitate comparison with the built-in *FFT* later on.

Exercise 12.2-1

Show that we could also enter simply

```
kzmax = Pi/dz //N
```

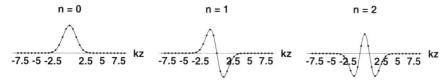

Figure 12.2-1. Lattice representations (dots) of the first three HO momentum wavefunctions as a function of `kz`. The solid curves are for comparison plots of the momentum wavefunctions as a function of the continuous variable `kz`.

We can now compute HO momentum-lattice wavefunctions simply by substituting `kzGrid` for the argument of the HO momentum wavefunctions `phiHOkz[n,kz]` that we defined in Section 11.4. We thus compute the first three states according to

```
phiHOkzGrid[0] = phiHOkz[0,kz] /.kz->kzGrid //N;
phiHOkzGrid[1] = phiHOkz[1,kz] /.kz->kzGrid //N;
phiHOkzGrid[2] = phiHOkz[2,kz] /.kz->kzGrid //N;
```

Here we've used the replacement `/. kz -> kzGrid` to set up the wavefunctions first before substituting the lattice. Otherwise, the system could bog down calculating the Fourier transform for each lattice point. These representations are plotted in Figure 12.2-1 along with the original momentum wavefunctions as a function of the continuous variable `kz`. Note that for **n** odd `Im[phiHOkzGrid[n]]` has to be plotted, as with `phiHOkz[n,kz]` (cf. Figure 11.4-1).

The strong similarity of the coordinate and momentum wavefunctions in Figures 12.1-3 and 12.2-1 is a result of having used **w = 1** (as well as **h = m = 1**) but is largely a property peculiar to gaussian functions. This is of course familiar from our study of coherent and squeezed states in Chapter 8 and Section 11.11.

We thus see that when choosing **L** and **nmax** we also have to take into consideration the momentum wavefunctions as well and try to strike a balance in the quality of the coordinate and momentum sampling for the level of numerical effort set by **nmax**. We see that **nmax** also determines the maximum allowed value of the momentum **kzmax** and hence of the kinetic energy. A large kinetic energy occurs when the total energy is large compared with the potential energy. Likewise, a large value of **L** determines how small the lattice spacing **dkz** will be. Again, the reciprocity of the coordinate and momentum representations is evident, and we see that **dz** and **dkz** define an uncertainty relation of sorts, namely, **dz dkz == 2Pi/nmax**.

Note that if **kzmax** and therefore **nmax** are too small, the momentum wavefunctions will have their tails chopped off too abruptly. Like their coordinate-space counterparts, the tails fall off exponentially for large **kz** and of course will always get chopped some on a finite lattice. In practice, however, this can be used to check whether or not the coordinate wavefunction has been competently sampled with a sufficient number **nmax** of small steps **dz**. We simply look to see whether the Fourier transforms of the wavefunctions of interest, i.e. the momentum wavefunctions, approach zero satisfactorily as **kz** -> ±**kzmax**.

Exercise 12.2-2

(a) Redo Figures 12.1-3 and 12.2-1 for different values of **L** and **nmax** and observe how the lattice representation changes in both spaces. For **L = 20** and **nmax = 32** for example, note that the round-off errors in the momentum matrix become noticable. Explain.

(b) Redo Part **(a)** for two different values of **w** ≠ **1**. Refer to Figures 8.3-2 and 11.11-1.

As with the coordinate lattice, we can approximate momentum overlap integrals and matrix elements simply by list dot products. For example, we verify orthonormality according to

```
Table[
    Conjugate[phiHOkzGrid[qp]].phiHOkzGrid[q] dkz,
    {qp,0,2},{q,0,2}
] //Chop //MatrixForm
```

```
1.    0    0

0     1.   0

0     0    1.
```

and matrix elements of **kz** according to

```
Table[
    Conjugate[phiHOkzGrid[qp]].(kzGrid phiHOkzGrid[q]) dkz,
    {qp,0,2},{q,0,2}
] //Chop //MatrixForm
```

```
0                  -0.707107 I    0

0.707107 I   0                    -1. I

0            1. I                 0
```

which agrees with the matrix **pMatrix** from Section 9.2 (again since **h** = **m** = **w** = **1** and **1/Sqrt[2]** ∼ **0.707107**).

Exercise 12.2-3

Verify the results of Problem 11.1-1 in the lattice representation using the functions **psiHOzGrid[n]** and **phiHOkzGrid[q]**. Refer to Exercise 12.1-3 above.

Discrete Fourier Transforms   12.3

We now go about setting up a direct and general connection between the coordinate and momentum *lattice* representations. This leads us to the notion of a Fourier transform on a lattice, the so-called *discrete Fourier transform*. The great utililty of this connection derives from the fact that discrete Fourier transforms can be computed quickly and efficiently for an arbitrary waveform with the *fast Fourier transform (FFT) algorithm*. An *FFT* package is included with almost all scientific subroutine libraries and has been used to define the built-in functions **Fourier** and **InverseFourier**. A good general reference to discrete and fast Fourier transforms is *Numerical Recipes* [54], Chapter 12. A thorough and readable introduction, including the connection with Fourier series, is given by Thompson [61], Section A5.

The first step is to approximate the Fourier integrals **FT** and **InvFT** defined at the beginning of Section 11.1 by discrete lattice sums. The reduction is relatively direct, and we define in this way the following two discrete transform functions **FTGrid** and **InvFTGrid**:

```
FTGrid[psizGrid_] :=
    Table[
        Sum[
            E^(-I kzGrid[[q]] zGrid[[n]]) psizGrid[[n]],
            {n,1,nmax}
        ] dz/Sqrt[2Pi] //N,
        {q,1,nmax}
    ]

InvFTGrid[phikzGrid_] :=
    Table[
        Sum[
            E^( I kzGrid[[q]] zGrid[[n]]) phikzGrid[[q]],
            {q,1,nmax}
        ] dkz/Sqrt[2Pi] //N,
        {n,1,nmax}
    ]
```

Here we've approximated **Integrate[f[x],{x,-Infinity, Infinity}]** by **Sum[fGrid[[n]],{n,1,nmax}]** and **Integrate[g[k], {k,-Infinity,Infinity}]** by **Sum[gGrid[[q]], {q,1,nmax}]**. The **Table** functions evaluate the sums at each point of the resulting lattice, a **q** lattice in the case of **FTGrid** and an **n** lattice in the case of **InvFTGrid**. The other replacements are self-evident.

We thus evaluate, for example, the discrete Fourier transform of the HO ground-state lattice wavefunction directly according to

```
FTGrid[psiHOzGrid[0]] //Chop
```

$\{-1.40184\ 10^{-9},\ 1.40647\ 10^{-9},\ -1.41861\ 10^{-9},\ 1.5084\ 10^{-9},$

$5.34474\ 10^{-10},\ 4.85399\ 10^{-8},\ 8.34997\ 10^{-7},\ 0.0000113154,$

$0.000116318,\ 0.000909155,\ 0.00540201,\ 0.0244011,\ 0.0837911,$

$0.218737,\ 0.434094,\ 0.654908,\ 0.751126,\ 0.654908,\ 0.434094,$

$0.218737,\ 0.0837911,\ 0.0244011,\ 0.00540201,\ 0.000909155,$

$0.000116318,\ 0.0000113154,\ 8.34997\ 10^{-7},\ 4.85399\ 10^{-8},$

$5.34474\ 10^{-10},\ 1.5084\ 10^{-9},\ -1.41861\ 10^{-9},\ 1.40647\ 10^{-9}\}$

which we can compare with the *exact* lattice momentum wavefunction:

```
phiHOkzGrid[0]
```

$\{4.31999\ 10^{-16},\ 3.02677\ 10^{-14},\ 1.61217\ 10^{-12},\ 6.52799\ 10^{-11},$

$2.00948\ 10^{-9},\ 4.70243\ 10^{-8},\ 8.36561\ 10^{-7},\ 0.0000113138,$

$0.00011632,\ 0.000909154,\ 0.00540201,\ 0.0244011,\ 0.0837911,$

$0.218737,\ 0.434094,\ 0.654908,\ 0.751126,\ 0.654908,\ 0.434094,$

$0.218737,\ 0.0837911,\ 0.0244011,\ 0.00540201,\ 0.000909154,$

$0.00011632,\ 0.0000113138,\ 8.36561\ 10^{-7},\ 4.70243\ 10^{-8},$

$2.00948\ 10^{-9},\ 6.52799\ 10^{-11},\ 1.61217\ 10^{-12},\ 3.02677\ 10^{-14}\}$

Clearly, these two lists are almost equal, especially towards their middles around the origin **kz = 0**. The small differences correspond of course to round-off errors introduced by the discrete Fourier transform. Applying **Chop** with a somewhat milder tolerence to the difference of the two lists, we thus obtain

```
% - %% //Chop[#,10^-8]&
```

$\{0,\ 0,\ 0,\ 0,\ 0,\ 0,\ 0,\ 0,\ 0,\ 0,\ 0,\ 0,\ 0,\ 0,\ 0,\ 0,\ 0,\ 0,\ 0,\ 0,$

$0,\ 0,\ 0,\ 0,\ 0,\ 0,\ 0,\ 0,\ 0,\ 0,\ 0,\ 0\}$

which shows that the round-off errors are no larger than **10^-8**. You can verify in Exercise 12.3-1 that these errors can be reduced by increasing **nmax**.

It turns out that the discrete Fourier transform has the general and usually undesirable property that the tails of the transformed function lying outside the range **{kz, -kzmax, kzmax}** get spuriously moved into the range. Consequently, the tails of a momentum wavefunction computed by a discrete Fourier transform of a coordinate wavefunction will always be somewhat larger than the tails of the true momentum wavefunction. This phenomenon is known as *aliasing* and is evident in our previous example (i.e. the tails of `phiHOkzGrid[0]` are smaller than those of `FTGrid[psiHOzGrid[0]]`).

As we pointed out at the end of Section 12.2, we can check to see whether our wavefunctions have been competently sampled by making sure that their Fourier transforms approach zero satisfactorily as `kz -> ±kzmax`. In other words, we can check to make sure that aliasing is minimized. As a general rule, the tails of the Fourier transforms will be satisfactorily small if the maximum lattice energy `kzmax^2/2 + Vmax` is larger than the any of the energies of the wavefunctions we're trying to represent, a bound we can often ensure on physical grounds.

We should mention in this connection that the maximum wavenumber `kzmax = Pi/dz` (see Exercise 12.2-1) determines what is also known as the *Nyquist critical frequency* (frequency in a time-frequency domain, but a wavenumber for us [see Section 11.3]). It's easy to see that a pure `Sin` wave with this wavenumber would be represented by two points per cycle (per wavelength) on the lattice. Thus, another way to ensure competent sampling is to set up a coordinate lattice dense enough that the highest planewave component present of any wavefunction under consideration is represented by a minimum of two lattice points per cycle. Refer to *Numerical Recipes* [54], Chapter 12, and also to Hamming [31], Chapter 34, for a fuller discussion. In practice, as long as the wavepacket doesn't develop too many nodes and remains smooth, one often ignores these somewhat subtle points and simply sets **nmax** to be as large as computer time allows.

The theorems and identities on Fourier transforms we introduced in Sections 11.6-11.10 carry over into the discrete representation. We easily verify for example the Fourier integral theorem for the ground state by

```
psiHOzGrid[0] - InvFTGrid[ FTGrid[psiHOzGrid[0]] ] // Chop

   {0, 0, 0, 0, 0, 0, 0, 0, 0, 0, 0, 0, 0, 0, 0, 0, 0, 0, 0,
    0, 0, 0, 0, 0, 0, 0, 0, 0, 0, 0, 0}
```

Likewise, we can take the derivative of a function simply by multiplying the Fourier transform by `I kGrid`. For example, with the gaussian `E^(-z^2)`, we check that

```
InvFTGrid[ I kzGrid FTGrid[E^(-zGrid^2)] ] -
   (D[E^(-z^2),z] /.z->zGrid) //N //Chop[#,10^-7]&

   {0, 0, 0, 0, 0, 0, 0, 0, 0, 0, 0, 0, 0, 0, 0, 0, 0, 0, 0,
    0, 0, 0, 0, 0, 0, 0, 0, 0, 0, 0, 0}
```

In the second input line we substituted the lattice after we calculated the derivative analytically. If we multiply this identity by `-I` we obtain an expression for the momentum operator in the coordinate-lattice representation (cf. Section 11.6 and Exercise 12.4-1 below).

Finally, there is of course a close connection between discrete Fourier transforms and ordinary Fourier series involving functions of the continuous variable **z**, which we investigate in Problems 12.3-1 and 12.3-2.

Exercise 12.3-1

Show with the functions **psiHOzGrid[n]** and **phiHOkzGrid[q]** that the accuracy of **FTGrid** and **InvFTGrid** increases if **nmax** is increased.

Exercise 12.3-2

In the lattice representation our definitions **FTGrid** and **InvFTGrid** are not dependent on an integration package. Check the Fourier integral theorem with the lorentzian function **1/(a^2 + z^2)** for a few different values of **a** and compare with Exercise 11.1-1.

Problem 12.3-1

Discrete Fourier transforms and ordinary (complex-valued) Fourier partial sums are closely related. Both representations are based on expansions in planewaves. The assumed periodicity of the function being expanded restricts the wavenumbers to the discrete values **kzGrid** *in both cases*, and we obtain completeness in the limit **nmax -> Infinity**. In a Fourier series the coordinate **z** is taken to be continuous, whereas in the full lattice representation administered by discrete Fourier transforms, **z** is replaced by the discrete values **zGrid**. It follows that the lattice elements of the discrete Fourier transform of a function approximate to within round-off errors the Fourier-series expansion coefficients of the function.

Let's take a closer look at this correspondence. We can obtain the usual expression for a Fourier (complex-valued) partial sum (refer to Park [50], Appendix 2) directly from our definition **InvFTGrid** by replacing the coordinate lattice variable **zGrid[[n]]** by the continuous coordinate **z**. That is, we can define the **nmax***th* partial sum according to

```
fourierPS[phikzGrid_] :=
    Sum[
        E^(I kzGrid[[q]] z) phikzGrid[[q]],
        {q,1,nmax}
    ] dkz/Sqrt[2Pi] //Chop
```

and approximate the usual Fourier-series expansion coefficients by the discrete Fourier transform **FTGrid** of the function being expanded.

Consider for example the gaussian **E^(-z^2)**. Since this function is real and even, we should be able to express its Fourier series as an even-parity **Cos** series. To that end, it's appropriate to switch to coordinate and momentum lattices which include the origins **z = 0** and **kz = 0**, respectively, but which are symmetric about the origins with equal numbers of positive and negative points. This requires an odd number of lattice points **nmax**.

Set up new lattices with **nmax = 33** and estimate the Fourier-series expansion coefficients according to

```
phiGauss = FTGrid[E^(-zGrid^2)]
```

Show that the Fourier partial sum transforms into the following `Cos` series:

```
0.147704 + 0.27584 Cos[0.523599 z] + 0.224574 Cos[1.0472 z] +

   0.159415 Cos[1.5708 z] + 0.0986658 Cos[2.0944 z] +

   0.0532441 Cos[2.61799 z] + 0.0250522 Cos[3.14159 z] +

   0.0102775 Cos[3.66519 z] + 0.00367616 Cos[4.18879 z] +

   0.00114649 Cos[4.71239 z] + 0.000311757 Cos[5.23599 z] +

   0.0000739143 Cos[5.75959 z] + 0.0000152795 Cos[6.28319 z] +
                -6                               -7
   2.75396 10    Cos[6.80678 z] + 4.32853 10    Cos[7.33038 z] +
                -8                               -8
   6.00383 10    Cos[7.85398 z] + 1.4169 10    Cos[8.37758 z]
```

Compare a plot of the partial sum with one of the original gaussian as a function of the continuous variable `z`. Derive the expansion coefficients analytically using the orthogonality of the original planewave components with periodicity `L` and compare a table of their values with `phiGauss`.

Problem 12.3-2

Redo Problem 12.3-1 for the rectangular wave we studied in Section 2.4. Compute the rectangular wave using

```
rect[z_] := 0           /;    0 <= z <    L/4
rect[z_] := Sqrt[2/L] /;   L/4 <= z <= 3L/4
rect[z_] := 0           /; 3L/4 <   z <=  L
```

with `L = 1` and with a new `zGrid` for the interval `{z,0,L}`. Using your lattice from the previous problem but with `nmax` = *even*, you should obtain the `Sin` series we derived in Section 2.4, albeit with approximate expansion coefficients. Plot your results as a function of `z` and compare with Figure 2.4-3.

Finally, improve convergence by including *Lanczos sigma factors* of the form

```
lanczos[0]   := 1
lanczos[q_] := Sin[q Pi/(nmax/2)]/(q Pi/(nmax/2))
```

directly in the complex-valued Fourier partial sum. Compare your results with Problem 2.4-3.

Local Energy   12.4

We can also compute the ground-state local energy in complete analogy with our calculation in Section 11.7. With `h = m = w = 1`, we want the value `0.5` across the lattice:

```
ListPlot[
    Elocal[0],
    AxesLabel -> {" n","Elocal[0]"}, PlotRange -> {-.1,1.1}
];
```

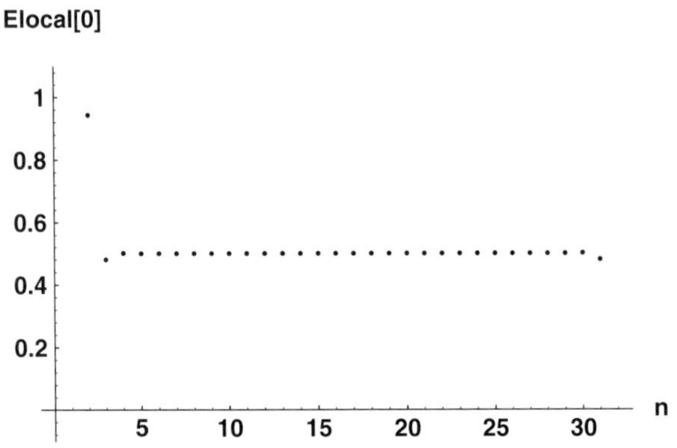

Figure 12.4-1.   HO ground-state local energy as a function of the lattice index n.

```
Elocal[0] =
    InvFTGrid[kzGrid^2/2 FTGrid[psiHOzGrid[0]]]/
    psiHOzGrid[0] + zGrid^2/2 //Chop
{-27.3477, 0.943684, 0.480801, 0.50139, 0.499859, 0.500019,
    0.499997, 0.500001, 0.5, 0.5, 0.5, 0.5, 0.5, 0.5, 0.5, 0.5,
    0.5, 0.5, 0.5, 0.5, 0.5, 0.5, 0.5, 0.5, 0.5, 0.500001,
    0.499997, 0.500019, 0.499859, 0.50139, 0.480801, 0.943684}
```

The deviations from the desired value near the beginning and end of the lattice, seen in Figure 12.4-1, are another indication of the round-off errors. We nevertheless still obtain an excellent ground-state expectation value because the wavefunction rapidly falls off to zero in these regions. That is,

```
(psiHOzGrid[0]^2).Elocal[0] dz
    0.5
```

which is very close to the exact result:

```
% - 1/2
                -16
    -2.26896 10
```

Exercise 12.4-1

Adapting the identity from the end of Section 12.3 for the derivative of a gaussian function, we define (cf. Section 11.6) a lattice momentum operator in the coordinate representation according to

```
pz @ psiGrid_ := InvFTGrid[ kzGrid FTGrid[psiGrid] ]
```

(for **h = 1**). Use this result to recompute the local energy of the HO ground-state.

Exercise 12.4-2

Extend our lattice computations to **n = 6** and check the corresponding HO energies. Note that the local energy for **n** = *odd* is infinite at the origin because of the node in the wavefunction there. Implement a work-around so that you can nevertheless obtain the correct energy expectation value from the local energy. Finally, compute the full energy matrix for these states.

Problem 12.4-1

Redo the variational calculation of Section 7.3 for the ground state of our model hamiltonian in the lattice representation. Hint: Compute **eTrial[0]** for several values of the variational parameter **a** and interpolate your results (refer to Exercise 3.5, Appendix III) in order to use **FindMinimum**, as in Section 7.3.

Problem 12.4-2

Estimate the ground-state energy of the HO variationally with the lorentzian function **1/(a^2 + z^2)** from Exercise 12.3-2 above. Compare with Problem 7.1-1 and with the previous problem. Hint: Be sure to normalize this trial function on the lattice or divide the energy expectation value by the norm.

FFT 12.5

We can thus apply an almost arbitrary hamiltonian to an equally arbitrary wavepacket, albeit only to the accuracy of a finite lattice representation. In analogy with our calculations in Section 11.10, we can also perform similar computations with more complicated operators, in particular, with the time-development operator. We thus obtain a very powerful, even if approximate, method for solving the time-dependent Schrödinger equation for an almost arbitrary initial state. We shall investigate the technique shortly.

What makes the approach particularly noteworthy is that the evaluation of the discrete Fourier transform can be considerably optimized. This is the origin of the *FFT* algorithm and of the built-in functions **Fourier** and **InverseFourier**. Basically, a straightforward evaluation of a discrete transform, like the one **FTGrid** performs, takes a list **psiGrid** and converts it into another list **phiGrid** by a matrix multiplication (see Exercise 12.5-1 below). Given list dimension **nmax**, this requires at least **nmax^2** complex-valued multiplications. The *FFT*, on the other hand, rearranges the sum and introduces some clever bookkeeping to reduce the number of multiplications to roughly **nmax Log[nmax,2]** (refer to *Numerical Recipes* [54], Chapter 12, and references therein). The result even for relatively small values of **nmax** can be an enormous savings in computer time.

Exercise 12.5-1

Redefine **FTGrid** and **InvFTGrid** using matrix multiplication. Thus, define a *unitary* square matrix **FTmatrix** of dimension **nmax** such that its inverse is the complex conjugate of its transpose (cf. Section 10.6). Then, verify by example the following relations:

```
   FTGrid[nList] == L/Sqrt[2Pi nmax] FTmatrix.nList
InvFTGrid[qList] == Sqrt[2Pi nmax]/L Conjugate[Transpose[
                         FTmatrix]].qList
```

See Problem 12.5-1.

Exercise 12.5-2

Compare values of **n^2** and **n Log[n, 2]** as a function of **n** up to **n = 2^15**.

Let's introduce the built-in *FFT* functions **Fourier** and **InverseFourier** by checking the Fourier integral theorem as we did above with **FTGrid** and **InvFTGrid**. Using the HO ground-state, we thus verify that

```
psiHOzGrid[0] - InverseFourier[ Fourier[psiHOzGrid[0]] ] //
    Chop

   {0, 0, 0, 0, 0, 0, 0, 0, 0, 0, 0, 0, 0, 0, 0, 0, 0, 0, 0, 0,

    0, 0, 0, 0, 0, 0, 0, 0, 0, 0, 0, 0}
```

We should not conclude, however, that these built-in transforms are directly equivalent to our functions **FTGrid** and **InvFTGrid**. There are in fact important differences. For example, **Fourier** and **InverseFourier** take the lattice origin of their argument list to be at the left end of the lattice, i.e. at the first element of the argument list. Our lattices **zGrid** and **kzGrid** and hence our wavefunctions, on the other hand, have been set up with origins in the middle of the lattice. Therefore, in order to be compatible with the built-in functions, we need to shift or *rotate* our lattice wavefunctions by half a lattice length. If, however, we're applying an operator, we can also simply shift the lattice representation of the operator instead (see Problem 12.5-2).

We can demonstrate how this works with the local energy. We thus define a shifted momentum lattice according to

```
kzpGrid = kzGrid //RotateLeft[#,nmax/2]&
   {0., 0.523599, 1.0472, 1.5708, 2.0944, 2.61799, 3.14159,

    3.66519, 4.18879, 4.71239, 5.23599, 5.75959, 6.28319,

    6.80678, 7.33038, 7.85398, -8.37758, -7.85398, -7.33038,

    -6.80678, -6.28319, -5.75959,-5.23599, -4.71239,

    -4.18879, -3.66519, -3.14159, -2.61799, -2.0944,-1.5708,

    -1.0472, -0.523599}
```

where here the built-in function **RotateLeft** performs the required shift by half a lattice length (Refer to **?RotateLeft**. Also, we could have just as well used the built-in function **RotateRight**.) This alternative lattice is plotted in Figure 12.5-1 along with the original momentum lattice, which makes clear that the origin **kz = 0** has been moved to the left end of the lattice. The left half of **kzpGrid** now contains the positive momenta with maximum **kzmax** at (nearly) the middle of the lattice while its right half now contains the negative momenta starting near **-kzmax** in the middle and ending near **kz = 0** on the right end. This structure corresponds exactly to that of **Fourier** and **InverseFourier**.

We can thus compute the local energy as we did before, but now with the built-in functions. For example, for the HO ground-state, we obtain using **kzpGrid**

```
ElocalFFT[0] =
    InverseFourier[kzpGrid^2/2 Fourier[psiHOzGrid[0]]]/
        psiHOzGrid[0] + zGrid^2/2 //Chop
                            -8                        -9
  {-27.3477 - 3.32368 10    I, 0.943684 + 1.54429 10    I,

                        -10
    0.480801 - 1.63716 10     I, 0.50139, 0.499859, 0.500019,

    0.499997, 0.500001, 0.5, 0.5, 0.5, 0.5, 0.5, 0.5, 0.5,

    0.5, 0.5, 0.5, 0.5, 0.5, 0.5, 0.5, 0.5, 0.5, 0.500001,

    0.499997, 0.500019, 0.499859, 0.50139,

                        -10                         -9
    0.480801 - 2.02956 10     I, 0.943684 + 3.82668 10    I}
```

which we see is very close to the quantity **Elocal[0]** we computed in Section 12.4. We therefore obtain the ground-state energy expectation value as before according to

```
(psiHOzGrid[0]^2).ElocalFFT[0] dz //Chop

    0.5
```

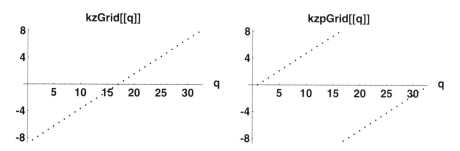

Figure 12.5-1. Our original momentum lattice on the left and the new shifted lattice on the right, both as functions of the lattice index q.

As we shall see in the problems below, the remaining differences in our discrete transform functions **FTGrid** and **InvFTGrid** and the built-in functions **Fourier** and **InverseFourier** are mostly reciprocal. That is, for most tasks requiring successive application of these functions, the differences tend to cancel out and can be ignored.

Exercise 12.5-3

Check the HO energies corresponding to the $n = 1$ and 2 lattice wavefunctions using the built-in functions **Fourier** and **InverseFourier**. Refer to Exercise 12.4-2 above.

Problem 12.5-1

Derive the connection between our discrete transform definitions **FTGrid** and **InvFT-Grid** and the built-in functions **Fourier** and **InverseFourier**.

(a) The built-in functions **Fourier** and **InverseFourier** are defined by the sums (see Wolfram [62], Section 3.8.3)

```
fourier[nList_] :=
    Table[
        Sum[
            E^( 2Pi I (q-1)(n-1)/nmax) nList[[n]],
            {n,1,nmax}
        ]/Sqrt[nmax],
        {q,1,nmax}
    ] //N //Chop

inversefourier[qList_] :=
    Table[
        Sum[
            E^(-2Pi I (q-1)(n-1)/nmax) qList[[q]],
            {q,1,nmax}
        ]/Sqrt[nmax],
        {n,1,nmax}
    ] //N //Chop
```

The built-in functions are computed, however, according to the *FFT* algorithm using compiled internal routines and therefore evaluate considerably faster than the above sums. These definitions of discrete Fourier transforms are essentially conventional (see *Numerical Recipes* [54], Chapter 12). Demonstrate by example the equivalence of the sums **fourier** and **inversefourier** with the built-in functions.

(b) Now relate our functions **FTGrid** and **InvFTGrid** to the sums in Part **(a)**, using paper and pencil if necessary, and show that

```
ftGrid[nList_] :=
    L/Sqrt[2Pi nmax] *
    Table[ (-1)^(q-1) *
```

```
        Sum[
            (-1)^(n-1) E^(-2Pi I (q-1)(n-1)/nmax) *
                nList[[n]],
            {n,1,nmax}
        ]/Sqrt[nmax],
        {q,1,nmax}
    ] //N //Chop

invftGrid[qList_] :=
    Sqrt[2Pi nmax]/L *
    Table[ (-1)^(n-1) *
        Sum[
            (-1)^(q-1) E^( 2Pi I (q-1)(n-1)/nmax) *
                qList[[q]],
            {q,1,nmax}
        ]/Sqrt[nmax],
        {n,1,nmax}
    ] //N //Chop
```

Verify these definitions by comparison with `FTGrid` and `InvFTGrid`.

Note that the coefficients `L/Sqrt[2Pi nmax]` and `Sqrt[2Pi nmax]/L` are reciprocal and appeared in Exercise 12.5-1. Note especially that the signs of the exponents in `ftGrid` and `invftGrid` are just the reverse of those in `fourier` and `inversefourier` in Part (a). Thus, `FTGrid` is proportional to `InverseFourier` and `InvFTGrid` to `Fourier`. (See Part (c). Also, you might keep in mind Exercise 11.1-2.)

(c) Show that the factors `(-1)^(n-1)` and `(-1)^(q-1)` in Part (b) arise from a momentum boost and a coordinate displacement, respectively, by half the lengths of the momentum and coordinate lattices (refer to Section 11.10) and are equivalent to multiplication by the lists

```
boost    = E^(I kzmax zGrid) //Chop
displace = E^(-I kzGrid L/2) //Chop
```

Thus, combine (a) and (b) to show that

```
FTGrid[nList] ==
    displace L/Sqrt[2Pi nmax] InverseFourier[boost nList]
InvFTGrid[qList] ==
    boost Sqrt[2Pi nmax]/L Fourier[displace qList]
```

Verify these relations by example.

Try interchanging `Fourier` and `InverseFourier` in these definitions but be sure to test both even and odd lattice functions. For quantities such as the kinetic energy which are computed as even powers of `kzGrid`, `Fourier` and `InverseFourier` can be interchanged (see Problems 12.5-2 and 12.5-3).

(d) Finally, show that the boosts and displacements in Part (c) are equivalent to rotations of the lattice functions by half a lattice length `nmax/2`. Thus, verify by example that

```
FTGrid[nList] ==
    L/Sqrt[2Pi nmax] *
    RotateLeft[
        InverseFourier[ RotateLeft[nList,nmax/2] ],
        nmax/2
    ] //N

InvFTGrid[qList] ==
    Sqrt[2Pi nmax]/L *
    RotateLeft[
        Fourier[ RotateLeft[qList,nmax/2] ],
        nmax/2
    ] //N
```

Problem 12.5-2

Compute the HO ground-state local energy **Elocal[0]** and the energy expectation value as in Section 12.4 with our original momentum lattice **kzGrid** but using the built-in functions **Fourier** and **InverseFourier**. Do this first by introducing the **boost** and **displace** operators from Problem 12.5-1, Part **(c)**, and then by using lattice rotations *of the functions* as in Problem 12.5-1, Part **(d)**. This latter rotation computation makes clear the origin of the rotated momentum lattice **kzpGrid**.

   Show by example that since the kinetic energy is computed as an even power of **kzGrid** or **kzpGrid**, **Fourier** and **InverseFourier** can be interchanged.

Problem 12.5-3

Compute the derivatives of some sample functions as we did at the end of Section 12.3 with our original momentum lattice **kzGrid** and the built-in functions **Fourier** and **InverseFourier**. Refer to Problem 12.5-2. Do this also with the rotated lattice **kzpGrid**. Note that in any case the order of application of **Fourier** and **InverseFourier** must correspond to **FTGrid** and **InvFTGrid** as in Problem 12.5-1.

Problem 12.5-4

Recompute the local energy **Elocal[0]** and the energy expectation value as in Problem 12.5-2 but using the *momentum* lattice representation and **phiHOkzGrid[0]**. Thus, first use the **boost** and **displace** operators from Problem 12.5-1, Part **(c)**, and then set up and use a rotated coordinated lattice **zpGrid** analogous to **kzpGrid**. Refer to Exercise 11.7-1.

12.6 Wavepacket Propagation

We are now ready to implement the ideas we introduced at the end of Section 11.10 and propagate an arbitrary lattice wavepacket in time. We shall thus apply the time-development operator (1.4-r1) directly to an initial wavepacket by taking a succession of small time steps **dt**. The short-step approximation allows us to split the time-development operator into kinetic- and potential-energy components that we can apply simply as lattice phase factors by

Fourier transforming back and forth between the coordinate and momentum lattices. The phase factors and therefore the time-development operator can thus be applied with straightforward lattice or list multiplication.

This "split-operator" technique was introduced by Fleck et al. [23], and has been nicely reviewed in the context of molecular dynamics by Kosloff [38]. The method has enjoyed enormous success in studies of femtosecond laser excitation of molecules. See for example Engel et al. [17]. We might also note that other propagation schemes of the time-dependent Schrödinger equation have been used in a variety of applications. Consult for example the special issue of Computer Physics Communications [39] and also the comparison study by Leforestier et al. [42].

To demonstrate the method's generality, let's drop a moving wavepacket into the model potential we introduced in Section 7.3 and diagonalized in Chapter 10. If we set up a coordinate lattice with length **L = 20** and **nmax = 128** points, we obtain a faithful representation of the potential on the lattice (see Figure 12.6-1). As in Sections 12.1 and 12.2, we then define the coordinate and momentum step sizes and maximum momentum according to

```
{nmax = 2^7, L = 20.0,
 dz = L/nmax, dkz = N[2Pi/L], kzmax = N[Pi/dz]}

  {128, 20., 0.15625, 0.314159, 20.1062}
```

and set up lattices as before:

```
zGrid   = Table[-L/2 + (n-1) dz,{n,1,nmax}];
kzGrid  = Table[-kzmax + (q-1) dkz,{q,1,nmax}];
kzpGrid = kzGrid //RotateLeft[#,nmax/2]&;
```

Here **kzpGrid** is the shifted momentum lattice we introduced in Section 12.5 for use with the built-in functions **Fourier** and **InverseFourier** (see Problem 12.5-2). Throughout most of this section and the next, we shall suppress output of lattice quantities with a semicolon for conciseness. When **nmax** is large, it's almost always better to **ListPlot** these quantities than to examine whole lists.

We thus define and evaluate the potential and kinetic energies on the coordinate and momentum lattices, respectively:

```
Vmodel[z_] = VHO[z] E^(-b z^2);     b  =  0.1;
VHO[z_] = Vo + z^2/2;               Vo = -2.0;

V = Vmodel[zGrid];        K = kzpGrid^2/2;
```

These quantities are plotted in Figure 12.6-1.

We perform the time development on a time grid in steps **dt** for a total of **ntmax** steps. Choosing a total propagation time **T**, we fix the step size according to

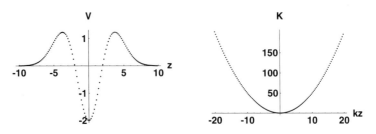

Figure 12.6-1. Potential and kinetic energies on the coordinate and momentum lattices, `zGrid` and `kzGrid`, respectively.

```
{ntmax = 2000, T = 10.0, dt = T/ntmax}
```

```
{2000, 10., 0.005}
```

We now have all we need to set up the time-development operator. Assuming the time step **dt** to be sufficiently small and using the ideas we introduced in Section 11.10, we split the time-development operator into potential- and kinetic-energy components according to

```
U[dt_] @ psi_ := UV[dt] InverseFourier[ UK[dt] Fourier[psi] ]
```

where **psi** is the wavepacket at time **t** and

```
UV[dt] = E^(-I V dt);        UK[dt] = E^(-I K dt);
```

i.e. coordinate- and momentum-space lists, respectively, and therefore simply lattice phase factors. Here, **Fourier[psi]** transforms **psi** to the momentum lattice allowing us to multiply by **UK[dt]**, while **InverseFourier** returns us to the coordinate lattice for a multiplication by **UV[dt]**. We thus apply the time-development operator in analogy with our computation of **ElocalFFT[0]** in the previous section.

We propagate the wavepacket forward **nt** steps from an initial wavepacket **psi0** by applying **U[dt]** repeatedly **nt** times. We can do this in a compact and efficient way with the built-in function **Nest** by defining (refer to Exercise 3.3, Appendix III).

```
psi[nt_,psi0_] := Nest[ U[dt][#]&, psi0, nt ]
```

If the wavepacket is competently sampled at every time step, then the transformation of the initial wavepacket to the one at time **t** preserves normalization, and the transformation is said to be *unitary*. This is because the component operators **UV[dt]** and especially **UK[dt]** are computed as unitary exponential operators (actually unitary matrices; see Exercise 12.5-1) and the product of two unitary operators is also unitary (see Section 15.3 and Exercise 15.3-1). That is, even the approximate time-development operator **UK[dt] UV[dt]** is unitary and preserves normalization exactly. This property greatly contributes to the stability of the resulting time-propagation scheme since there is no spurious amplification of wavepacket components. The split-operator approximation instead manifests itself in the wavepacket's time-dependent *phase* and in (slight) non-conservation of the wavepacket's energy (see Kosloff [38] and also Leforestier et al. [42]).

It will be convenient in the following to examine the energy expectation value **eExp** and energy uncertainty **dele** in the state **psi**, which we define according to

```
eExp[psi_] := Conjugate[psi].H @ psi/Conjugate[psi].psi
dele[psi_] := Conjugate[psi].H @ H @ psi/
              Conjugate[psi].psi - eExp[psi]^2 //Sqrt
H @ psi_   := InverseFourier[K Fourier[psi]] + V psi
```

where **H** is the lattice hamiltonian. Note that the ordering of the built-in functions **Fourier** and **InverseFourier** here and in the time-development operator is unimportant because the kinetic energy **K** and therefore **UK[dt]** are even functions of **kz** (see Problems 12.5-2 and 12.5-3). An example for which the ordering is important is the momentum expectation value, which we will compute in a moment.

It's convenient to take the initial wavepacket to be a gaussian (cf. Section 8.0), which we shall locate at the origin **z = 0** with width parameter **wPo = 2** and initial momentum **kPo = 1**. (It is important, of course, we do not take **kPo** bigger than **kzmax**.) Substituting **zGrid**, we thus enter

```
psi[0] = E^(I kPo z) E^(-wPo z^2/2) (wPo/Pi)^(1/4) /.
         {kPo -> 1.0, wPo -> 2.0} /.
         z -> zGrid //N;
```

and check the normalization, initial position and momentum according to

```
{Conjugate[psi[0]].psi[0] dz,
 Conjugate[psi[0]].(zGrid psi[0]) dz,
 Conjugate[psi[0]].Fourier[kzpGrid InverseFourier[
         psi[0]]] dz}//Chop
```

```
  {1., 0, 1.}
```

The ordering of **Fourier** and **InverseFourier** is important here in order to obtain the correct sign for the momentum expectation value, unlike for the kinetic-energy and time-development operators above.

Likewise, the energy expectation and uncertainty values are found to be

```
{eExp[psi[0]], dele[psi[0]]} //Chop
```

```
  {-0.835622, 1.11319}
```

We easily check that the energy expectation value agrees with the perturbation estimate we made in Section 7.4 for **wPo = 1** (see also Problem 12.4-1 above). These values tell us that the initial packet has peak energy between the ground- and first-excited states of the well (see Section 10.2) and spread in energy reaching into the continuum but much below the tops of the barriers. Nevertheless, we can expect these positive energy components to leak out of the barriers and propagate towards the ends of the lattice, if given enough time.

Exercise 12.6-1

Compute the kinetic-energy expectation value of the initial packet and compare with the analytic result of Exercise 4.2-1.

The method is numerically efficient because the *FFT* is. Although we will not pursue it, the technique is readily extended to two and even three dimensions, but after that requires truly supercomputing efforts. The computations we shall perform here are quite modest. Nevertheless, run times can be relatively long compared to other tasks we perform in this book. Therefore, you should have patience with the following sections and keep in mind that the run times given refer to a 33 MHz, 16Mb (NeXT Cube) computer. Quite adequate results, however, have been obtained in an afternoon on a 16 MHz, 8 Mb (Macintosh) computer by scaling down grid sizes by factors of two or four.

Despite *Mathematica*'s sluggishness, *Mathematica* code based on the built-in functions **Fourier and InverseFourier** is extremely compact. The present approach is therefore a good candidate for linking to fast remote machines and external *FFT* routines.

Time Development

We propagate the initial wavepacket out to a time **nt dt** by running **psi[nt, psi[0]]**. If we want to save intermediate wavepackets at each time step **dt**, we can replace **Nest** by **NestList** in the definition of **psi**. Such a computation, however, can become quite demanding of computer memory, so it is convenient to skip several steps before saving the wavepacket. This will minimize memory requirements and still allow us to obtain say a smooth animation.

We thus use the following **Do** loop to take snapshots of the wavepacket every **10** time steps and save **200** of the **ntmax = 2000** wavepackets computed. This takes less than 7 minutes to run.

```
Do[ psi[nt] = psi[10,psi[nt-10]], {nt,10,ntmax,10} ]
```

The resulting coordinate- and momentum-space probability densities **Abs[psi[nt]]^2** and **Abs[Fourier[psi[nt]]]^2** are shown in Figure 12.6-2 for selected times. The motion is of course analogous to that of a squeezed state in a pure HO potential, as shown in Figure 11.11-1. Preservation of normalization is easily checked.

In Figure 12.6-2, we plot the potential in the background for perspective using the **Epilog** option for **ListPlot** by including

```
Vbkgrd := Epilog -> {Line[Thread[{zGrid,V}]]}
```

A similar function gives the *FFT* of the potential for the momentum-space plots. (See also Figure 14.1-1.)

Exercise 12.6-2

Check that the normalization of the wavepacket is preserved. Make the same check in the momentum representation by defining the Fourier transform of the wavepacket according to

```
L/Sqrt[2Pi nmax] InverseFourier[psi]
```

from Problem 12.5-1. To what extent is energy conserved? Try and improve energy conservation by reducing the time step by a factor of 2 or 4.

Figure 12.6-2 makes clear that we can work with rather arbitrary wavepackets and potentials. Nevertheless, we rely here on lattice approximations, and we should always check our results and conclusions in the limit of decreasing lattice step size in both space and time, viz. of increasing lattice lengths and number of lattice points. Moreover, the finiteness and periodicity of the lattice places an unrealistic bound on the positive-energy continuum components of the wavepacket, which in reality are free to escape to infinity. Since the lattice confines these components forever, they will eventually overlap and spuriously interfere.

The following problems explore these effects and limitations and examine possible work-arounds. Problem 12.6-2 introduces an algorithm for absorbing wavepacket components which approach the ends of the lattice. Problem

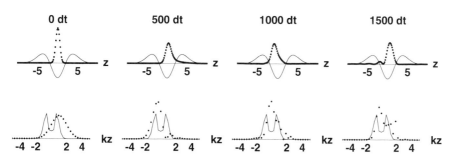

Figure 12.6-2. Time development of the wavepacket's probability density on the coordinate (top row) and momentum (bottom row) lattices. The solid curves show, respectively, the potential and its Fourier transform.

12.6-4 shows that an initial wavepacket can be reconstructed from the time-developed wavepacket using time reversal, an interesting result to observe which also provides a powerful check on the accuracy of the time propagation. We note in this regard that the more accurate `O[dt]^3` split of the time-development operator we derived in Problem 11.10-1 can be easily implemented (see Problem 12.7-1 below).

We should point out that we shall restrict our investigations here to potentials that are regular enough that the wavepacket has in principle periodic and continuous derivatives to all orders. Otherwise, the discrete Fourier transform or *FFT*, which forms the backbone of the approach, can develop severe convergence difficulties and uncontrollable errors. A notable pathological example is the Coulomb `-1/r` potential, which is infinite at the origin. Its ground-state wavefunction is of the form `r E^-r` for `{r,0,Infinity}` (see Section 20.2) and thus has discontinuous derivatives in the lattice length `L`, in the sense that its derivatives essentially vanish at the end of the lattice `rmax` but not at the origin. One possible way of handling the difficulty directly is to introduce *Lanczos sigma factors*, as in Problems 2.4-3 and 12.3-2.

Exercise 12.6-3

Animate `psi[t]` and its Fourier transform. Recompute the wavepacket with a somewhat increased energy (however, keep in mind the limit imposed by `kzmax`) and look for components which leak out or pass over the barriers.

Exercise 12.6-4

Change to the pure HO potential `z^2/2` and propagate the squeezed states shown in Figure 11.11-1.

Problem 12.6-1

Propagate an initially gaussian wavepacket displaced to one side of a deep rectangular well with a thin rectangular barrier in the middle and animate the motion. Refer to Chapter 2 and to Problems 4.2-4 and 10.6-5. You will know if the well is deep enough when penetration of the wavepacket into the walls is relatively small. You might find it convenient to make the wavepacket fairly fast by giving it a large initial momentum.

Adjust the barrier height to study the three cases corresponding to an energy above, equal to, and below the barrier. The penetration of the wavepacket through the barrier is an example of *quantum tunneling*, which we shall study in more detail in Chapter 14. Examine this effect as a function of wavepacket energy and barrier thickness. Compute the energy uncertainty to keep track of how much of the packet has energy above the barrier.

Problem 12.6-2

(a) Propagate a free-particle wavepacket and with animation compare the time-dependent probability density with the analytic result from Chapter 4. Check that the computed momentum distribution is shifted by the initial momentum but is otherwise independent of time. Check that the uncertainty product has the time dependence calculated in Problem 4.2-2. In order to minimize effects of a finite lattice, take a large enough initial momentum so that the wavepacket is able to outrun its own spreading before it reaches the ends of the lattice.

(b) Because the lattice is finite and the *FFT* intrinsically periodic in the lattice length L, a free wavepacket moving to the right automatically reappears on the left end of the lattice as it exits the right end. Examine this effect by propagating a wavepacket for a time L/kPo, the time of passage of a classical particle across the lattice, which should bring the centroid of the packet back to its starting point.

For many applications, it is useful to eliminate this effect by introducing some mechanism for absorbing or destroying wavepacket components that approach the ends of the lattice. One possibility is simply to introduce a damping factor in the time-development algorithm such as the following:

```
SetAttributes[Absorb,Listable]
Absorb[z_] := 1 - E^(-(z-zo)^2/a) /; z >= 0
Absorb[z_] := 1 - E^(-(z+zo)^2/a) /; z <  0
Absorption = Absorb[zGrid];

psiAbsorb[nt_,psi0_] := Nest[Absorption U[dt][#] &, psi0, nt]
```

Here the attribute **Listable** allows us to use the list **zGrid** as the argument of **Absorb[z]**, although it is defined with the conditionals /;. Clearly, the absorption process does not conserve probability, and this time-development algorithm is not unitary. Rather, it can be thought of as being generated by a *complex or optical potential* (see Exercise 5.0-6 and refer to D. Neuhauser and M. Baer, J. Chem. Phys. **90**, 4351 (1989)).

The position parameter **zo** can be adjusted to make **Absorption** vanish at the ends of the lattice **zGrid**. The range parameter **a** can be adjusted to control the interval over which the absorption takes place. If **a** is too small and hence the interval too short, **Absorption** itself will create an "impedance mismatch" and generate unwanted reflections of the wavepacket back onto the lattice. A large value of **a** could make the nonabsorbing part of the lattice too short to study anything useful. Examine plots of **Absorption** for various values of **a** and **zo**.

Use **psiAbsorb** to propagate the free-particle wavepacket in part **(a)** with **zo = 10.0** and **a = 5.0** and animate the motion. Note the tell-tale reflections of the wavepacket from the end of the lattice.

Problem 12.6-3

Redo Problem 12.6-2 but for the finite step potential defined by

```
SetAttributes[VStep,Listable]
VStep[z_] := Vo /; z >= 0;    Vo = 70;
VStep[z_] := 0  /; z <  0
V = VStep[zGrid];
```

to simulate scattering of a particle by a center of force.

Study three cases corresponding to an incident energy above, equal to, and below the step. In each case, examine the time development of the momentum distribution and explain. (The transmitted waves have shifted wavenumbers.) Compute the energy uncertainty to keep track of how much of each packet has energy above the step. You might want to compare with conventional discussions of this problem, as for example in Park [50], Chapter 4, or in Merzbacher [45], Chapter 5.

The splitting of the wavepacket into transmitted and reflected components is a nonclassical effect which reinforces the probability interpretation of the wavepacket (see Section 14.1).

The reflection and transmission coupled with the periodicity of the time development in the lattice length **L** quickly degenerates into a great deal of spurious interference across the lattice. Show that the absorption algorithm introduced in Problem 12.6-2 will eliminate most of these unphysical effects and generate a more realistic simulation of the time development.

Finally, for an incident energy above the step, compare scattering of packets incident from the left and right of the step.

Problem 12.6-4

Redo Problems 12.6-2 and 12.6-3 but propagate time-reversed wavepackets with initial state `Conjugate[psi[to>0]]` from the future of `psi[t]`. Refer to Exercises 5.0-6 and 11.10-1.

You want to avoid the absorption algorithm here since it destroys time-reversal symmetry. Therefore, you will have to make sure **to** is a time before the leading edges of the packets reach the ends of the lattice. For more accuracy with the potential step, you might want to bump up **nmax** and take smaller time steps. (Be sure to recompute **UV[dt]** and **UK[dt]** if you do.) Or you might want to use the **O[dt]^3** time-development operator we define in Problem 12.7-1 below.

Correlation Function

The propagation in time of an arbitrary initial wavepacket can also provide a powerful technique for determining the energy spectrum of the hamiltonian. The key is to introduce the so-called *correlation function* defined as the overlap integral of the propagated wavepacket **psi[t]** with the initial wavepacket **psi[0]**, viz.

```
c[t] ==
     Integrate[Conjugate[psi[0]] psi[t],{z,-Infinity,Infinity}]
```
$$(12.6\text{-}r1)$$

In the lattice representation, we can compute this simply as

```
c[Cpsi0_,psit_] := Cpsi0.psit dz
```

where **Cpsi0** is the complex **Conjugate** of the initial wavepacket. It is convenient to set up a procedure to collect together and evaluate this and similar quantities, such as expectation values, while propagating the wavepacket as a function of time. Thus, consider

```
TimeDevelopment[ntmax_,psi0_] :=
    Module[
        {psit = psi0, Cpsi0 = Conjugate[psi0]},
         c[t] = {};
         zExp = {}; zsqExp  = {}; kzExp = {}; kzsqExp = {};

        Do[
            AppendTo[c[t],
                c[Cpsi0,psit]];
            AppendTo[zExp,
                dz Conjugate[psit].(zGrid psit)];
            AppendTo[zsqExp,
                dz Conjugate[psit].(zGrid^2 psit)];
            AppendTo[kzExp,
                dz Conjugate[psit].
                    Fourier[kzpGrid InverseFourier[psit]]
            ];
            AppendTo[kzsqExp,
                dz Conjugate[psit].
                    Fourier[kzpGrid^2 InverseFourier[psit]]
            ];

            psit = U[dt] @ psit,
            {ntmax}
        ]
    ]
```

Here the correlation function `c[t]` and the desired expectation values, `zExp`, etc. are first initialized as empty lists `{}`. Then, each of these quantities is computed at the current time step and the result appended to the end of its respective list with the built-in function `AppendTo`. The `Do` loop permits a single evaluation of the wavepacket `psit` at time `t` before updating `psit` with another application of the time-development operator `U[dt]`. We avoid, in particular, having to store intermediate wavepackets and thus eliminate additional demand for computer memory regardless of the propagation time `T`.

We can run `TimeDevelopment` using our previous gaussian wavepacket `psi[0]`, although the choice of initial state is mostly unimportant (see also Problem 12.6-5 below). The results will be more interesting, however, if we increase the propagation time `T` and the number of time steps `ntmax`; of course, we have to recompute the time-development factors (still using the model potential) for the new time step `dt`. For plotting purposes, it is also useful to introduce a time grid `tGrid`.

```
{ntmax = 2^11, T = 60.0, dt = T/ntmax}

    {2048, 60., 0.0292969}

UV[dt] = E^(-I V dt);          UK[dt] = E^(-I K dt);

tGrid = Table[ (nt-1) dt, {nt,1,ntmax} ];
```

Here **tGrid** has been constructed in analogy with **zGrid** but starting at **t = 0**. It thus takes just over 20 minutes to compute **ntmax** time steps upon entering

```
TimeDevelopment[ntmax,psi[0]]
```

When this is finished, the correlation function and expectation values are represented by lists each containing **ntmax** elements, one for each time step. We extract, for example, the first two elements of the correlation-function list and pair them with the corresponding times according to

```
Thread[{tGrid,c[t]}] [[{1,2}]] //Chop
```

    {{0, 1.}, {0.0292969, 0.999169 + 0.0244734 I}}

The correlation function, the coordinate expectation value, and the coordinate-momentum uncertainty product are shown as **ListPlots** in a **GraphicsArray** in Figure 12.6-3 as functions of time. Although the motion is fairly complicated, the period **2Pi/w** of classical oscillation in an approximate HO

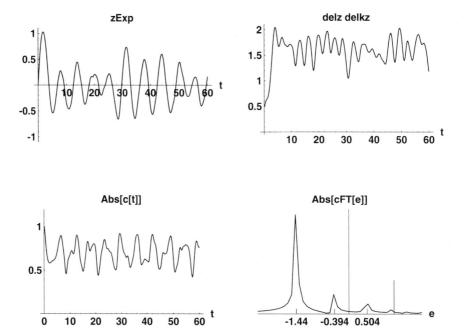

Figure 12.6-3. Shown are the coordinate expectation value (top left), the coordinate-momentum uncertainty product (top right), and the correlation function (bottom left), all as functions of time. The *FFT* of the correlation function is shown (bottom right) as a function of energy. Here the tick marks and their labels show energies we estimated in Section 10.2; also, the center vertical axis marks the continuum threshold **e = 0**, while the shorter right vertical axis marks the top of the potential barriers.

potential (with **w = 1**) is evident. To make this connection clear, we can redo our present analysis using a pure HO potential. The details are relegated to Problem 12.6-9, and Figure 12.6-4 summarizes the results. The correlation function thus also shows traces of a squeezing frequency **2w**, analogous to a pure squeezed state (see Section 8.3).

Exercise 12.6-5

If time and computer resources allow, increase **nmax** and **ntmax** and rerun **TimeDevelopment** to check that no new features appear in Figure 12.6-3.

Energy Spectrum

We can see the connection of the correlation function with the energy spectrum, if we consider a formal expansion of the state **psi[t]** in the desired energy eigenstates **E^(-I e[n] t) phi[n,z]** of the model hamiltonian, where the time-development phases are determined by the desired eigenenergies **e[n]**. We show in Exercise 12.6-6 that the correlation function can be formally expressed in terms of the expansion coefficients **a[n]** of the initial state **psi[0]** as

$$c[t] \ == \ Sum[Abs[a[n]]^2 \ E^{\char94}(-I \ e[n] \ t),\{n\}] \qquad (12.6\text{-}r2)$$

where here **{n}** is meant to symbolize a sum over the bound states and an integration over the continuum states.

We then show in Exercise 12.6-6 that the *nth* term in the Fourier transform of this expression with respect to time is a function of energy proportional to **Abs[a[n]]^2**. In particular, the discrete bound states appear as a sequence of Dirac delta functions centered on precisely the eigenenergies **e[n]**, while the continuum appears as a single broad curve with possible peaks corresponding to resonance states. Thus, the Fourier transform of the correlation function yields the desired energy spectrum, which is actually a frequency spectrum **e/h** although we've set **h = 1** here.

When we perform the time propagation on the computer in steps **dt**, we must estimate the infinite Fourier integral with a discrete Fourier transform on a finite interval of time **{t,0,T}**, for example with an *FFT*. We also see in Exercise 12.6-6 that this has the effect of replacing the delta functions in the spectrum by functions of finite width. In other words, the energy resolution of the method is limited in practice by the maximum propagation time **T** we can achieve (i.e. our computer allows). Clearly the method will work best for well-separated bound and resonance states.

Exercise 12.6-6

(a) Verify relation (12.6-r2) with paper and pencil if necessary.

(b) Estimate the Fourier transform of the correlation function with a finite time interval {t,0,T} by multiplying the sum (12.6-r2) by E^(I e t) and integrating over t. Show that the *nth* term of the resulting sum is proportional to

```
                2        T (e - e[n])
    2 Abs[a[n]]     Sin[-------------]
                                2
    _____
    
             e - e[n]
```

which for a bound state gives a peak of finite width positioned at the desired eigenenergy e[n]. Show that the peak becomes sharper as T increases by plotting Sin[T de]/de as a function of de for increasing values of T. Finally, use the result of Exercise 11.5-2 to show in the limit T -> Infinity that the *nth* term in the exact Fourier transform is proportional to Abs[a[n]]^2 DiracDelta[e-e[n]].

The Fourier transform of the correlation function is easily computed as a function of energy as

```
cFT[e] = Fourier[c[t]] //RotateLeft[#,ntmax/2]&;
```

where we rotate by half a lattice length to have the middle of the lattice correspond to zero energy. For reference and plotting, it is appropriate to construct an energy grid reciprocal to **tGrid** in the same way we set up **kzGrid** in Section 12.2. We thus define an energy step size **de**, maximum energy **emax**, and **eGrid** according to

```
{de = N[2Pi/T], emax = N[Pi/dt]}

  {0.10472, 107.233}
```

```
eGrid = Table[ -emax + (qe-1) de, {qe,1,ntmax} ];
```

The energy spectrum **cFT[e]** is shown in Figure 12.6-3 as a function of energy **e**. The abscissa marks the two bound states and first resonance we estimated by diagonalization in Chapter 10. The peaks in the spectrum match up fairly well with these previous estimates, especially the ground state. The spectrum clearly shows a second resonance *below* the tops of the potential barriers, marked in the figure by the right vertical line, and confirms our conclusions in Exercise 10.2-2 based on an improved diagonalization. Other weaker resonances are also discernable above the barriers. We shall analyze these continuum states and discuss their significance more in Chapter 14.

The widths of the peaks in the spectrum are controlled by the finite propagation time **T** (see Problem 12.6-6). This energy resolution can be increased by increasing **T** and therefore decreasing **de**. At the same time, however, we should increase **ntmax** in order to keep the time step **dt** small enough to maintain the accuracy of the time-development operator.

The method is readily generalized to obtain the energy eigenstates **phi[n, z]** of the model hamiltonian corresponding to the eigenenergies **e[n]**. The idea is to Fourier transform the propagated wavepacket *with respect to time* and tune the Fourier-transform energy to one of the eigenenergies, determined previously from the correlation-function spectrum. If the propagation time **T** is long enough, the Fourier transform will become proportional to the corresponding eigenfunction. The technique is referred to as *filtering* and is developed and applied in Problems 12.6-8 and 12.6-9. We should point out in this regard that these methods can be also be used to determine the widths or lifetimes of resonance states, quantities we shall define and discuss in Chapter 14 (refer to Engel et al. [18]).

Exercise 12.6-7

Show that the correlation function has the (time-reversal) symmetry `c[-t] == Conjugate[c[t]]`. What is the consequence of this symmetry on its Fourier transform?

Exercise 12.6-8

Show that the correlation function `c[2t]` is equivalent to the scalar product of `psi[t]` with `psi[-t]`, a state related to the time-reversed wavepacket `Conjugate[psi[-t]]` (cf. Exercises 5.0-6 and 12.6-6). Hence, show that the correlation function up to time `t` can be computed from a propagation of the wavepacket `psi[t]` just up to time `t/2`, in effect reducing computational effort by a factor of two. This useful trick was suggested by V. Engel, Chem. Phys. Lett. **189**, 76 (1992), who also showed how to circumvent the drawback that the correlation function is obtained only in steps `2 dt`, where `dt` is the time step of the propagation.

   Test the method out by recomputing the energy spectrum in Figure 12.6-3.

Exercise 12.6-9

We can compute the correlation function more directly using the built-in the function `NestList`, analogous to the propagation of `psi[t]` (see Exercise 3.3, Apendix III). Compare for example the procedure

```
correlation[ntmax_,psi0_] :=
   Map[
       dz Dot[Conjugate[psi0],#]&,
       NestList[U[dt][#]&, psi0, ntmax-1]
   ] //Chop
```

with that defined in **TimeDevelopment** to compute the correlation function `c[t]`. This algorithm has the nice feature that it eliminates the **Do** loop, but it also requires that every intermediate wavepacket be saved. Can you find a way around this and relieve the burden on computer memory without the **Do** loop?

Problem 12.6-5

We can pick out the even- or odd-parity spectrum of levels by propagating an even- or odd-parity initial packet. Explain, with paper and pencil if necessary, and demonstrate with a computation.

Problem 12.6-6

Demonstrate that the energy resolution is inversely proportional to the propagation time **T** by say doubling **T**.

Problem 12.6-7

Redo the infinite and semi-infinite rectangular wells from Chapter 2 and Problem 6.2-1 and generate Figures 12.6-2 and 12.6-3 for these systems.

Problem 12.6-8

The propagation of an arbitrary initial wavepacket can also be used to filter out energy eigenstates of the system (refer to M.D. Feit et al., J. Comp. Phys. **47**, 412 (1982)). To see how this works, consider again a formal expansion of the time-dependent wavepacket **psi[t]** in the desired energy eigenstates **E^(-I e[n] t) phi[n, z]**. Estimate now the Fourier time transform of this expansion by introducing a finite interval of time **{t, 0, T}**, as we did with the correlation function in Exercise 12.6-6, and show that the *nth* term of the resulting sum is proportional to

$$
\frac{2 \; \mathtt{phi[n, z]} \; \mathtt{Sin}[\dfrac{\mathtt{T \; (e - e[n])}}{2}]}{\mathtt{e - e[n]}}
$$

Thus, if **T** is long enough and we tune **e** to one of the eigenenergies **e[n]**, determined beforehand say from the correlation-function spectrum, then the sum is filtered to a single term proportional to the corresponding eigenfunction **phi[n, z]**. For finite **T**, the method clearly works best for well-separated bound and resonance states, analogous to extracting the eigenenergies from the correlation spectrum.

Use the method to try filtering the bound states of the model potential. Thus, consider the following slight modification of our procedure **psi[nt, psio]** for propagating **psi[t]**

```
phi[en_] :=
    Nest[# + E^(I en dt) Absorption U[dt][#] &, psi[0], ntmax]
```

in order to estimate the Fourier time transform of the wavepacket while simultaneously propagating the wavepacket on our finite time grid. Note that intermediate wavepackets are not saved here so the procedure is memory efficient. Since part of the initial wavepacket can leak out of the well and generate spurious interference effects, we include the factor **Absorption** introduced in Problem 12.6-2 above.

Refer to Exercise 10.2-2 and to the next section for better estimates of the energies. For convenience, use HO eigenfunctions for the initial wavepackets. Finally, try filtering the first few resonance states.

Problem 12.6-9

The harmonic oscillator provides as usual an important test case. The system is convenient for lattice studies because wavepackets are confined to the potential well for all energies and can be kept away from the ends of the lattice in a natural way. With **h = m = 1**, propagate a squeezed state in the HO potential **z^2/2** with **w = 1**. Try **wPo = 2** for the initial width parameter of the squeezed state and **zPo = 0** and

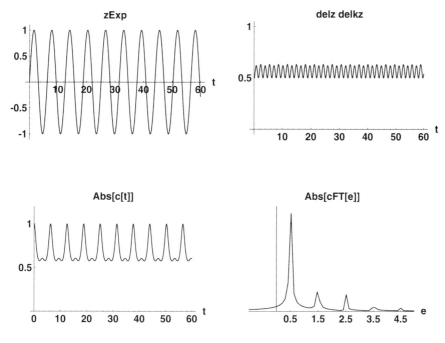

Figure 12.6-4. Figure 12.6-3 but for a pure HO squeezed state with $w = 1$ and $wPo = 2$.

$kPo = 1$ for its initial displacement and momentum. Refer to Chapter 8. Check the energy expectation value and uncertainty of the initial wavepacket and compare with the analytic value from Exercise 8.0-2.

(a) Run `TimeDevelopment[ntmax,psi[0]]` and generate Figure 12.6-4. Make separate plots of the `Re` and `Im` parts of the correlation function. Compare a plot of the uncertainty product with the analytic result from Problem 8.3-1. Estimate from the energy spectrum `cFT[e]` the relative magnitudes `Abs[c[n]]` of the expansion coefficients of the initial state and compare with the analytic values from Section 8.1. Redo Problem 12.6-5.

(b) Calculate the correlation function analytically using the result for the squeezed-state time dependence from Section 8.2 or Problem 8.2-2. Make plots of the correlation function and its `Re` and `Im` parts and compare with your plots of the lattice-propagated correlation in part **(a)**.

(c) Redo parts **(a)** and **(b)** for the coherent state defined by `wPo = w = 1`.

(d) Use the filtering method introduced in Problem 12.6-8 to obtain the first few HO energy eigenstates. Because the HO potential naturally confines the wavepacket away from the ends of the lattice, you may omit the factor `Absorption` from the time development. Normalize your results and check them by plotting with the exact HO eigenfunctions from Section 6.6.

Problem 12.6-10

Fearn and Lamb [21] have applied the present wavepacket propagation scheme to the problem of quantum measurement (see Section 2.6). They examined how wavepacket tunneling through the barrier of a model potential, defined by $V = -1.6725 \ x^2 + 0.0408 \ x^4$, is affected by position measurements on the packet.

Plot their potential energy and observe the symmetric double wells and central barrier. Compute the correlation function for this potential and use it to estimate the energy spectrum of a particle of mass $m = 1$ confined to the potential with $h = 1$. You should find that there are just four states below the barrier, in particular, two pairs of nearly degenerate levels. The members of each pair are distinguished by their even- and odd-parity, and hence their energies can be estimated separately by propagating both even and odd parity initial wavepackets (cf. Problem 12.6-5). Use your energies to estimate the tunneling time through the barrier of a wavepacket which is a linear superposition of just the ground and first-excited states (cf. Exercise 2.5-2).

Fearn and Lamb included coupling to a meter in order to simulate measurement of the particle's position. Introducing a rapid succession of measurements, they found an enhanced tunneling rate through the barrier and thus little evidence for the *quantum Zeno effect*. That is, a suppressed tunneling rate would support the idea that rapid repeated measurements of the particle's position should restrain the particle to its initial position and hence the correlation function to a value near unity. See also Home and Whitaker [33] and Fearn and Lamb [22].

12.7 **Quantum Diffusion**

It may seem somewhat paradoxical or at least surprising that a time-propagation scheme can be used to obtain eigenenergies and eigenfunctions of a system, i.e. solutions of the *time-independent* Schrödinger equation. Nevertheless, we've seen in the previous section how we can extract eigenenergies from the Fourier transform of the correlation function and in Problem 12.6-8 how we can filter the corresponding eigenfunctions.

If it's just the eigenenergies and eigenfunctions we seek, especially the ground state, there is a simple way to improve convergence and obtain very accurate results. The trick is to replace the time t by the time parameter $tC = I \ t$ and solve in effect the Schrödinger wave equation (1.1-r1) as a classical *diffusion equation*, namely,

$$-D[psi[z,tC],tC] == hamiltonian[V] \ @ \ psi[z,tC] \qquad (12.7\text{-}r1)$$

To see how this works, consider again a formal expansion of the solution $psi[z,t]$ of the wave equation in the desired energy eigenstates $E^{\wedge}(-I \ e[n]t)phi[n,z]$ (cf. Exercise 12.6-6). If we now replace t by the time parameter tC, $psi[z,tC]$ becomes a solution of (12.7-r1). If we introduce tC into the formal expansion in energy eigenstates, the time-dependent phases become strictly decaying exponentials $E^{\wedge}(-(e[n] - e[0])tC)$ relative to the ground state with decay times $1/(e[n] - e[0])$. Therefore, by propagating $psi[z,tC]$ for long times tC, the excited states will decay out each in turn starting with the highest and leaving just the ground state, like the diffusion of mixed gases through a porous membrane. In other words, `psi[z,tC -> Infinity]` becomes proportional to `phi[0,z]`. The method is analogous to the filtering technique we developed in Problem 12.6-8 and is referred to as *diffusion filtering* or *quantum diffusion*.

Exercise 12.7-1

Expand `psi[z,tC]` with paper and pencil in energy eigenfunctions `phi[n,z]` and show that it is a solution of the diffusion equation (12.7-r1). Evaluate the normalization and show that `phi[0,z]` is obtained from `psi[z,tC -> Infinity]` after renormalization.

In principle, `psi[z,tC]` converges to the exact ground state only if the propagation time is long enough and the time step size small enough. The method nevertheless improves on variational and diagonalization techniques, which only provide upper bounds on the exact energies. In fact, diffusion filtering is mostly independent of the initial state, although convergence is greatly enhanced if the *initial* and hence *trial wavefunction* is close to the desired eigenfunction. Moreover, if as in a variational calculation we require `psi[z,tC]` to be orthogonal to the first `n` excited states, the diffusion will converge to the *(n+1)th* excited state.

We easily implement diffusion filtering using our original time-propagation scheme simply by redefining the time-development factors `UV[dt]` and `UK[dt]` with the substitution `dt -> dtC = I dt`. Setting up a shorter time grid, because convergence is faster, we thus enter

```
{ntmax = 400, T = 4.0, dtC = T/ntmax}

  {400, 4., 0.01}

tCGrid = Table[(nt-1) dtC, {nt,1,ntmax}];

UV[dtC] = E^(-V dtC);    UK[dtC] = E^(-K dtC);
```

Clearly, the decaying (non-unitary) nature of `UV[dtC]` and `UK[dtC]` and therefore of the time-development operator means that the time propagation will no longer preserve normalization. Hence, `phi[0,z]` is obtained from `psi[z,tC -> Infinity]` after renormalization (see Exercise 12.7-1 above).

To demonstrate how well this works, let's use *white noise* for the initial wavefunction which we generate as a list of **nmax** random numbers between 0 and 1. That is, let's take simply (see **?Random**)

```
psi[0] = Table[Random[],{nmax}];
```

This unnormalized trial wavefunction is plotted in Figure 12.7-1 with **List-Plot** and the option **PlotJoined -> True**.

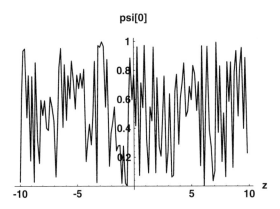

Figure 12.7-1.   "White-noise" trial wavefunction (not normalized).

We can propagate this initial wavefunction with the same procedure we used in real time. Thus, we can again enter

```
Do[ psi[nt] = psi[10,psi[nt-10]], {nt,10,ntmax,10} ]
```

Figure 12.7-2 shows snapshots of the resulting normalized probability distributions converging to the ground state. When normalized, **psi[nt]** becomes in the long-time limit the ground-state eigenfunction **phi[0,zGrid]** of the model potential.

Exercise 12.7-2

Normalize **psi[nt]** and animate the convergence of the initial state to the ground state shown in Figure 12.7-2.

We verify the rapid convergence to the ground-state energy by making a **ListPlot** of energy expectation values **eExp[psi[nt]]** as in Figure 12.7-

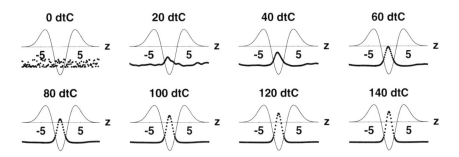

Figure 12.7-2.   Evolution of the normalized time-dependent probability density **Abs[psi[nt]]^2** from white noise to the ground state of the model potential.

```
ListPlot[
    Table[
        {tCGrid[[nt]],eExp[psi[nt]]},
        {nt,10,ntmax,10}
    ] //Chop,
    AxesLabel -> {" tC",""}, PlotLabel -> "eExp[tC]",
    PlotRange -> {-2,1}
];
```

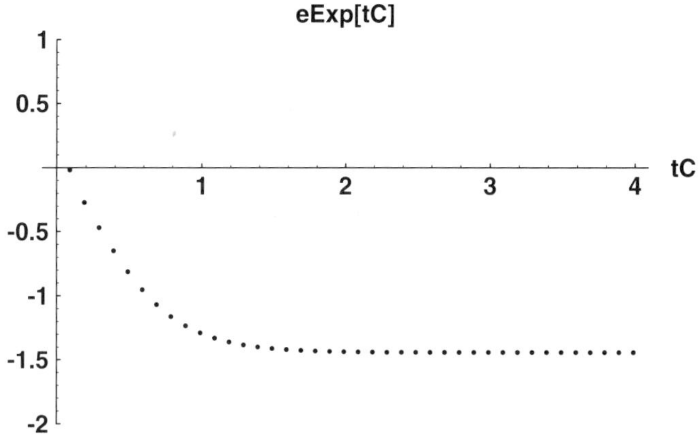

Figure 12.7-3. Convergence of the energy expectation value to the ground-state energy.

3. Also, we can examine a table of the energy expectation values in the long-time limit:

```
Table[eExp[psi[nt]],{nt,ntmax-40,ntmax,10}] //Chop

    {-1.44144, -1.44145, -1.44146, -1.44147, -1.44148}
```

These values are very close to our diagonalization estimate of the ground-state energy from Section 10.2. Here, the last value -1.44148 agrees well with the improved estimate from Exercise 10.2-2. In the limit of a very long propagation time and very short time steps, the energy expectation value will approach the true ground-state energy. (Note that you will obtain slightly different results if you recompute **psi[0]** and thereby introduce another set of random numbers.)

We can make a strong check of the filtered eigenfunction by plotting the local energy as in Figure 12.7-4. We see that this quantity is as desired nearly constant at least over the range of the ground-state eigenfunction (Figure 12.7-2) and equals the asymptotic energy expectation value from Figure 12.7-3. We show in the following exercises and problem that we can improve this situation dramatically with better convergence.

We point out in passing that diffusion filtering is often computed with *Monte Carlo techniques*, rather than with a uniform lattice, by introducing a

```
ListPlot[
    Thread[{zGrid,H @ psi[ntmax]/psi[ntmax]}] //Chop,
    AxesLabel -> {" z",""}, PlotLabel -> "eLocal",
    PlotRange -> {-2,2}, Vbkgrd
];
```

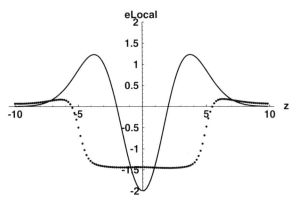

Figure 12.7-4. Ground-state local energy after **ntmax** time steps. The solid curve shows the potential.

set of random walkers and evaluating the time propagation as a *path integral*. The general approach is called *Quantum Monte Carlo*. It has the advantage of being efficient even for systems with many degrees of freedom such as molecules or quantum liquids but has the severe drawback that it is not known how to extract excited states generally. (Refer to Koonin [37] and to Beckmann and Feagin [5] and references therein.) Finally, we note that Quantum Monte Carlo is closely related to *Greens Function Monte Carlo*, which filters with a time-independent iteration scheme and thus eliminates the time-step approximation. (See Kalos and Whitlock [36] and references therein.)

Exercise 12.7-3

Show that if the initial trial function is a stationary gaussian convergence improves. Examine the local energy.

Exercise 12.7-4

Increase the propagation time **T** in order to obtain a better estimate of the ground-state *local* energy and hence of the ground-state energy.

Problem 12.7-1

Implement the **O[dt]^3** time-development operator from Problem 11.10-1 in a new procedure for propagating wavepackets. Thus, consider the following function

```
psiBigStep[nt_,psi0_] :=
    Fourier[ UK[dt/2] *
```

```
        InverseFourier[
            psi[nt-1,
                UV[dt] Fourier[UK[dt/2] InverseFourier[psi0]]
            ]
        ]
    ]
```

in order to obtain a better estimate of the ground-state local energy by increasing the propagation time T but keeping the total number of time steps `ntmax` fixed.

Excited States

Finally, let's see how we can achieve convergence to the first excited state. The key is to define a projection operator which will keep the propagated wavefunction orthogonal at all times to the ground state. This will ensure that a ground-state component will not develop in the propagated wavefunction and that convergence is to the next highest or first-excited state.

Exercise 12.7-5

Redo Exercise 12.7-1 but include the constraint that `psi[z,tC]` be orthogonal to the ground state `phi[0,z]`. Thus, show that the first-excited state `phi[1,z]` is obtained in the limit `tC -> Infinity`.

Consider therefore the function

```
project[psi_,phi_] := psi - phi Conjugate[phi].psi dz
```

to remove the component of a *normalized* lattice function `phi` from another lattice function `psi`. (The second term on the right-hand side proportional to `phi` is actually the projection operator acting on `psi`.) It is easy to see that `project[psi,phi]` is orthogonal to `phi`.

Exercise 12.7-6

Check the properties of `project` with some examples. Note that we implicitly assume `phi` is normalized. How should we generalize `project` if this isn't the case?

Now we can define a time-development procedure for the first-excited state simply by wrapping `project[..., psiGrd]` around each time step, where `psiGrd` is the ground-state wavefunction. Thus, we modify our original procedure `psi[nt,psi0]` by defining

```
psi1st[nt_,psi0_] :=
    Nest[ project[U[dtC][#], psiGrd]&, psi0, nt]
```

We can use our previous result `psi[ntmax]` for the ground-state projection function. Normalizing, we therefore define

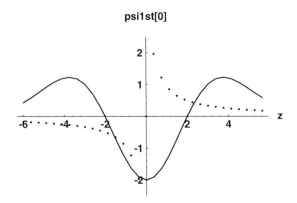

Figure 12.7-5. Trial wavefunction for the first-excited state. The solid curve shows the potential.

```
psiGrd = psi[ntmax]/Sqrt[psi[ntmax].psi[ntmax] dz];
```

Let's take for our initial trial state a lorentzian function with one node and odd parity and thus define

```
psi1st[0] = zGrid/(0.05+zGrid^2);
```

This function is plotted in Figure 12.7-5. We propagate this initial state in the usual way

```
Do[psi1st[nt] = psi1st[10,psi1st[nt-10]],{nt,10,ntmax,10}]
```

and obtain a table of the energy expectation values in the long-time limit according to

```
Table[eExp[psi1st[nt]],{nt,ntmax-40,ntmax,10}] //Chop
```

```
{-0.397963, -0.397986, -0.398007, -0.398025, -0.39804}
```

Here the last value -0.39804 is better (i.e. lower) than our diagonalization estimate from Section 10.2 and again agrees well with the improved estimate from Exercise 10.2-2. (Refer also to Figure 12.6-3.) We show in Exercise 12.7-7 and Problem 12.7-2 that convergence and accuracy improves if a better trial function is used. Figure 12.7-6 compares convergence of the energy expectation values of the ground and first-excited states.

Clearly, the method can be generalized to higher excited states by projecting out all previously computed states in turn. We can accomplish this by wrapping **project** around each time step once for each eigenfunction previously computed.

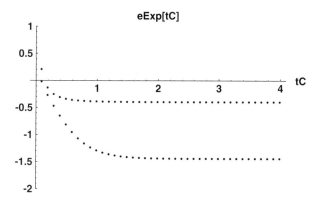

Figure 12.7-6.   Time convergence of the energy expectation value to the ground-state (bottom curve from Figure 12.7-3) and first-excited-state (top curve) energies.

Exercise 12.7-7

Plot the convergence of the normalized wavefunction to the first excited state as in Figure 12.7-2 and then animate the motion. Redo the computation with a gaussian initial state with the correct parity and show that convergence improves.

Problem 12.7-2

Redo the previous exercise by implementing the higher-order time-development operator with the procedure **psiBigStep** from Problem 12.7-1 modified to include **project[psi,psiGrd]**. Plot the local energy and compare your best energy expectation value with your best diagonalization estimate from Exercise 10.2-2.

Problem 12.7-3

Use quantum diffusion to compute the first few energy levels of the harmonic oscillator.

Problem 12.7-4

Use quantum diffusion to compute the bound-state energy spectrum of the finite rectangular well from Problem 6.2-1.

Problem 12.7-5

Check the results of Problem 7.3-2 for the anharmonic oscillator by quantum diffusion.

Morse Oscillator

Long ago Morse introduced the potential function

```
V = -De + De(1 - E^(-b x))^2 //Expand
```

$$\frac{De}{E^{2\,b\,x}} - \frac{2\,De}{E^{b\,x}}$$

to model the vibrational energy of a diatomic molecule with an analytical solution of the Schrödinger equation. (See Morse [48] and Morse and Feshbach [49].) This potential is shown in Figure 13.0-1 as a function of the dimensionless coordinate $y = b\,x$ and scaled by the depth parameter **De**. Because the model is practical and found to be a very good approximation, especially for low-energy vibrations, the Morse oscillator is often referred to in molecular applications. Here we mostly consider it as a system closely related to the harmonic oscillator, but one with useful differences which nevertheless affords a full solution in terms of confluent hypergeometric functions.

We see that the potential approaches zero for large positive **x**, has the minimum value **De** at **x = 0**, and quickly becomes large and positive for large negative **x**. The overall form is what one expects for a diatomic molecule when **x = r - ro** measures the separation of the two nuclei relative to an equilibrium separation **ro**. For large separations **r > ro**, the neutral atoms don't interact, whereas for small separations **r < ro** the nuclear Coulomb repulsion pushes the two atoms apart. Equilibrium occurs for **r ~ ro** as a balance between the nuclear repulsion and the electronic binding. Evidently, for negative energies the Morse oscillator is bounded and the spectrum of negative eigenenergies is discrete. On the other hand, the spectrum of positive eigenenergies is continuous. At the continuum threshold **Energy = 0**,

```
Plot[
    V/De /. x -> y/b,
    {y,-1,4}, PlotRange -> {-1.25,1},
    AxesLabel->{" b x","V[x]/De"}
];
```

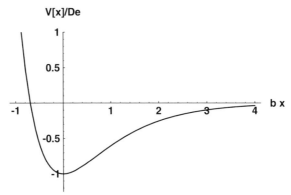

Figure 13.0-1.   The Morse potential.

the molecule dissociates into two neutral atoms, and **De** is also the *dissociation energy* of the molecule relative to the potential minimum.

The Morse potential is asymmetric and doesn't have a definite parity. Dynamically, the restoring force is softer on the right for large **x** than on the left. Hence, parity arguments won't enter here, and the Morse oscillator is often referred to as an *asymmetric oscillator*. The width parameter **b** is therefore also called the *asymmetry parameter*.

13.1 Kummer's Equation

We begin by transforming the Schrödinger equation for a particle of mass **m** oscillating in a Morse potential to the confluent hypergeometric equation introduced in Section 6.5. We shall follow the general approach we used to obtain a scaled Schrödinger equation for the harmonic oscillator. Introducing the Morse potential into the Schrödinger differential operator from Section 1.3, the equation of motion takes the form

```
schroedingerD[V_] @ psi_ :=
    -h^2/(2m) D[psi,{x,2}] + V psi - Energy psi

schroedingerD[V] @ psi[x] == 0
```

$$
(\frac{De}{E^{2 b x}} - \frac{2 De}{E^{b x}})\ psi[x]\ -\ Energy\ psi[x]\ -\ \frac{h^{2}\ psi''[x]}{2\ m}\ ==\ 0
$$

We can eliminate the exponentials here by introducing a new independent variable $x \to z = a E^{\wedge}(-b x)$. Hence, we enter

```
psi[x] = phi[z] /. z -> a E^(-b x);
```

```
schroedingerD[V] @ psi[x] //Expand
```

$$\frac{De\ phi[\frac{a}{E^{b\ x}}]}{E^{2\ b\ x}} - \frac{2\ De\ phi[\frac{a}{E^{b\ x}}]}{E^{b\ x}} - Energy\ phi[\frac{a}{E^{b\ x}}] -$$

$$\frac{a^2\ b^2\ h^2\ phi'[\frac{a}{E^{b\ x}}]}{2\ E^{b\ x}\ m} - \frac{a^2\ b^2\ h^2\ phi''[\frac{a}{E^{b\ x}}]}{2\ E^{2\ b\ x}\ m}$$

which, when equated to zero, gives the new Schrödinger equation. This expression can be cleaned up with pattern replacements and by dividing out some constants:

```
-2m/(b h z)^2 % /.
    E^(n_?Negative b x) :> (z/a)^-n //Expand
```

$$-\frac{2\ De\ m\ phi[z]}{a^2\ b^2\ h^2} + \frac{2\ Energy\ m\ phi[z]}{b^2\ h^2\ z^2} + \frac{4\ De\ m\ phi[z]}{a\ b^2\ h^2\ z} +$$

$$\frac{phi'[z]}{z} + phi''[z]$$

We see from the first term that setting $a \to \mathtt{Sqrt[2De\ m]/(b\ h)}$ will further simplify things. It is convenient to do this by introducing a new parameter $A = \mathtt{Sqrt[2m]/(b\ h)}$ and setting $a \to A\ \mathtt{Sqrt[De]}$. Since we are considering negative energies, it's also convenient to define the *positive* parameter $p = \mathtt{Sqrt[-Energy]}$. We thus obtain

```
% /. a -> A Sqrt[De] /.
    m/(b^2 h^2) -> A^2/2 /. Energy -> -p^2
```

$$-phi[z] - \frac{A^2\ p^2\ phi[z]}{2} + \frac{2\ A\ Sqrt[De]\ phi[z]}{z} +$$

$$\frac{phi'[z]}{z} + phi''[z]$$

Since $\mathtt{-Infinity} < \mathtt{x} < \mathtt{Infinity}$, we have that $0 \le \mathtt{z} < \mathtt{Infinity}$, and we seek well-behaved solutions subject to the boundary conditions $\mathtt{phi[z = 0]} = \mathtt{phi[z \to Infinity]} = 0$. In the next exercise you are asked to show that **phi** $\sim$ **E^-z** aysmptotically for large **z** and that **phi** $\sim$ **z^(A p)** for small **z**. This suggests we introduce a new function **G[z]** according to

```
phiG[z_] := z^(A p) E^-z G[z]
```

Exercise 13.1-1

Determine the large- and small-**z** solutions **phi[z]** which are everywhere finite. Assume **A** and **p** positive.

Exercise 13.1-2

Transform the Schrödinger equation for the Morse potential to the dimensionless form

$$
A^2 \, De \, (-E^{-2y} + \frac{2}{E^y}) \, phi[y] + A^2 \, Energy \, phi[y] + phi''[y] == 0
$$

where **y = b x**. This result shows the problem reduces to two parameters **A** and **De** and is convenient for numerical integration. What are the asymptotic solutions for **y -> Infinity**?

However, the resulting equation is not quite Kummer's equation. Combining our previous steps into one, we find instead

```
psi[x] = phiG[z] /. z -> a E^(-b x);

Expand[
    schroedingerD[V] @ psi[x] *
        (-2m/(b h z)^2 z^(-A p + 1) E^z)
] /.
    a -> A Sqrt[De] /. m/(b^2 h^2) -> A^2/2 /.
    Energy -> -p^2 //.
    E^(n_?Negative b x) -> (z/(A Sqrt[De]))^-n //
    Expand //Collect[#,{G[z],G'[z]}]&

(-1 + 2 A Sqrt[De] - 2 A p) G[z] +

    (1 + 2 A p - 2 z) G'[z] + z G''[z]
```

(Note **//.** (**ReplaceRepeated**) is needed here to remove all of the exponentials; cf. Exercise 2.2, Appendix III.) The coefficient of **G'[z]** hints we should introduce a function of **2z** to obtain Kummer's equation. We thus replace **G[z]** with **F[2z]** and try again:

```
psi[x] = phiG[z] /. G[z] -> F[2z] /. z -> a E^(-b x);

Expand[
    schroedingerD[V] @ psi[x] *
```

```
            (-m/(b h z)^2 z^(-A p + 1) E^z)
   ] /.
      a -> A Sqrt[De] /. m/(b^2 h^2) -> A^2/2 /.
      Energy -> -p^2 //.
      E^(n_?Negative b x) -> (z/(A Sqrt[De]))^-n //
      Expand //Collect[#,{F[2z],F'[2z]}]&

      1
   (-(-) + A Sqrt[De] - A p) F[2 z] +
      2

            (1 + 2 A p - 2 z) F'[2 z] + 2 z F''[2 z]
```

When equated to zero, this result is Kummer's equation, as desired.

Eigenenergies   13.2

We conclude that **G[z] = 1F1[1/2 - A Sqrt[De] + A p, 1 + 2A p, 2z]**. We thus reject any contribution from the other independent solution **U[a, b, 2z]** Kummer's equation. Recall that **U[a,b,2z]** behaves like **z^(1-b)** and therefore like **z^(-2A p)** in the present case (see Section 6.5 and Exercise 6.5-2). Hence, it's irregular (infinite) at the origin **z = 0**, since **p** is positive, and therefore can be dismissed as a probability amplitude (see Section 1.3).

Now in order to ensure that **phiG[z]** vanishes as **z -> Infinity**, we require the series **G[z]** to truncate to a polynomial of degree **n** in the usual way. (Note **2z** in the argument of **1F1** is crucial in this regard. See Problem 6.5-2.) We therefore require that

```
   1/2 - A Sqrt[De] + A p == -n;
```

is a negative integer or zero, i.e. **n = 0, 1, 2,** ... This in turn restricts the parameter **p** to the discrete values

```
   p[n_] = p /. Solve[%, p][[1,1]] //Collect[#,A]&

                   1
                 -(-) - n
                   2
      Sqrt[De] + --------
                    A
```

which of course quantizes the bound-state energy spectrum since **-p^2 -> Energy**. With some rearrangement (see Exercise 2.1, Appendix III) we find that

```
   -p[n]^2 //Expand[#,A]& //Map[Factor,#]&

                                          2
            Sqrt[De] (1 + 2 n)   (1 + 2 n)
      -De + ------------------ - ----------
                   A                  2
                                   4 A
```

and thus define the eigenenergies as

```
e[n_] = % /. 1+2n -> 2(n+1/2)
```

$$-De + \frac{2\ Sqrt[De]\ (\frac{1}{2} + n)}{A} - \frac{(\frac{1}{2} + n)^2}{A^2}$$

We recognize the first two terms of this result to be the spectrum **h w (n + 1/2)** of a harmonic oscillator shifted by **-De**. The last term proportional to **(n + 1/2)^2** is a correction which derives from the asymmetric deviation of the Morse potential from a HO potential.

Unlike the harmonic oscillator spectrum, the number of bound states is finite and the energy levels unevenly spaced. Since we require **p[n]** to be positive for all **n** in order for the wavefunction to be regular, we require that **A Sqrt[De] > nmax + 1/2**, where **nmax** labels the last bound state. Since **nmax $\geq$ 0**, we thus require **A Sqrt[De] > 1/2**. Hence, **nmax + 1** gives the total number of bound states, which we see is the largest integer less than **1/2 + A Sqrt[De]** (see also Exercise 13.2-2 below). In particular, if **A Sqrt[De] < 1/2** there are no bound states.

Because of the potential's asymmetry, the equilibrium separation of the nuclei rapidly increases with increasing energy of the molecule. Moreover, because the potential is softer (less steep) for large **x** in the Morse oscillator compared with the symmetric harmonic oscillator, the energy level separation decreases as **n** increases. If there happen to be many bound levels, those just below the continuum threshold will form a quasi-continuum.

Exercise 13.2-1

Show that the first two terms of **e[n]** are given by first-order perturbation theory (see Section 7.4) when the Morse potential is expanded in powers of **b** and HO wavefunctions with frequency

$$w \ -> \ \frac{2\ Sqrt[De]}{A\ h}$$

are retained in zeroth order.

Exercise 13.2-2

To facilitate comparison with molecular spectroscopic data, one usually converts **Energy** to wavenumbers, i.e. inverse wavelengths **2Pi/λ**, according to the Planck relationship **Energy -> 2Pi h c/λ**. One thus defines a spectroscopic energy formula **G[n]** (in cm^-1) according to **G[n] = (e[n] + De)/(2Pi h c)**, where c is the speed of light. (Recall that our **h** is Planck's constant divided by **2Pi**). This function is conventionally written

```
G[n_] := we (n+1/2) - wexe (n+1/2)^2
```

where **we** and **wexe** are positive constants defined by **A** and **De**, which can be obtained by fitting to experimental data. One generally finds for diatomic molecules that **wexe << we**.

(a) Show for the Morse potential that the ratio **we^2/(4 wexe)** determines the dissociation energy **De** of the molecule (in **cm^-1**) relative to the potential minimum. Show that a small ratio **wexe/we** implies that the potential's asymmetry parameter **b** is small, or that the depth parameter **De** is large, or both. In such cases, the bottom of the potential well strongly resembles a HO potential and therefore has a similar energy spectrum (see Exercise 13.2-1). Make some potential plots to check this.

(b) Here is another way to determine **nmax** and hence the maximum number of bound states **nmax + 1**. Consider the level separation

```
delG[n_] := G[n+1] - G[n]
```

and show that it decreases as **n** increases until it reaches zero at the continuum threshold **Energy = 0**. Show that

$$
nmax \; \to \; -1 \; + \; \frac{we}{2\ wexe}
$$

Compare this result with the one obtained in the text and with tables of values of **G[n]** and **delG[n]** for **A = 1** and **De = 9, 25** and **30**.

Eigenfunctions 13.3

Collecting results, we thus define the bound-state wavefunctions as a function of the scaled coordinate **z** according to

```
phiF[n_,z_] := phiF[n,z] = norm[n] *
    z^(A p[n]) E^-z Hypergeometric1F1[-n,1+2A p[n],2z]
```

where **norm[n]** is the normalization constant, which we will determine in a moment. The ground state, for example, is found to be

```
phiF[0,z] //ExpandAll
```

$$
\frac{z^{-(1/2)\ +\ A\ Sqrt[De]}\ norm[0]}{E^{z}}
$$

Finally, we transform back to the coordinate **x** and obtain the bound-state probability amplitudes for displacement of the oscillator in real space:

```
psi[n_,x_] := psi[n,x] =
    phiF[n,z] /. z -> A Sqrt[De] E^(-b x)
```

Now the ground-state wavefunction becomes

```
psi[0,x] //ExpandAll
```

$$
\frac{\left(\cfrac{\text{A Sqrt}[\text{De}]}{E}\right)^{-(1/2) \ + \ \text{A Sqrt}[\text{De}]} \text{norm}[0]}{E^{(\text{A Sqrt}[\text{De}])/E^{b \ x}}}
$$

It follows (see Exercise 6.5-4 and Problem 6.4-1) that these functions form a complete, orthogonal set.

13.4 Normalization

We require the integral over all **x** of the wavefunction **psi[n,x]** squared to be unity. It's convenient to perform this integral with the coordinate **z = A Sqrt[De]E^(-b x)**, i.e. with a change of variables. Since **dz = -b z dx**, we thus require

```
1 == Integrate[ phiF[n,z]^2/(b z), {z,0,Infinity} ]
```
(13.4-r1)

For specific **n** values, these integrals are easily evaluated if we load our own function **integExp**, specialized for integrating combinations of exponentials and powers. Defined in the package **Quantum'integExp'**, this function is analogous to our gaussian integration rules **integGauss** (see also Exercise 2.7, Appendix III). Let's load the package and examine its usage statement:

```
<<Quantum'integExp'
```

```
?integExp
```

```
    integExp[integrand,{r,0,Infinity}] integrates linear
      combinations of patterns of the form r^n E^(-a r)
      for Re[a] > 0. WARNING: The requirement Re[a] > 0
      must be enforced by the user.
```

Thus, for the ground-state we find

```
1 == integExp[phiF[0,z]^2/(b z),{z,0,Infinity}] //
      ExpandAll
```

$$
1 == \frac{2 \ (-2 + 2 \ \text{A Sqrt}[\text{De}])! \ \text{norm}[0]^2}{2^{2 \ \text{A Sqrt}[\text{De}]} \ b}
$$

and hence

```
norm[0] = norm[0] /. Solve[%, norm[0]][[1]]
```

$$\frac{-(1/2) + A \; Sqrt[De]}{Sqrt[(-2 + 2 \; A \; Sqrt[De])!]} \quad Sqrt[b]$$

Note that the built-in factorial function evaluates as `z! == Gamma[z+1]` for nonintegral arguments `z` (see Problem 6.5-4). We derive a general expression for the normalization in the next section.

Exercise 13.4-1

(a) Show that for `n < 4` the `psi[n,x]` are mutually orthogonal. Note that parity arguments don't apply here, since the Morse potential does not have a definite parity. Hint: You will need to introduce the property `q! == (q+1)!/(q+1)`. For example, you might try attaching the rule

```
//.(q_+2A p_)! :> (q+1+2A p)!/(q+1+2A p) /; q < -2
```

See also the built-in function `SimplifyGamma`.

(b) Evaluate `norm[n < 4]`.

Exercise 13.4-2

The eigenfunction factor `1F1[-n,1 + 2A p,2z]` is actually proportional to a common orthogonal polynomial known as an *associated Laguerre polynomial*. These functions, which are also built-in as `LaguerreL[q,p,x]`, satisfy Laguerre's differential equation

```
q L[q,p,y] + (1 + p - y) L'[q,p,y] + y L''[q,p,y] == 0
```

In practice, various normalizations of the Laguerre polynomials are found. One possibility is to relate them to the confluent hypergeometric functions, which are normalized such that the first term in the hypergeometric series is unity. Show that the relation used by *Mathematica* is

```
LaguerreL[q,p,y]==(q+p)!/(q!p!) Hypergeometric1F1[-q,p+1,y]
```

Express the *normalized* eigenfunction `phiF[n,z]` in terms of the Laguerre polynomial and compare with our results in the text. Laguerre polynomials are familiar from the description of the bound states of the hydrogen atom (see Section 20.2).

We can plot the wavefunctions along with the potential if we give the parameters some numerical values. Except that the normalization introduces the scale parameter **b** explicitly, the wavefunction is determined entirely by **A** and **De** (see Exercise 13.1-2). (Of course, **A** depends on **m**, **h** and **b**.) Thus, for tasks such as plotting, it is convenient to introduce the scaled coordinate **y = b x** and work with `psi[n,y/b]/Sqrt[b]`. You can see that this function is normalized with respect to **y**.

Let's substitute for example **A = 1** and **De = 9** and evaluate the ground-state wavefunction. We thus obtain

```
psi[0,y/b]/Sqrt[b] /. {A -> 1, De -> 9}
```

$$
\frac{18 \ (E^{-y})^{5/2}}{E^{3/E^{y}}}
$$

First, let's determine the total number **nmax + 1** of bound states the potential can support given these parameter values. From our discussion above (see also Exercise 13.2-2 above), this is the largest integer less than **1/2 + A Sqrt[De]**. We thus see there are

```
Floor[1/2 + A Sqrt[De]] /. {A -> 1, De -> 9}
```

```
3
```

bound states and that **nmax = 2** (see **?Floor**). We can verify this by computing a table of **p[n]**, which we require to be positive in order for the wavefunction to be everywhere finite.

```
Table[p[n],{n,0,4}] /. {A -> 1, De -> 9}
```

$$
\{\frac{5}{2}, \ \frac{3}{2}, \ \frac{1}{2}, \ -(\frac{1}{2}), \ -(\frac{3}{2})\}
$$

Exercise 13.4-3

Try evaluating **psi[n, x]** for n > 3 with {**A -> 1, De -> 9**}.

The bound-state eigenenergies are therefore

```
Table[e[n],{n,0,2}] /. {A -> 1, De -> 9}
```

$$
\{-(\frac{25}{4}), \ -(\frac{9}{4}), \ -(\frac{1}{4})\}
$$

These three states (with their normalizations from Exercise 13.4-1) are plotted in Figure 13.4-1, shifted by their eigenenergy **e[n]**. The plots exhibit all the familiar features. For example, the quantum number **n** specifies the number of nodes, starting with **n** equal to zero since the ground state has no node. The inflection points are seen to coincide with the classical turning points in correspondence with the vanishing of the classical kinetic energy there. Also, the probability is seen to be smaller at the left turning point where the potential is steepest and the classical velocity highest (cf. Problem 6.6-4).

psi[n, y/b]/Sqrt[b]

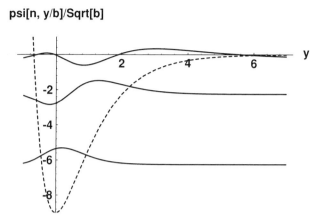

Figure 13.4-1. Bound states of the Morse potential as a function of **y = b x** for **A = 1** and **De = 9**.

Problem 13.4-1

(a) Generate Figure 13.4-1. Verify the orthogonality of these wavefunctions for **n < 3** directly with **NIntegrate**.

(b) Compute the wavefunctions in Figure 13.4-1 using the shooting method and the scaled Schrödinger equation from Exercise 13.1-2 above. Normalize your results with **NIntegrate** and compare with the analytic functions.

Problem 13.4-2

(a) Verify that **psi[0,x]** is the ground-state wavefunction with eigenenergy **e[0]** by calculating the local energy **schroedingerD[V]@ psi[0,x]/psi[0,x]** directly and simplifying.

(b) Repeat part **(a)** for the two excited bound-states for **h = m = A = 1** and **De = 9**. Apply **//N** to simplify.

Problem 13.4-3

Animate the time development of a simple wavepacket which is some linear combination of the three bound states in Figure 13.4-1. Refer to Section 2.5.

Problem 13.4-4

Estimate the energies of the bound states in Figure 13.4-1 *variationally* using gaussian wavefunctions. Discuss your results and try improving them by improving your trial functions. Refer to Chapter 7.

Problem 13.4-5

Redo the Problem 13.4-4 as a partial exact diagonalization with an HO basis set. Refer to Chapter 10.

Problem 13.4-6

Redo Problem 13.4-4 by propagating a wavepacket in the Morse potential with the *FFT* method of Section 12.6. Hence, compute the correlation function and its Fourier transform. Also, estimate the bound-state energies by the quantum diffusion method of Section 12.7.

Problem 13.4-7

Determine the bound states and eigenenergies of a particle moving in the potential **V = -Vo/Cosh[d x]^2**. Introduce **z = Tanh[d x]** and show that the solution of the Schrödinger equation which is finite as **x -> +Infinity** is **(1-z^2)^(p/2) 2F1[a,b,c,(1-z)/2]**, where **p** is proportional to **Sqrt[-Energy]**. (Note this is a hypergeometric function and not Kummer's function. Use the connection formulas for the hypergeometric function from Problem 6.5-5 to apply the boundary conditions.)

Show that, if the wavefunction is to remain finite as **x -> -Infinity**, one must require **a = -n** with **n = 0, 1, 2, ...** so that **2F1** is a polynomial of degree **n**. Thus, show that

$$
\text{Energy} \rightarrow \frac{-(d^2\ h^2\ (-1\ -\ 2\ n\ +\ \text{Sqrt}[1\ +\ \dfrac{8\ m\ \text{Vo}}{d^2\ h^2}])^2\ )}{8\ m}
$$

The polynomial solutions which arise here happen to be *Legendre polynomials* (cf. Problem 19.1-2).

13.5 Hypergeometric Integrals

Finally, let's go back and determine the bound-state normalization generally. Since relatively compact expressions have been tabulated for integrals involving combinations of hypergeometric functions, powers, and exponentials, we will follow the procedure we used to derive the normalization of the harmonic oscillator wavefunctions (see Section 6.6). This example also serves to exemplify the great utility of hypergeometric functions.

Recall that we require

```
Clear[norm]
phiF[n,z]^2/(b z)  //ExpandAll

    -2 + 2 A Sqrt[De] - 2 n
   (z                          Hypergeometric1F1[-n,

                                    2              2        2 z
         2 A Sqrt[De] - 2 n, 2 z]   norm[n]  ) / (b E    )
```

integrated over **z** for **0 ≤ z < Infinity** to be unity. Here we cleared **norm** in order to prevent *Mathematica* from substituting values we've already determined. We thus have an integrand of the form **E^(-k z) z^nu 1F1[-n,g,**

k z]^2, and given an expression for the integral, we want to construct a replacement rule of the form **/. integrand -> *integral*.** For the present class of integrals, this is much more efficient than calling on the built-in function **Integrate**.

We shall use the results derived in the *Mathematical Appendices d, e* and *f* of Landau and Lifshitz [41]. They express the integral we need for integer **n** and **Re[nu] > 0** in terms of a function **Jnu[k,nu+1,n,g]**, which they evaluate as a finite sum of combinations of factorial functions. Hence, we define the replacement rule

```
expressJnu :=
    z_^nu_. E^(k_?Negative f_. z_) *
    Hypergeometric1F1[-n_, g_, kp_. f_. z_]^2  :>
        Jnu[-k f,nu+1,n,g]     /; kp == -k
```

in order to implement equation *f.6* of Landau and Lifshitz [41]. The syntax of this rule is similar to that of replacement rules we use in other packages (see Exercises 2.2-2.7, Appendix III). For example, the factor **f_.** is included to cover patterns such as **-2k z** in addition to **-k z**. The normalization integrand we entered above can thus be integrated according to

```
%% /. expressJnu
```

$$\frac{Jnu[2, -1 + 2 A Sqrt[De] - 2 n, n, 2 A Sqrt[De] - 2 n] \, norm[n]^2}{b}$$

which, when equated to unity, can be solved for **norm[n]**:

```
norm[n_] = norm[n] /. Solve[ % == 1, norm[n] ][[1]]
```

$$\frac{Sqrt[b]}{Sqrt[Jnu[2, -1 + 2 A Sqrt[De] - 2 n, n, 2 A Sqrt[De] - 2 n]]}$$

Now all we need is a replacement rule for **Jnu**. We thus enter Equation *f.7* of Landau and Lifshitz [41], re-expressed in terms of Pochhammer symbols for compactness and computational efficiency (see Problem 6.5-4):

```
JnuEval = Jnu[k_,nu_,n_,g_] :>
    n! Gamma[nu] k^-nu/Pochhammer[g, n] *
    (1 + Sum[
            Pochhammer[n-s,s+1] *
            Pochhammer[g-nu-s-1,2s+2]/
            ((s+1)!^2 Pochhammer[g,s+1]),
            {s,0,n-1}
        ]);
```

Here the sum will evaluate when **n** is given a value. For example, for **n = 0**, we obtain with the relation **Gamma[n] -> (n-1)!** (see Problem 6.5-4)

```
norm[0] /. JnuEval /. Gamma[n_] -> (n-1)! //
      PowerExpand //Expand
```

$$\frac{2^{-(1/2)\,+\,A\,Sqrt[De]}\;Sqrt[b]}{Sqrt[(-2\,+\,2\,A\,Sqrt[De])!]}$$

in agreement with our previous calculation.

Many integrals involving hypergeometric functions can be implemented in a similar fashion, as for example the very general result on page 862 of Gradshteyn and Ryzhik [28], *Tables of Integrals, Series and Products.* We thus have an efficient and powerful technique for evaluating integrals and matrix elements which might arise in a wide variety of applications.

Problem 13.5-1

(a) Make a table of **norm[n]** for **n < 6** and simplify. Apply **PowerExpand** and **PowerContract**.

(b) From your results in part **(a)**, deduce and verify the following simpler expression for **norm[n]**:

```
2^(A p[n]) Sqrt[b] Sqrt[Product[q+2A p[n],{q,n}]/
   (n!(-1+2A p[n])!)]
```


Problem 13.5-2

Landau and Lifshitz [41] express integrals of **E^(-1 z) z^(g-1) 1F1[a,g,k z] 1F1[ap, g, kp z]** for **{z, 0, Infinity}** in terms of functions **J[k, kp, a, ap, l, g]** which they evaluate with hypergeometric functions **2F1**. Refer to their Equations *f.9-11*. Their results can be summarized by the following two replacement rules:

```
expressJ :=
    {z_^gp_. E^(l_?Negative f_. z_) *
        Hypergeometric1F1[a_, g_,k_.  z_] *
        Hypergeometric1F1[ap_,g_,kp_. z_] :>
            J[k,kp,a,ap,-l f,g] /; gp == g-1,

     z_^gp_. E^(l_?Negative f_. z_) *
        Hypergeometric1F1[a_,g_,k_. z_]^2  :>
            J[k,k,a,a,-l f,g] /; gp == g-1}

JEval = J[k_,kp_,a_,ap_,l_,g_] :>
    Module[{w,wp},
        j1wwp =
            Expand[
                (g-1)! l^(a+ap-g) w^-a wp^-ap *
                Hypergeometric2F1[a,ap,g,k kp/(w wp)]
            ];
        j1wwp /.{w -> 1-k, wp -> 1-kp}
    ];
```

The local variables **w** and **wp** introduced in **JEval** are necessary in order to evaluate the **2F1** function properly and avoid indeterminate results when **l** = **k** or **kp** and when **a** or **ap** is a negative integer. Use these rules to verify the normalization relation (6.4-r3) of the Hermite polynomials.

14
Potential Scattering

We return to the model potential energy we studied in Chapter 10 and Section 12.6 and compute its stationary-state continuum spectrum by integrating the time-independent Schrödinger equation numerically. The basic importance of this solution is in formulating a description of scattering of a particle of mass **m** by a potential and in developing an understanding of resonances and metastable states. We especially want to follow this through for a system that has no analytical solution.

We work here within a *time-independent* framework, although we will construct moving wavepackets in Section 14.6 by forming a time-dependent superposition of stationary continuum states. We connect in this way with our discussion of wavepacket propagation in Section 12.6. Neither approach outplaces the other; on the contrary, the two points of view are complementary and contribute to a full understanding of continuum states.

We thus re-enter the potential energy from Section 10.0:

```
V   = VHO E^(-b z^2);     b   =  0.1;
VHO = Vo + z^2/2;         Vo  = -2.0;
```

For simplicity, we continue to work with the scaled coordinate **z** and in effect set **m = h = 1**. It is convenient to write the Schrödinger equation as **D[psi,{z,2}]+ksq[e,z]psi == 0** and thus introduce the *square* of the particle's wavenumber

```
ksq[e_,z_] = 2(e - V);
k[e_] = N[Sqrt[2e]];
```

where **e** is the particle's energy. Here **k[e]** is the asymptotic wavenumber defined far from the center of force **z = 0**. Hence, the particle's asymptotic

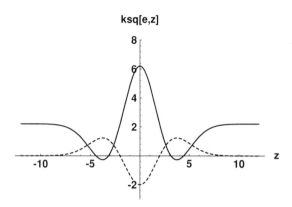

Figure 14.0-1. Square of the local momentum for **e = 1.1**. The potential is shown by the dotted curve.

deBroglie wavelength is **2Pi/k[e]**. The wavenumber is also the particle's local momentum, since **h = 1**, where local means the value of the momentum at the position **z**. The wavenumber squared is therefore twice the kinetic energy and is plotted in Figure 14.0-1 for a particle incident with (total) energy slightly below the tops of the potential barriers. The figure shows that the kinetic energy goes to zero at the classical turning points and becomes negative under the barriers in the classically forbidden regions, as expected.

We shall generate a numerical solution with the built-in function **NDSolve**. We thus need to specify a range of integration **-zo < z < zo** and values for the wavefunction and its derivative at the left or right end of the range, e.g. **psi[zo]** and **psi'[zo]**. (The choice of a symmetric range of **z** is simply a matter of convenience.) In order to include the free-particle motion far from the center of force, **zo** should be at least a few free-particle deBroglie wavelengths beyond the range of the potential. The value **zo = 15** is sufficient for energies above about **e = 0.4**. For a fixed energy **e**, our stationary-state solution will be the corresponding continuum energy eigenfunction.

Exercise 14.0-1

Plot the asymptotic deBroglie wavelength of the particle for **0.1 < e < 2.0**.

14.1 Numerical Solution

Let us assume we have a particle incident from the left for **z < 0** moving to the right towards **z > 0**. The potential scatters the corresponding incident wave, generating a reflected component to the left and a transmitted or outgoing component to the right for **z > 0**. The computed wavefunction is of course a superposition of all three components. However, whereas the wavefunction contains both incident and reflected components to the left of the potential, it contains only the transmitted component to the right, which asymptotically for **z >> 0** must be an outgoing free-particle planewave **E^ (I k z)**. Therefore, it is convenient to integrate in from **z = zo** right to left, starting with **psi[zo] == E^ (I k zo)** and thus with **psi'[zo] == I k[e]psi[zo]**.

The partial reflection and transmission and hence splitting of the wavefunction is a notable nonclassical effect which reinforces the probability interpretation of the wavefunction. The squares of the amplitudes of the respective wavefunction components give the probabilities that reflection or transmission occurs. We reject in particular any other interpretation which would imply a physical breakup of the particle and therefore contradict the fact that electrons for example are always detected as whole entities.

The following function **psiNDSolve[e]** calls **NDSolve** to integrate the Schrödinger equation subject to these boundary conditions. We give it an optional triple-blank argument **opts** (see Exercise 1.10, Appendix III) to pass on requests to **NDSolve** for extra precision if needed, for example if changes to **MaxSteps**, **AccuracyGoal**, and **PrecisionGoal** are needed. (See **?NDSolve**.)

```
psiNDSolve[e_,opts___] := psi[z] /.
    NDSolve[
        {psi''[z] + ksq[e,z] psi[z] == 0,
         psi [zo] ==          E^(I k[e] zo),
         psi'[zo] == I k[e] E^(I k[e] zo)},
         psi[z], {z,-zo,zo}, opts
    ][[1]]
```

Here the subscript **[[1]]** picks out the inner list from the list of lists returned by **NDSolve** so that **psi[z]** is not a **{**_list_**}** (see Exercise 1.3, Appendix III).

In order to extract the amplitudes of the asymptotic incident, reflected and transmitted planewaves, one can evaluate the interpolated result **psi[z]** from **NDSolve** on a grid or lattice in steps **dz**. The amplitudes can then be obtained directly with the built-in function **Fit** and linear combinations of **E^ (I k z)** and **E^ (-I k z)**, as will be demonstrated in the next section. Introduction of a lattice also has the advantage that we could replace **NDSolve**, with only minor modification of the remaining discussion in this chapter, by a custom-tailored function, for example, an integration routine based on the efficient Numerov algorithm (see Johnson [35] and also Koonin [37]). We thus define a table of coordinate and wavefunction values with

```
psi0[e_,opts___] :=
    Table[
        Evaluate[{z, psiNDSolve[e,opts]}],
        {z,-zo,zo-dz,dz}
    ]
```

Here **Evaluate** forces **NDSolve** to compute before construction of the **Table** begins. For variety, we assume that the lattice step size **dz** is given.

In Chapter 12 we specified instead the number of lattice points **nmax**, in addition to the lattice length **L**. The connection of course is **nmax = 2 zo/dz** and **L = 2 zo**.

Let's try out **psiNDSolve[e]** for an energy **e** slightly below the tops of the barriers, and plot the resulting eigenfunction **psi0[e]**. For perspective, we will display the potential in the background by setting up an **Epilog** for **ListPlot**. We can do this once and for all by defining the function

```
bkgrd[e_,zo_:15] := Epilog ->
    {{Dashing[{.01,.01}], Line[{{-zo,e},{zo,e}}]},
       Line[Table[{z,V},{z,-zo,zo,.05}]]}
```

which also draws a dashed line to mark the energy (cf. the analogous function **Vbkgrd** defined in Section 12.6).

We set **dz = 0.1**, which is small compared to a deBroglie wavelength, and compute **psi0[e]** and save the result for reuse later on as **psi0e**. Its real part is shown in Figure 14.1-1 and demonstrates that even though the energy is below the barriers, the wavefunction tunnels through the barriers, a quantum wave effect which is classically forbidden. A powerful check of the wavefunction is a computation of its *local energy* (cf. Sections 10.4 and 12.7). We verify in Exercise 14.1-2 and in Figure 14.1-2 that the local energy is satisfactorily flat, as desired. As we mentioned already in Section 2.2, numerical integration is the most efficient way of *computing* (time-independent) wavefunctions in *1D*.

```
e = 1.1;     zo = 15;     dz = 0.1;
psi0e = psi0[e];

ListPlot[ Re[psi0e], bkgrd[e],
    PlotRange -> {-3,3}, Axes -> {Automatic,None},
    AxesLabel -> {" z",""}
];
```

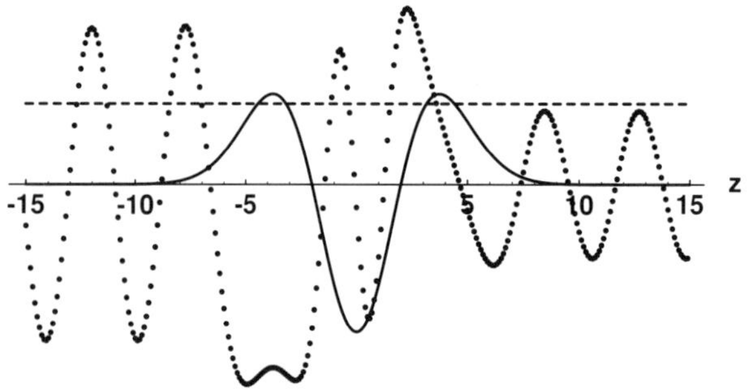

Figure 14.1-1. Real part of the continuum eigenfunction **psi0[e = 1.1]**. The model potential is shown in the background by the solid curve and the energy by the dashed straight line.

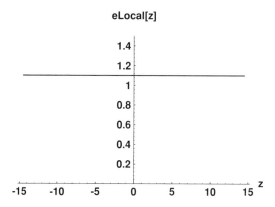

Figure 14.1-2. Check of the continuum-eigenfunction local energy for **e = 1.1**.

Exercise 14.1-1

Plot the imaginary part of the wavefunction shown in Figure 14.1-1. Compute **psi0[e]** for some other energies and plot your results. Try looking for the resonances we predicted by diagonalization in Section 10.2 and Exercise 10.2-2. How will you recognize a resonance? Although we shall examine this question in detail shortly, be warned that the first resonance is very narrow in energy and therefore difficult to find.

Exercise 14.1-2

We saw back in Section 10.4 that the flatness of the local energy of an eigenfunction **psi[e,z]** as a function of **z**, defined by **hamiltonian[V] @ psi[e,z]/psi[e,z]**, provides a very powerful check of the eigenfunction's numerical accuracy. If the calculation is exact, the local energy will be a constant equal to the eigenvalue **e** (independent of the wavefunction's normalization). See Figure 10.4-1.

   Compute and plot the local energy of our numerical wavefunction for **e = 1.1**, shown in Figure 14.1-1. In order to compute the derivatives of the eigenfunction and hence **hamiltonian[V] @ psi[e,z]**, use the interpolated solution **psiNDSolve[e]** directly. Check your results by generating Figure 14.1-2. You can get a better idea of the numerical accuracy, however, by replotting with the option **PlotRange -> All**.

Scattering Amplitudes   14.2

Since the Schrödinger equation is homogenous, its solutions are arbitrary to within a multiplicative constant, and the value we assign to the wavefunction to start the integration is unimportant. Ordinarily, we would fix the value of the constant by normalizing the wavefunction after computation. Continuum wavefunctions, however, extend over all space and the integral of their square over all space is infinite. Analysis shows that these infinities can be described by Dirac delta functions, and one usually employs a delta-function normalization (see Section 20.4 and also Merzbacher [45], Sec. 6.3). An alternative method is to assume that the system is enclosed in a large but finite box (see Park [50], p. 273), an approach that is appropriate for numerical wavefunctions since computers confine us to a finite lattice anyway. Although we could then compute the normalization by numerical integration, it's sufficient to assign the constant a convenient value.

We shall simply normalize the amplitude of the incoming planewave from the left for **z << 0** to unity. Then the squares of the amplitudes of the asymptotic reflected and transmitted planewaves determine the reflection and transmission coefficients or probabilities **R** and **T**, respectively, such that **1 == R + T**. This relation expresses of course conservation of probability by the scattering process. (Dirac-delta-function, or any other, normalization can then be obtained by multiplication by a suitable factor.)

We therefore proceed as follows. Numerical integration with **psiND-Solve** generates **psi0 -> E^(I k z)** asymptotically for **z >> 0**. Because of small numerical errors, however, we actually obtain **psi0 -> cT0 E^(I k z) + dR0 E^(-I k z)**, where **dR0** is a spurious but small reflection amplitude and **cT0** a transmission amplitude close to unity. The deviations from **dR0 = 0** and **cT0 = 1** are a measure of the accuracy of **NDSolve**. Likewise, **psi0 -> aI0 E^(I k z) + bR0 E^(-I k z)** asymptotically for **z << 0**. As we shall see, the *scattering amplitudes* **{aI0, bR0, cT0, dR0}** are easily extracted by fitting these asymptotic forms to **psi0[e]**. Then, dividing **psi0[e]** by **aI0**, we *renormalize* **psi0[e]** to **psi[e]** and therefore to an incoming planewave with unit amplitude for **z << 0**. The renormalized wavefunction **psi[e]** thus takes the asymptotic form

```
{E^(I k z) + bR  E^(-I k z),
 cT  E^(I k z) + dR  E^(-I k z)}
```
(14.2-r1)

for **z << 0** and **z >> 0**, respectively, with the renormalized amplitudes **{aI = 1, bR = bR0/aI0, cT = cT0/aI0, dR = dR0/aI0}**. The reflection and transmission coefficients are then given by **R = Abs[bR]^2** and **T = Abs[cT]^2**, respectively. The spurious reflection amplitude **dR** is a measure of the numerical error and should remain small.

This solution is actually more general than it might at first appear. The reason is degenerate eigenfunctions with any desired asymptotic form, e.g. incoming waves from the right, can be obtained directly from this solution by symmetry arguments (since the potential is symmetric) and a suitable linear superposition *without another numerical integration*. For example, this idea can be used to derive a fundamental connection between the phases of the reflection and transmission amplitudes, or the so-called *scattering phase shifts* (see Problem 14.2-1).

Exercise 14.2-1

Construct degenerate states of definite even and odd parity from **psi0e** for **e = 1.1** and plot the real and imaginary parts as in Figure 14.1-1. (See also Problem 14.2-1.)

We note in passing that if we were to connect the scattering amplitudes by a (scattering) matrix, we would require the matrix to be unitary in the usual way in order for probability to be conserved. The statement of conservation of probability $1 == R + T$ is therefore sometimes referred to as a *unitarity relation*. Its verification provides a useful check on the integrity of any numerical solution.

We define now fitting functions to extract the reflection and transmission amplitudes from the first and last **nFit** points of **psi0[e]**. For accuracy, we set this up with ten-point fits by default using **nFit_:10** (cf. Exercise 1.12, Appendix III), although it is often done simply with two-point fits (cf. Koonin [37]). Here **fitIR** determines the incident and reflected waves for **z << 0** while **fitT** determines the transmitted wave (and the spurious reflection) for **z >> 0**.

```
fitIR[e_,psi_,nFit_:10] :=
    Fit[
        Take[psi,nFit],
        {E^(-I k[e] z), E^(I k[e] z)}, z
    ]
fitT[e_,psi_,nFit_:10] :=
    Fit[
        Take[psi,-nFit],
        {E^(-I k[e] z), E^(I k[e] z)}, z
    ]
```

We can test these functions out on **psi0[e]** from Figure 14.1-1:

```
{fitIR[e,psi0e], fitT[e,psi0e]}
```

$$\{(-0.000043583 + 0.953813\ I)\ E^{-1.48324\ I\ z} +$$
$$(1.06186 - 0.884388\ I)\ E^{1.48324\ I\ z},$$
$$(7.3547\ 10^{-7} + 7.65244\ 10^{-7}\ I)\ E^{-1.48324\ I\ z} +$$
$$(0.999997 - 2.6595\ 10^{-6}\ I)\ E^{1.48324\ I\ z}\}$$

The results are of the form **b E^(-I k z) + a E^(I k z)**, and we can pick off the amplitudes **a** and **b** by attaching the appropriate subscripts **[[i,j]]** to **fitIR** and **fitT**. We thus define the scattering amplitudes and the reflection and transmission coefficients. Dividing by **aI0**, we define the renormalized wavefunction **psi[e]** from **psi0[e]**.

```
aI[e_,psi_] := fitIR[e,psi][[2,1]]
bR[e_,psi_] := fitIR[e,psi][[1,1]];
R[e_,psi_]  := Abs[bR[e,psi]]^2
```

```
cT[e_,psi_]  :=  fitT[e,psi][[2,1]]
dR[e_,psi_]  :=  fitT[e,psi][[1,1]];
T[e_,psi_]   :=  Abs[cT[e,psi]]^2

psi[e_,psi0_,aI0_]  :=
   Table[{psi0[[n,1]],psi0[[n,2]]/aI0},{n,Length[psi0]}]
```

Keep in mind that **psi0** is a list of pair lists with the first element of each pair a grid point **z** and the second the corresponding value of the wavefunction. Therefore, we can perform this last renormalization step more compactly using **Map** (**/@**) according to (cf. Exercise 3.2, Appendix III)

```
psi[e_,psi0_,aI0_]  :=  {#[[1]],#[[2]]/aI0}& /@ psi0
```

We extract the (unnormalized) scattering amplitudes **{aI0,bR0,cT0, dR0}** by applying these functions to **psi0[e]**:

```
{aI[e,psi0e], bR[e,psi0e], cT[e,psi0e], dR[e,psi0e]}
```

$$\{1.06186 - 0.884388\ I,\ -0.000043583 + 0.953813\ I,$$
$$0.999997 - 2.6595\ 10^{-6}\ I,\ 7.3547\ 10^{-7} + 7.65244\ 10^{-7}\ I\}$$

which are of course just the coefficients from **fitIR** and **fitT** computed above. As desired, **cT0** is very close to unity and **dR0** nearly equals zero. Dividing by **aI0**, we obtain the renormalized scattering amplitudes with

```
%/aI[e,psi0e] //Chop
```

$$\{1.,\ -0.441739 + 0.530336\ I,\ 0.556038 + 0.463102\ I,$$
$$5.45613\ 10^{-8} + 7.66104\ 10^{-7}\ I\}$$

These values can also be extracted directly, however, from the renormalized wavefunction **psi[e]**:

```
psie = psi[e, psi0e, aI[e,psi0e]];

{aI[e,psie], bR[e,psie], cT[e,psie], dR[e,psie]} //
   Chop
```

$$\{1.,\ -0.441739 + 0.530336\ I,\ 0.556038 + 0.463102\ I,$$
$$5.45613\ 10^{-8} + 7.66104\ 10^{-7}\ I\}$$

Now **aI = 1**, as desired. As we've already observed in connection with Figure 14.1-1, the transmission coefficient **T = Abs[cT]^2** is nonvanishing, a situation which is classically forbidden for incident energies below the barriers. We check, however, that the sum of the reflection and transmission coefficients equals unity to within numerical accuracy, as required:

```
{R[e,psie], T[e,psie], R[e,psie] + T[e,psie]}

   {0.47639, 0.523641, 1.00003}
```

This verifies that our computation satisfactorily conserves probability (here to better than five significant digits).

Problem 14.2-1

Scattering phase shifts describe the phases of the scattering solution relative to the incident plane wave. These quantities are thus the phases or arguments of the complex-valued reflection and transmission amplitudes defined by **bR == Abs[bR]E^(I argR)** and **cT == Abs[cT]E^(I argT)**. The phase shifts can therefore be computed as

```
argT[e_,psi_] := -ArcTan[Im[cT[e,psi]]/Re[cT[e,psi]]]
argR[e_,psi_] := -ArcTan[Im[bR[e,psi]]/Re[bR[e,psi]]]
```

The minus signs are simply a matter of choice.

Use symmetry arguments (time reversal and parity) and superposition to derive generally the following relations among the scattering amplitudes:

$$\{Abs[cT]^2 == 1 - Abs[bR]^2, \frac{Conjugate[bR]}{bR} == -(\frac{Conjugate[cT]}{cT})\}$$

Refer to Chapter 5 and to Exercise 5.0-6. The first relation is just conservation of probability **T == 1 - R**. The second connects the reflection and transmission phase shifts. Show that the second relation is equivalent to **argR == argT ± nOdd Pi/2** where **nOdd** is an odd integer. Verify these relations with our computed results for **e = 1.1**.

Problem 14.2-2

Starting with the asymptotic forms

```
{E^(I k z) + bR E^(-I k z), cT E^(I k z)}
```

of **psi[e]** for **z << 0** and **z >> 0**, derive the asymptotic forms of degenerate states with even and odd parity. Hint: Use the results of Problem 14.2-1 and show that **Abs[bR ± cT] == 1**.

Problem 14.2-3

Compare plots of the real and imaginary parts of **psi[e]** and of a planewave **E^(I k[e] z)** for **e = 1.1** and estimate graphically the transmission phase shift **argT**. Refer to the definitions and results from Problem 14.2-1.

Finally, for plotting purposes it's helpful to define functions which compute the real and imaginary parts of the wavefunction and the corresponding probability density and also shift the results by the energy **e** and scale them by an arbitrary factor **scale**. We can do this in analogy with our above renormalization of the wavefunction **psi0[e]** and the definition of **psi[e]**. Thus, we define

```
Repsi[e_,scale_,psi_] :=
    {#[[1]], e + scale  Re[#[[2]]]  }& /@ psi
Impsi[e_,scale_,psi_] :=
    {#[[1]], e + scale  Im[#[[2]]]  }& /@ psi
psisq[e_,scale_,psi_] :=
    {#[[1]], e + scale Abs[#[[2]]]^2}& /@ psi
```

These functions can be passed directly to **ListPlot** and are used in a **GraphicsArray** in Figure 14.2-1 to display three representations of the renormalized wavefunction **psi[e]**.

Exercise 14.2-2

Reproduce Figure 14.2-1 for three energies **e = 0.1, 1.1** and **10.0**.

Figure 14.2-1 makes clear several expected results. Beyond the range of the potential, the real and imaginary parts of the wavefunction oscillate with the deBroglie wavelength of the incident particle **2Pi/k[e]**. The particle's wavelength shortens between the potential barriers as the potential deepens corresponding to an increase in the kinetic energy **ksq[e,z]/2** there. The incident and reflected waves interfere and give rise to a standing wave in the probability distribution for **z << 0**. The transmitted wave, however, departs the scattering region alone and therefore without interference so that the probability distribution for **z >> 0** is a constant with magnitude **T**. The probability distribution has three approximate nodes inside the well, indicating that **e = 1.1** is near the second resonance (see Exercise 10.2-2). The wavefunction decays exponentially under the barriers and thus describes the *tunneling* of the particle through the barriers, while the probability density maximizes in the well near the classical turning points corresponding to a classical particle trapped there (see Problem 6.6-4).

Exercise 14.2-3

Calculate the minima and maxima and hence the amplitude of the standing wave in the probability distribution for **x << 0** and relate it to **Sqrt[R] = Abs[bR]**. Compare values of these quantities for **e = 1.1** with a detailed plot and a table of values of **Abs[psi[e]]^2**.

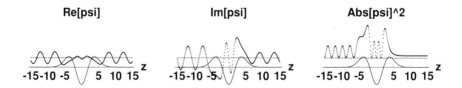

Figure 14.2-1. The renormalized wavefunction **psi[e = 1.1]**. Here the **Re** and **Im** parts are scaled by **0.9** and the probability density by **0.5**.

Resonance Hunting **14.3**

We now have all the tools we need to investigate the continuum spectrum. We begin by computing the energy dependence of the scattering amplitudes as a function of the energy **e**. Of particular interest are the transmission resonances for which the transmission coefficient **T** is unity or equivalently the reflection coefficient **R** vanishes. Unless we have performed a partial-exact diagonalization or made some other determination (see Problem 14.3-1), we simply have to search for the resonances by trial and error.

If the resonances are sharp and therefore occur in a narrow range of energies, we will have to compute wavefunctions on a fine enough energy grid to avoid missing a resonance. This will happen whenever the barriers are high and thick, so that transmission through a single barrier is unlikely. Nevertheless, for just the right energies, 100% transmission occurs when the incident and reflected waves to the left of the potential interfere destructively and only the transmitted wave to the right survives. This remarkable fact means that although a single barrier can have a very low transmission coefficient two such barriers in a row can be completely transparent. Such a phenomenon also reminds us of the operation of certain optical interferometers.

Problem 14.3-1

The semiclassical WKB approximation is prominent historically for studying tunneling and scattering resonances. In the case of a symmetric potential with barriers, one derives that the transmission coefficient reaches its maximum value unity whenever the *classical action variable* defined by

```
L[e_] := 2 NIntegrate[ Evaluate[Sqrt[ksq[e,z]]], {z,0,a[e]} ]
```

equals **(2n+1) Pi/2**, where **n** is the number of nodes in the wavefunction between the barriers. The range of integration **a[e]** is an inner classical turning point and therefore a root of **e == V**. (Refer to Merzbacher [45], Sec. 7.4 and to Bohm [10], Chap. 12.)

Define a function using **FindRoot** to compute **a[e]**, and then evaluate **L[e]** as a function of **e** to estimate the resonance energies for our model potential.

Let's look for the first resonance. In Exercise 10.2-2, we estimated it to be near **e = 0.48,** and from the variational principle we know this estimate provides an upper limit on the true resonance energy. We thus compute the following table of scattering amplitudes in the narrow range **{e, 0.47515, 0.48445}** in steps of **de = 0.0003** to obtain thirty or so data points. We introduce the **Module** with the local variables **psi0e** and **psie** for computational efficiency to avoid recomputing the wavefunction unneccessarily. (Nevertheless, this might take 35 minutes to compute.)

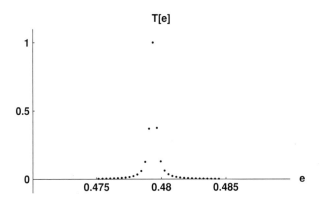

Figure 14.3-1. The first transmission resonance as a function of energy.

```
zo = 15;     dz = 0.1;

amplitudes =
    Table[
        Module[{psi0e,psie},
            psi0e = psi0[e];
            psie  = psi[e, psi0e, aI[e,psi0e]];
          {e, cT[e,psie], bR[e,psie]}
          ],
          {e, 0.47515, 0.48445, 0.0003}
        ];
```

The output has been suppressed with a semicolon, since it's rather long. We easily obtain a plot of the results, however, by making a table of, for example, the transmission coefficients, i.e. the elements **amplitudes[[n,2]]** squared, versus energy, i.e. the elements **amplitudes[[n,1]]**. Thus, (cf. Exercise 3.2, Appendix III)

```
Tvalues = {#[[1]], Abs[#[[2]]]^2}& /@ amplitudes;
```

A **ListPlot** of these values is shown in Figure 14.3-1.

We see that the fifteenth data point, corresponding to **e = 0.47935**, is very close to resonance, which we verify by computing **T** and **R** with this element:

```
amplitudes //{Abs[#[[15,2]]]^2, Abs[#[[15,3]]]^2}&
```

```
{1.00001, 0.0000349723}
```

The "full width at half-maximum" (FWHM) or just the "width" **g** of the resonance peak (usually denoted Γ) is also seen to be less than **0.0006** (i.e less than **2 de**). The width happens to be an important quantity, and we shall determine it accurately in the next section. Because the barriers are relatively thick at this energy, the transmission coefficient quickly approaches zero off resonance.

Exercise 14.3-1

Check the unitarity and symmetry relations from Problem 14.2-1 with our computed scattering amplitudes near the first resonance. **ListPlot** is convenient for this. Also, plot the reflection coefficient **R** as a function of energy and show that **R == 1 - T**.

Exercise 14.3-2

Compute more points nearer to resonance and estimate better the FWHM of the transmission peak in Figure 14.3-1. You might have to increase the precision of **NDSolve** and use for example

```
psi0e =
    psi0[e,MaxSteps->3000,AccuracyGoal->15,PrecisionGoal->15];
```

We can now compute the particle's coordinate probability distribution at resonance and plot it along with the real and imaginary parts of the wave-function, as we did in Figure 14.2-1. The result is shown in Figure 14.3-2.

```
e = 0.47935;     zo = 15;     dz = 0.05;
psi0e = psi0[e];
psie  = psi[e, psi0e, aI[e,psi0e]];
```

Since our model potential supports two bound states, the first resonant wavefunction is the third eigenfunction in the sequence and thus has two nodes, which are clearly visible in the figure (refer to Section 10.3). The wavefunction between the barriers is seen to be large (note the scale factors in the figure caption), corresponding to an intense wave trapped inside the well constructively interfering with itself and tunneling out only weakly. The situation is of course analogous to standing waves on a string which are resonantly excited by a small periodic vibration applied to one end of the string. The smaller the energy losses on the string, the larger the wave amplitude and the sharper the mechanical resonance. In the case of the quantum resonance, the incident wave from the left acts as the forcing term. If it has the same frequency as the wave reflecting back and forth between the

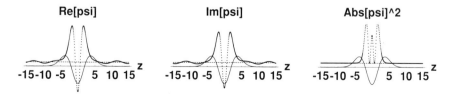

Figure 14.3-2. First resonant wavefunction and probability distribution. Here the **Re** and **Im** parts are scaled by **0.15** and the probability density by **0.0035**.

barriers, a large wave builds up inside the well. The smaller the transmission through the barriers, the larger the wave inside and the sharper the resonance.

The resonant wavefunction between the barriers is thus seen to be analogous to a bound-state wavefunction, and one which can be approximated for example by the pseudo-continuum states we discussed in Sections 10.3 and 10.5 (compare with Figure 10.3-1). As we shall see shortly, a time-dependent wavepacket which enters the well stays for a long time, relative to the passage of a free-particle wavepacket, and only slowly leaks out through the barriers. One thus speaks of *the lifetime of the resonance*, a quantity clearly related to the transmission probability and the width of the resonance. If the barrier is thick, the resonance is sharp and the lifetime long. These concepts account for the label *metastable* state and find application in all areas of physics ranging from lasing mechanisms to the stability of elementary particles.

14.4 Radial Wavefunctions

Our entire approach here is easily modified to the computation of *radial* continuum eigenfunctions which arise when an expansion of the potential energy in spherical harmonics, a so-called *partial-wave analysis*, is introduced. (The formal solution of the radial wave equation and its relation to the partial-wave analysis of the scattering problem is found in most books on quantum mechanics. See also Chapter 19.) The *1D* Schrödinger equation for the *reduced* radial wavefunction `u[r]` is identical in form to the equation we have solved here for `psi[z]`. The difference is that the range of the radial coordinate is restricted to positive values `{0,r,Infinity}`. The origin therefore acts like an infinite potential barrier, and we require that `u[0] = 0` there. This means there is no transmitted wave, only an incident wave moving towards the origin and a reflected wave moving away, and it's convenient in this case to integrate outward from the origin to a large value of `r`. The computational procedure is otherwise the same as long as the potential has finite range. (In the case of long-range interactions, as for example Coulomb forces, more elaborate modifications are required. Refer to Section 20.4 on continuum eigenfunctions of the hydrogen atom.)

These constraints on a radial solution are mimicked by any *1D* potential which has an infinite barrier on one side, such as the Morse potential (refer to Chapter 13 and the problems at the end of Section 10.6). In fact, Morse designed the potential for describing the radial motion of a vibrating diatomic molecule. An acceptable solution under the barrier in the classically forbidden region must attenuate such that `psi[z] -> 0` there. Physically, the barrier blocks the transmitted wave, just like the origin does in the radial problem. You can explore this solution in Problem 14.4-3 below for the case of the Morse potential and also in Problem 14.6-5 at the end of this chapter.

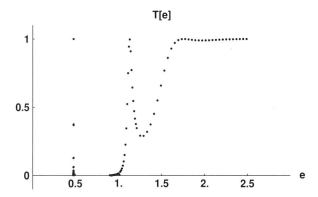

Figure 14.4-1.   The first three transmission resonances as a function of energy.

Problem 14.4-1

Extend our computations to include the second and third resonances near the tops of the barriers, one below and one above. Thus reproduce Figure 14.4-1, which includes our data for `T[e]` from the first resonance considerably "squeezed" because of the larger energy scale.

   Above the first resonance, fewer points are needed to reveal the structure of `T[e]` since the resonances are much wider. Near the second resonance, this is because the barriers are narrower and hence the transmission is larger over a wider range of energies. Also, above the barriers the transmission coefficient is close to unity anyway. This occurs because the wavelength of the particle is shorter at higher energies compared to variations in the potential, so that the "impedance mismatch" and therefore the reflections are small. Moreover, because the second resonance is close to the tops of the barriers, `T[e]` never reaches zero again as `e` increases above resonance.

   Plot the resonance wavefunctions as in Figure 14.3-2. Estimate the FWHM of the second resonance.

Problem 14.4-2

Consider scattering from a rectangular potential well and from a rectangular potential barrier defined by

```
vB[z_,w_,vo_] := 0   /;          z <  -w
vB[z_,w_,vo_] := vo  /; -w <= z <=  w
vB[z_,w_,vo_] := 0   /;   w <  z
```

where `w` is the well or barrier width in `z`. For the well, take `vo < 0` for the depth; for the barrier, take `vo > 0` for the height. (Refer to Park [50], Section 4.2 for a full discussion and analytical details. See also Problem 6.2-1.)

(a) For a well with `vo = -1` and `w = 2`, compute renormalized continuum levels `psi[e]` and transmission coefficients `T` for incident energies in the range `0 < e < -10 vo`. For simplicity, take `m = h = 1`. Use `ListPlot` to display a `Table` of pair values `{e,T}`. You should find that the transmission coefficient is mostly less than unity, which is another wave effect. Reflections at the edges of the well are generated by the abrupt changes (*impedance mismatch*) in the momentum at the edges, although transmission resonances with **T = 1** do occur. Compare your numerical results with the analytical expression

$$\frac{1}{T} == 1 + \frac{vo^2 \; Sin[2 \; kp \; w]^2}{4 \; e \; (e - vo)}$$

where the wavenumber **kp = Sqrt[2(e - vo)]**. Determine the energies of the transmission resonances for **T = 1**.

(b) For a barrier with **vo = +1** and **w = 2**, compute renormalized continuum levels **psi[e]** and the transmission coefficients **T** for incident energies **e > vo** *above* the barrier. Again, reflections occur and the transmission coefficient is mostly less than unity. The analytical expression for **1/T** in Part **(a)** still applies. Compare its values with your numerical results.

Finally, extend your results to incident energies **e < vo** *below* the barrier. Show that the expression for **1/T** in Part **(a)** still applies if **kp** is replaced by **I Kp == I Sqrt[2(vo-e)]**. Note that **Sin[2kp w]** becomes **Sinh[2Kp w]**, which is a consequence of the tunneling of the particle through the barrier and the corresponding exponential decay of the wavefunction in the classically forbidden region.

Plot your wavefunctions as in Figure 14.3-2.

Problem 14.4-3

Compute several continuum states of a Morse oscillator for **A = 1** and **De = 9**. Refer to Exercise 13.1-2 and Problem 13.4-1.

The infinite barrier in the potential for **x < 0** requires that the asymptotic limits on the wavefunction be reconsidered. The component in the classically forbidden region under the barrier must attenuate if the solution is to be physically acceptable. We can think of this component as the transmitted wave. The incident wave then comes in from the right (**x > 0**) towards the origin and the reflected wave goes out to the right away from the origin.

It's thus appropriate to begin the numerical integration at **x = -xo** under the barrier with **psi[-xo]= 0**. This will also help to minimize exponentially diverging solutions which might be triggered by numerical errors. Try generating solutions both close to and far away from the continuum threshold **e = 0**. Extract the amplitudes of the incident and reflected asymptotic planewaves and verify that reflection is 100% to within numerical accuracy. Plot your wavefunctions as in Figure 14.3-2 and check them by computing their local energy as in Figure 14.1-2.

14.5 Resonance Parameterization

In the case of a sharp, isolated resonance, the scattering amplitudes near resonance can be conveniently parameterized in terms of the energy and the FWHM of the resonance, **er** and **g**, respectively. The basic idea is that when **e ~ er**, the transmission amplitude **cT** is large and nearly constant and the reflection amplitude **bR** small and proportional to **e - er**. The amplitudes, however, must also satisfy the unitarity and symmetry relations from Problem 14.2-1 and therefore in the neighborhood of a resonance must be of the general form

```
{cT[e_] = E^(I d) I(g/2)/((e-er) + I g/2),
 bR[e_] = E^(I d) (e-er)/((e-er) + I g/2)}
```

$$\left\{ \frac{-\dfrac{I}{2}\, E^{I\,d}\, g}{e - er + \dfrac{I}{2}\, g}, \ \frac{E^{I\,d}\,(e - er)}{e - er + \dfrac{I}{2}\, g} \right\}$$

where **d** is a constant and common phase shift (see Problem 14.5-1). Since these expressions both approach zero off resonance for **Abs[e-er] >> g/2**, they're more appropriate at energies for which the barriers are high and thick, as in the case of the first resonance. We can check them directly with our symbolic **Conjugate** rule defined in the package

```
Needs["Quantum`QuickReIm`"]
```

Thus, we verify that the reflection and transmission probabilities sum to unity

```
Conjugate[bR[e]] bR[e] + Conjugate[cT[e]] cT[e] //
    Together
```

> 1

while the transmission and reflection amplitudes are related by the general relation from Problem 14.2-1:

```
Conjugate[bR[e]]/bR[e] == -Conjugate[cT[e]]/cT[e] //
    ExpandAll
```

> True

Problem 14.5-1

Starting with the two properties

$$\left\{ \mathrm{Abs}[cT]^2 == 1 - \mathrm{Abs}[bR]^2, \ \frac{\mathrm{Conjugate}[bR]}{bR} == -\left(\frac{\mathrm{Conjugate}[cT]}{cT} \right) \right\}$$

from Problem 14.2-1, show that **cT** and **bR** can be expressed in terms of three real parameters **x**, **y** and **d** as

$$\left\{ cT == \frac{I\, E^{I\,d}\, y}{x + I\, y}, \ bR == \frac{E^{I\,d}\, x}{x + I\, y} \right\}$$

These expressions take the appropriate (and conventional) form with the replacements **x = e - er** and **y = g/2**. (See Merzbacher [45], Sec. 7.4.)

The approximation **cT[e]** for the transmission amplitude also provides us with a simple expression for the transmission coefficient near resonance

```
T[e_] = Conjugate[cT[e]] cT[e] /.
          ((x_ + I/2 y_)(x_ - I/2 y_))^-1 -> (x^2 + y^2/4)^-1
```

$$\frac{g^2}{4\left((e - er)^2 + \frac{g^2}{4}\right)}$$

which we recognize to be a familiar *lorentzian* function. This result is also known as the *Breit-Wigner one-level formula* after the two physicists who first introduced it long ago for describing nuclear reactions. (It's discussed in detail in this context in Chapter 8 of Blatt and Weisskopf [8].) Its form makes clear the designation full width at half maximum (FWHM) for **g**. Namely, if we solve for the energies for which the transmission coefficient falls to half its maximum value unity, we obtain

```
Solve[T[e] == 1/2, e] //ExpandAll
```

$$\left\{\left\{e \to er + \frac{g}{2}\right\}, \left\{e \to er - \frac{g}{2}\right\}\right\}$$

Hence, the FWHM is just the difference in these two energies, **2 (g/2)**.

We could determine **g** by fitting the lorentzian form to our computed transmission coefficients shown in Figure 14.3-1. However, since this would require a nonlinear fit, let's look for a simpler linear function of **e**. Evidently, if we normalize **cT[e]** by its value on resonance, **cT[er] == E^(I d)**, we can use the ratio of its **Im** and **Re** parts:

```
Im[cT[e]/cT[er]]/Re[cT[e]/cT[er]] //Together
```

$$\frac{2 (e - er)}{g}$$

Let's see what our computed scattering amplitudes give for this quantity. Since the fifteenth data point in Figure 14.3-1 is very near resonance, we define **er** and **cT[er]** by

```
{er = amplitudes[[15,1]], cT[er] = amplitudes[[15,2]]}
```

```
{0.47935, 0.158358 + 0.987388 I}
```

and hence compute values of the desired ratio near resonance according to (cf. the computation of **Tvalues** in Section 14.3)

```
cTratio =
    {#[[1]], Im[#[[2]]/cT[er]]/Re[#[[2]]/cT[er]]}& /@
        amplitudes;
```

These values are plotted in Figure 14.5-1 and display, as desired, a near linear dependence on **e** in the vicinity of **er**.

Dropping the first **10** and the last **12** (nonlinear) points, we thus compute a linear fit of **cTratio** (cf. Exercise 3.4, Appendix III):

```
cTratioFit = Fit[ Take[cTratio,{11,20}], {1, e}, e ]
```

```
  -1979.95 + 4130.54 e
```

When equated to the approximate expression **2(e-er)/g** from above, we obtain

```
Solve[
    cTratioFit == 2(e-erFit)/gFit,
    {erFit,gFit}, e
][[2]] //N
```

```
  {erFit -> 0.479343, gFit -> 0.000484198}
```

Here the syntax **Solve[equation,variables,e]** has been used to eliminate the variable **e**. Because of the approximate nature of the fit, the value **erFit** is only close to our original value **er = 0.47935**, which, nevertheless, gives us an idea of the accuracy of **gFit**. The "goodness of fit" is indicated by Figure 14.5-2.

Now we can include the lorentzian approximation for **T[e]** and the value **gFit** in a plot of our original numerical data from Figure 14.3-1. The result is shown in Figure 14.5-3, and, as expected from Figure 14.5-2, the fit is quite reasonable.

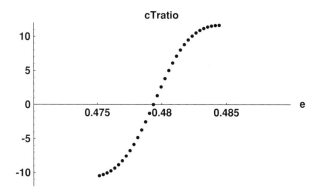

Figure 14.5-1.   The ratio **Im[cT[e]/cT[er]]/Re[cT[e]/cT[er]]** near the first resonance.

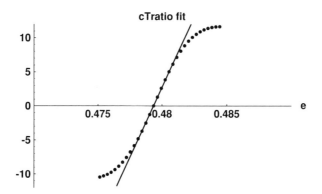

Figure 14.5-2. Linear fit to the ratio `Im[cT/cTr]/Re[cT/cTr]` near the first resonance.

We shall relate the width **g** to the lifetime of the resonance in the next section when we examine the scattering and delay of a wavepacket by the potential.

Although our parameterization has been based specifically on potential scattering, our conclusions are largely independent of the forces governing the reaction. Indeed, resonance theory provides a means of classifying reactions for which the forces are unknown or simply too complicated to allow a full solution of the Schrödinger equation. Thus, resonance theory has a broad and distinguished history in nuclear and elementary particle physics and as well in the interpretation of complex spectra of atoms, molecules and solids.

Problem 14.5-2

Redetermine **gFit** with the computed *reflection* amplitudes **amplitudes[[n,3]]**. Make plots of **bRratio** and of **R[e]** analogous to Figures 14.5-2 and 14.5-3.

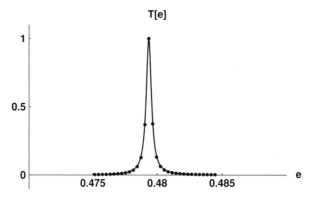

Figure 14.5-3. Lorentzian fit to the computed transmission coefficient near the first resonance.

Problem 14.5-3

Estimate the width of the second resonance using the computed transmission amplitudes from Problem 14.4-1 and Figure 14.4-1. To obtain reasonable results, you'll have to exclude values of **cT** above resonance, since the transmission coefficient doesn't go to zero above resonance and our approximate formulas really don't apply there. Make plots of **cTratio** and of **T[e]** as in Figures 14.5-2 and 14.5-3.

Wavepacket Impact   14.6

As a final application, we construct a time-dependent scattering wavepacket as a linear superposition of continuum eigenfunctions belonging to a small range of eigenenergies with an average value **eP**. Since continuum eigenfunctions extend over all space, our intent is to represent a particle which is more localized in space and incident on the well. Because the eigenenergies are continuous, the superposition we need is a Fourier integral. Of course, since our solutions are fully numerical, we have to approximate the integral by a discrete sum over a grid of discrete energies. However, if we don't require our wavepacket to be too localized in space, we can achieve quite satisfactory results by summing over just a few eigenfunctions. We've already seen how this works in Chapter 3 and in Problem 4.2-3.

We also extend in this way our previous *FFT* investigation of wavepacket propagation from Section 12.6. With a suitable superposition we can now study very narrow wavepackets in energy and therefore broad wavepackets in space. In particular, the spatial width of a wavepacket narrow in energy may exceed the length of the lattice, a situation not permitted with the *FFT*.

We get a good idea of the linear combination we want by constructing first a free-particle wavepacket. Thus, consider the time-dependent superposition of planewaves defined by

```
freePacket[t_,dz_:1.0,zP_:-50] :=
    Table[
        Evaluate[
            {z,
             Sum[
                 f[e] E^(-I e t) E^(I k[e] (z-zP)),
                 {e,emin,emax,de}
             ]}
        ],
         {z,-zo,zo,dz}
    ]
```

where the phase **E^(-I k[e] zP)** shifts the **t = 0** peak of the packet to **zP**, which we take to be **zP = -50** by default far to the left of the potential (see Section 4.2). A value **zP < -zo** just means that the peak of the packet is initially outside our "viewing window" defined by **-zo < z < zo** (compare with Figure 14.6-2 below). Since each planewave corresponds to a momentum along the positive z axis, the wavepacket moves with time to the right towards **z > 0**.

It is convenient to assign each planewave component a weight **f[e]** according to a gaussian distribution, as we did in Problem 4.2-3. (Here, however, we need a function of energy and not of wavenumber.) To be specific, let's tune to the second resonance, which according to Figure 14.4-1 is very close to **e = 1.13** (see also Problems 14.4-1 and 14.5-3), by defining

```
{eP = 1.13, emin = 1.01, emax = 1.25, de = 0.04};
```

This gives us a grid of seven energy values which includes **eP**. We thus compute the amplitudes according to

```
{f[e_] = E^(-(e-eP)^2/(2w))/nrm, w = 0.005, nrm = 1.76653};
```

The normalization constant **nrm** has been evaluated for this energy grid such that (see also Exercise 14.6-3 below)

```
Sum[f[e]^2,{e,emin,emax,de}]
```

```
    1.
```

These amplitudes are plotted in Figure 14.6-1 along with the transmission coefficient **T[e]** from Figure 14.4-1 for comparison. The width parameter **w** has been adjusted to make **f[e]** roughly as wide as the resonance peak in order to ensure that a large portion of the scattered wavepacket will be transmitted through the potential (refer again to Exercise 14.6-3 below).

The probability density of the resulting free-particle wavepacket is plotted in Figure 14.6-2 using our previously defined function **psisq** (see also below). The viewing window has been extended by setting **zo = 125** to

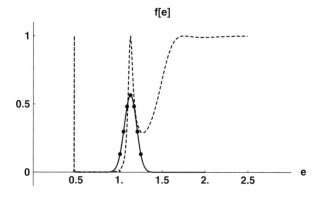

Figure 14.6-1. Amplitude function **f[e]** (solid curve) of the **freePacket[t]** planewave components. The seven points are the actual values used in the superposition. The dashed curve shows the transmission coefficient **T[e]** from Figure 14.4-1.

show the initial peak position at **zP = -50**, which however considerably "compresses" the potential.

Exercise 14.6-1

Verify the peak position at **t = 65** of the free-particle wavepacket in Figure 14.6-2 from the average speed of the wavepacket. Animate the motion of the packet.

Exercise 14.6-2

Because we approximate a Fourier integral by a discrete sum with just seven components, our wavepackets are periodic with a large but nevertheless finite period. This means that other peaks will eventually pass over the potential (see also Problem 12.6-2). Demonstrate this shortcoming by extending the viewing window and animating the free-particle wavepacket for a long enough time.

It's now straightforward to form a scattering wavepacket as a time-dependent linear combination of (renormalized) continuum eigenfunctions **psi[e]**. In analogy with **freePacket[t]** we thus define the superposition (recall that **psi[e][[n,1]]** gives the *nth* grid point in **z**)

```
psiPacket[t_,zP_:-50] :=
    Table[
        {psi[eP][[n,1]],
         Sum[
            f[e] E^(-I e t) E^(-I k[e] zP) psi[e][[n,2]],
            {e, emin, emax, de}
         ]},
         {n, Length[psi[eP]]}
    ]
```

using the energy grid and amplitudes **f[e]** from above. Again, the phase **E^(-I k[e] zP)** shifts the **t = 0** peak of the packet to **zP**. Since each component eigenfunction **psi[e]** corresponds asymptotically for **z << 0** to an incident planewave, the wavepacket moves initially to the right towards

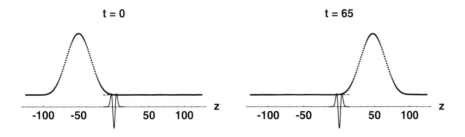

Figure 14.6-2. The probability density of a moving free-particle wavepacket at two different instances.

z > 0. However, because the eigenfunctions, unlike planewaves, contain transmitted and reflected components, the wavepacket splits into transmitted and reflected packets as the particle scatters off the potential.

Before we go any further, we have to compute the continuum eigenfunctions. The following **Do** loop is a convenient way to set this up with a syntax identical to **Table**. The value **zo = 30** increases our viewing window without making unreasonable computational demands, although the initial wavepacket will be mostly outside the viewing window. (The option **MaxSteps** on **NDSolve** has to be bumped up, nevertheless, to cover the longer interval. The computation takes about 15 minutes.)

```
zo = 30;     dz = 0.1;

Do[
    Module[{psi0e},
        psi0e = psi0[e, MaxSteps -> 1500];
        psi[e] = psi[e, psi0e, aI[e, psi0e]];
    ],
    {e, emin, emax, de}
]
```

Exercise 14.6-3

Compute the transmission and reflection coefficients **T[e]** and **R[e]** for these seven states **psi[e]**, and define and evaluate average coefficients according to

```
Tavg = Sum[ f[e]^2 T[e], {e, emin, emax, de}]
Ravg = Sum[ f[e]^2 R[e], {e, emin, emax, de}]
```

Show that the normalization of **f[e]** ensures that **Tavg + Ravg == 1**. Since the width of **f[e]** is close to the width **e** of the resonance, a large portion of the scattered wavepacket will be transmitted by the potential.

Now we're ready to examine the scattered wavepacket. We begin by setting up some functions to plot **psiPacket[t]** together with **freePacket[t]** as a function of time, and thus define

```
psiPlot[t_] :=
    ListPlot[
        Evaluate[psisq[eP,0.65,psiPacket[t]]],
        bkgrd[eP], PlotRange -> {-3,15},
        Axes -> {Automatic,None},
        PlotLabel -> "t = " <> ToString[t],
        AxesLabel -> {" z",""}, PlotJoined -> True,
        DisplayFunction -> Identity
    ];

freePlot[t_] :=
    ListPlot[
        Evaluate[psisq[eP,0.65,freePacket[t]]],
```

```
          bkgrd[eP], PlotRange -> {-3,15},
          Axes -> {Automatic,None},
          PlotLabel -> "t = " <> ToString[t],
          AxesLabel -> {" z",""},
          DisplayFunction -> Identity
      ];

  psifreePlot[t_] :=
      Show[
          psiPlot[t], freePlot[t],
          DisplayFunction -> Identity
      ];
```

We can then generate, for example, a snapshot of the scattered wavepacket at a given instant of time, as in Figure 14.6-3.

Figure 14.6-3 shows clearly the formation of the resonance state. The transmitted front portion of the scattered wavepacket has just passed through the potential well, as indicated by the finite probability which is already being leaked to the right of the well and by the position of the free-particle wavepacket. Also, the three nodes that characterize the second resonance are developing between the barriers. The rear portion of the scattered wavepacket which is still incoming is now interfering with its promptly reflected front portion, as indicated by the strong interference to the left of the well. Up to this point in time, the free-particle wavepacket still determines the overall form of the scattered wavepacket.

In order to follow the development more closely, we need to examine several snapshots of the wavepacket's motion, as in Figure 14.6-4 (which takes about 30 minutes to generate). Alternatively, we can make an animation as a function of time (see Exercise 14.6-4).

```
Show[ psifreePlot[30], DisplayFunction -> $DisplayFunction ];
```

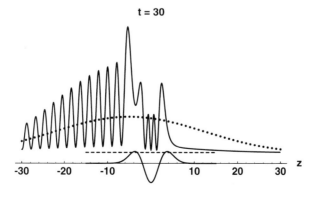

Figure 14.6-3. Probability density of the scattered wavepacket for $t = 30$. The dotted curve is the corresponding free-particle density.

The sequence of pictures in Figure 14.6-4 demonstrates the persistance of the metastable state long after the incident packet has passed through the scattering region. This delay of the wavepacket by the potential well defines the *resonance lifetime* Δt. We see that although the initial wavepacket is spatially long compared to the range of the potential, it is nevertheless narrow enough that it passes through in a time less than the delay we attribute to the lifetime. At later times after passage of the free-particle wavepacket, the metastable state between the barriers begins to resemble a bound state, although the energy is positive.

Close inspection of Figure 14.6-4 reveals that the probability density oscillates or "rattles" for a time between the barriers. However, with each reflection near an inner turning point, some of the probability tunnels through the barrier and a small wavepacket appears outside the well. Such transmitted and reflected packets are evident at $t = 60$ and $t = 80$ to the right and to the left of the potential, respectively. We can interpret $1/\Delta t$ as just the number of instances per unit time the particle trapped between the barriers strikes one of the barriers multiplied by the probability for tunneling through the barrier (see Bohm [10], Section 12.19). The process is evidently self damping. This effect is much more dramatic when the initial packet is spatially narrow compared with the range of the potential, a situation which can be efficiently studied with *FFT* propagation as in Section 12.6.

Detailed analysis shows that the decay in time of the metastable-state probability eventually follows an exponential law with a mean lifetime Δt. Moreover, Δt is inversely proportional to the width of resonance, i.e. $\Delta t =$ h/g (see Merzbacher [45], Sections 7.4 and 11.6). (Here we reinstate $h \neq 1$ for generality.) Therefore, if the resonance is sharp its lifetime is long, whereas if the resonance is broad its lifetime is short. Clearly, the higher resonance is much shorter lived than the lower one for our model potential.

We can also take the point of view that the decaying metastable state is a solution of the Schrödinger equation corresponding to the *complex eigenvalue* $er - I\,g/2$. We account in this way for the exponential decay since the time-development operator for the state becomes

```
E^(-I (er - I g/2) t/h) //ExpandAll

   (-I er t)/h - (g t)/(2 h)
  E
```

which when squared gives (again using our symbolic **Conjugate** rule)

```
Conjugate[%] %

   -((g t)/h)
  E
```

as required.

The interested reader will enjoy the beautiful reference on wavepacket scattering by Brandt and Dahmen [11], *The Picture Book of Quantum Mechanics*,

```
Show[
    GraphicsArray[
        Evaluate[
            Table[
                {psifreePlot[t],psifreePlot[t+20]},
                {t,0,80,40}
            ]
        ]
    ],
    GraphicsSpacing -> {.1,.2},
    DisplayFunction -> $DisplayFunction
];
```

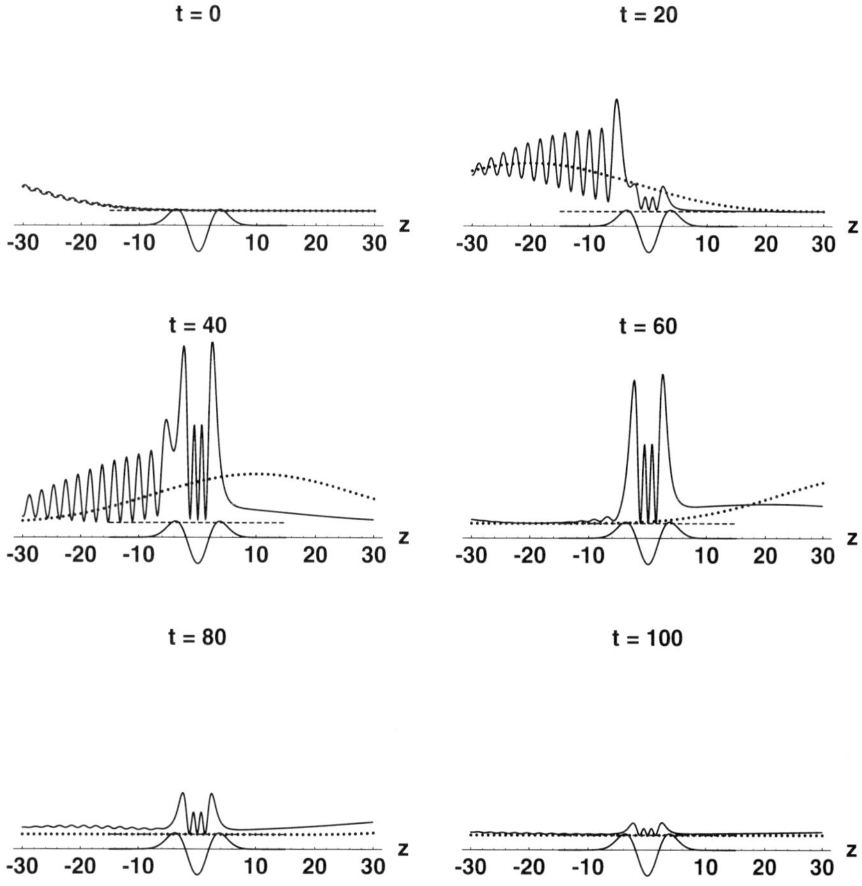

Figure 14.6-4.   Formation and decay of the second metastable state. The dotted curve shows the passage of the free-particle wavepacket.

a more sophisticated text than its title might imply. Indeed, a great deal of what we have done here was inspired by this book.

Exercise 14.6-4

Reproduce Figure 14.6-4 with finer time steps, or better as an animation, to reveal the release of the delayed transmitted and reflected wavepackets. Estimate the delays from your plots and the separations of the delayed packets with the free-particle and promptly reflected packets. Show that the delays are consistent with the width of resonance and hence the resonance time delay from Problem 14.5-3.

Exercise 14.6-5

Tune the wavepacket off the **er = 1.13** resonance by changing the mean energy **eP**. Try an energy between the first and second resonances and then one above the barriers.

Problem 14.6-1

Estimate the total probability between the barriers of the **er = 1.13** metastable state as a function of time by numerical integration. Attempt to estimate the width **g** of resonance from the slope of a plot of **Log[*probability*]** versus **t**.

Problem 14.6-2

Drop even- and odd-parity wavepackets with average energy **eP = 1.13** but with zero velocity into the middle of the well and propagate them with time. (Refer to Exercise 14.2-1 above.) Use first the narrow amplitude function **f[e]** in the text, and then one with a wider energy spread. Plot or animate the time evolution.

Problem 14.6-3

Generate a scattering wavepacket tuned to the first resonance and hence with mean energy **eP = 0.47935**. You thus want a new amplitude function **f[e]** with a width that roughly equals that of the transmission coefficient **T[e]** in Figure 14.5-3. However, you might also want to consider limiting the spatial width of your packet since your viewing window **-zo < z < zo** has to be limited. If this proves to be too demanding computationally, redo Problem 14.6-2 for this energy.

Animate your results for a wavepacket initially located to the left of the potential or generate a **GraphicsArray** of snapshots as in Figure 14.6-4.

Problem 14.6-4

Consider scattering from a pair of rectangular barriers defined by

```
vB[z_,w_,vo_] := 0   /;              z <  -1.3w
vB[z_,w_,vo_] := vo  /; -1.3w <= z <= -w
vB[z_,w_,vo_] := 0   /;    -w <  z <   w
vB[z_,w_,vo_] := vo  /;     w <= z <= 1.3w
vB[z_,w_,vo_] := 0   /;   1.3w <  z
```

for **w = 2** and **vo = 1**. For simplicity, take **m = h = 1**. Locate the resonances and determine their widths. Plot and discuss the resonance wavefunctions. Construct a scattering wavepacket which is initially located to the left of the potential and animate its time development or generate a **GraphicsArray** of snapshots as in Figure 14.6-4.

This problem is examined in detail by Brandt and Dahmen [11].

Problem 14.6-5

Redo the previous problem for the model potential with an infinite barrier defined by

```
V = z^2/2                 /; z <  0
V = z^2/2 E^(-b z^2)      /; z >= 0
```

with **b = 0.15** from problem 10.6-2. This and related potentials have been recently discussed in connection with molecular resonances by J. F. Babb and M. L. Du, Chem. Phys. Lett. **167**, 273 (1990).

PART II

Quantum Dynamics

Quantum Dynamics

We now pick up the pace somewhat and lift the restriction to life in one dimension. Our main task will be to develop computer methods for handling the noncommuting operators needed to describe quantum systems and the protracted algebra which often arises. We begin with a purely formal approach based on the built-in function `NonCommutativeMultiply`, and develop for example an algebraic solution of the harmonic oscillator problem and a formal representation of angular momentum, including angular momentum coupling.

We then turn to the representation of commutator algebra with derivative operators. As an application, we develop the cartesian representation of angular momentum and determine the hydrogen-atom energy spectrum algebraically. We also examine the transformation of angular momentum to curvilinear coordinates and solve the hydrogen atom problem in spherical and parabolic coordinates with emphasis on constructing a complete set of commuting observables. In addition, we establish the fundamental connection between the hydrogen atom and the $2D$ isotropic harmonic oscillator.

15
Quantum Operators

In quantum mechanics, the conjugate (x axis) components **x** and **px** of the position and momentum vectors **rvec** and **pvec** of a particle are taken to be *noncommuting* quantities or *operators* such that **x px** is not equal to **px x**. The same is assumed of the components **y** and **py** and **z** and **pz**, although independent variables must commute, i.e. **x y = y x**, **x py = py x**, etc. All physical quantities or *dynamical variables* describing a system are represented in a similar way by operators, called *observables*, generally constructed from components of **rvec** and **pvec**. One ensures in this way that the Heisenberg uncertainty principle is built into the mathematical foundation of quantum mechanics: measurements of noncommuting observables disturb each other; measurements of commuting observables don't.

The degree of noncommutivity is determined by Planck's constant through the *fundamental commutation relations*, for example, **x px - px x = I h**. This is of course an inherently nonclassical quantity. (As usual, **h** stands for Planck's constant divided by **2Pi**.) In practical terms, it means we have to keep track in general of the order in which operators are first applied when working with an expression, unless we explicitly change their order with a rule. Specifically, the product **x px** means that **x** has to stay to the left of **px**, unless we transform it using **x px -> px x + I h**. Thus, if **x** is the number of meters along the x axis from the origin, then **px** has to be a more complicated mathematical entity, of course with units of momentum (such as kg-m/s).

As we have seen already in Section 1.1, the commutation relations can be implemented, and usually are, by assuming that **px** is proportional to a derivative with respect to **x**. In more than one dimension, we make analogous assumptions about **py** and **pz** (free space is isotropic) and therefore take the *vector* operator **pvec** to be proportional to a *gradient* with respect to **rvec**. These derivatives operate on the wavefunction as a function of the coordinates **x**, **y**, **z**, and this is referred to as the *coordinate representation*. One also often works in the *momentum representation* with **x** proportional to a derivative with respect to **px** (see Section 11.8) and therefore with **rvec** proportional to a gradient with respect to **pvec** and a wavefunction as a function of **px**, **py**, and **pz**. Another familiar approach is to represent operators as matrices, since matrix multiplication is also noncommutative.

We will examine the coordinate and momentum representations later on beginning with Chapter 18 and turn first to a purely formal representation which also lends itself to the computer. We will investigate the matrix representation of angular momentum in Chapter 16.

15.1 Commutator Algebra

In *Mathematica*, we can represent the product of two noncommuting quantities **A** and **B** by the special built-in function **NonCommutativeMultiply[A,B] = A ** B**. This mathematical operation preserves the order of application of **A** and **B** and can be customized for quantum observables like **x** and **px**. Compare for example how ordinary and noncommutative multiplies are rearranged when entered:

```
{B A, B ** A}
```

```
    {A B, B ** A}
```

In particular, the *commutator* of **A** and **B**, defined by

```
Commutator[A_,B_] := A ** B - B ** A
```

does not automatically vanish as it would if ordinary multiplication (**Times**) were used:

```
{Commutator[A,B], A B - B A}
```

```
    {A ** B - B ** A, 0}
```

(In the text we'll usually write the commutator in the familiar way **[A, B]** for short.) Now in order to be able to simplify expressions involving both kinds of multiplies, we need to define rules for ******, which essentially has just the associative law built-in, for example

```
(A ** B) ** C == A ** (B ** C)
```

```
    True
```

In addition, however, since quantum operators are linear, we want to be able to invoke the distributive law of multiplication and ensure that ****** is independent of multiplication and *division* by ordinary numbers. As it stands, no simplification occurs:

```
(2 A + B/3) ** C

         B
 (2 A + -)  ** C
         3
```

We relgate the development of the necessary rules to Exercise 2.5 in Appendix III. Here, we simply implement the rules by loading (see Appendix IV)

```
Needs["Quantum'NonCommutativeMultiply'"]
```

and examine their properties. We thus now obtain as desired

```
(2 A + B/3) ** C

                B ** C
 2 A ** C +  ───────
                3
```

Moreover, ****** now has the desirable properties that multiplication by zero vanishes, that multiplication by one is the identity operation, and that factors like **Sqrt[2]** commute, i.e.

```
{0 ** A,  1 ** A,  (Sqrt[2] A) ** B}

 {0, A, Sqrt[2] A ** B}
```

Note that the package introduces the rules *globally* so that *Mathematica* automatically uses them on all expressions it evaluates involving ****** (see Maeder [43], Chapter 6). This means we don't have to explicitly call a function to get things done, which will greatly simplify our input in the following and help us to automate operator algebra. One drawback is that it's almost impossible to prevent the rules from being used. Should you want to do that, you might find it more useful to convert the package and define a function, say **ExpandNCM**, which applies the rules. See Exercise 2.5, Appendix III.

In addition, we can check that the commutator **Commutator[A,B]** obeys an algebra appropriate to quantum mechanics. Thus, we find that any operator commutes with itself and with multiplication (and division) by ordinary numbers:

```
c /: NumberQ[c] = True;

{Commutator[c A,A],
 Commutator[c A,B] == c Commutator[A,B]} //
   ExpandAll

 {0, True}
```

Likewise, the fundamental linearity of quantum operators means that

```
Commutator[A,B+C] ==
   Commutator[A,B] + Commutator[A,C]

 True
```

while noncommutivity further implies that

```
Commutator[A**B,C] ==
    A**Commutator[B,C] + Commutator[A,C]**B
```

```
    True
```

In this rule, operator ordering on the right-hand side is important since in general two operators **A** and **B** don't commute with their commutator **[A, B]**. Finally, it's easy to see that commutators are *antisymmetric* in the sense that

```
Commutator[A,B] == -Commutator[B,A]
```

```
    True
```

Although entirely adequate for a wide range of tasks in quantum mechanics, our rules aren't completely general and won't handle for example fractional powers or quotients of operators correctly, a generalization rarely required. Readers seeking more applicability or just interested in noncommutative algebra on the computer should refer to packages available from Wolfram Research. See for example the package **NCAlgebra** developed by J. W. Helton and R. L. Miller.

Exercise 15.1-1

Take a closer look at the above five basic commutators. Evaluate for example individual terms and examine the first identity without the **ExpandAll**. Try evaluating all the identities as a single expression.

Exercise 15.1-2

Verify **[A, [B, C]]** + **[B, [C, A]]** + **[C, [A, B]]** == 0, i.e. that the sum of the cyclic permutations of the double commutator of three operators vanishes. Hence, the commutator "product" is not associative. This rule together with the above linearity and antisymmetry rules define a *Lie algebra*, familiar in applications of group theory to quantum mechanics.

In the formal study of a physical system, our primary tools will be the fundamental commutation relations **[x , p]== I h**, etc., which we conveniently invoke with a rule such as **p ** x -> x ** p - I h**. For example,

```
Commutator[x,p] /. p**x :> x**p - I h
```

```
    I h
```

(In one dimension, we will write simply **p** instead of **px**.) In order to simplify more complicated expressions, it is useful to introduce a function which will apply this rule to an expression until no further change occurs. We thus define (refer to Exercises 2.2 - 2.5 in Appendix III)

```
xpCommute[expr_] := Expand[expr //. p ** x :> x ** p - I h]
```

and re-evaluate the previous expression as

```
Commutator[x,p] //xpCommute
```

```
I h
```

As an example, let's consider the commutator of **p/(-I h)** with powers of **x**. In order to simplify ******, we have to declare **h** to be a number.

```
h /: NumberQ[h] = True;
```

```
{Commutator[p/(-I h), x],
 Commutator[p/(-I h), x**x],
 Commutator[p/(-I h), x**x**x],
 Commutator[p/(-I h), x**x**x**x]} //xpCommute
```

```
{1, 2 x, 3 x ** x, 4 x ** x ** x}
```

We see that this process is tantamount to differentiation. We in fact deduce that for any *analytic* function **f[x]** which can be expanded in a power series in **x** the commutator **[p/(-I h),f[x]] == f'[x]** (see Problem 15.1-1). (This result can actually be established for any function whose derivative with respect to **x** is defined. See Exercise 18.2-2)

Exercise 15.1-3

Show that if an operator **a** commutes with the commutator **[a,b]**, then **[a^n,b]== n a^(n-1)[a,b]**.

Problem 15.1-1

(a) For some applications it is convenient to have a function which expands a power, such as **a^3** into **a**a**a**, while another function contracts the result back into **a^3**. Thus define functions **PowertoNCM[expr]** and **NCMtoPower[expr]** which will transform any occurrences of **a^n** into **a**a**** ... and **a**a**** ... into **a^n**, respectively, in **expr** for any positive integer **n**. (Hint: Use **Nest** and refer to Exercise 3.3 in Appendix III.)

(b) Show that if a function **f[x]** is analytic then **[p/(-I h), f[x]]== f'[x]**. (Hint: Expand **f[x]** and **f'[x]** in power series and use the functions in part **(a)**.)

Every physical quantity or dynamical variable in a system is represented by a hermitian operator, called an *observable*. It's easy to show that if two observables **P** and **Q** have a common complete set of eigenfunctions then their commutator **[P,Q]** must vanish. One can show the converse is also true, although the proof is more difficult: *if **P** and **Q** are commuting observables then a common set of eigenfunctions can be found for them.* We then refer to **P** and **Q** as *compatible observables* with the following experimental consequence. As we discussed in Section 2.6, if we make a measurement of one of the observables such as **P**, we obtain one of its eigenvalues **p**, and the system will be left in the corresponding eigenstate of **P**. If we make a second measurement now of **Q**, we obtain one of its eigenvalues **q** and the system will be left in the eigenstate of **Q**. When **P** and **Q** commute, these states are the same and the eigenstate of **P** won't be disturbed by the measurement of **Q**, so that a third measurement of **P** will give the same eigenvalue **p** as the first.

If two observables **P** and **Q** don't commute, the corresponding physical quantities can't be measured without disturbing one another, and sharply defined eigenvalues can't be defined for each quantity simultaneously. The inevitable degree of *fuzziness* in their values in an ensemble of measurements is specified by their commutator through the Heisenberg uncertainty relation (see for example Park [50], Section 3.5, or Merzbacher [45], Section 8.7):

$$delP \; delQ \; >= \; Abs[<I \; Commutator[P,Q]>]/2 \qquad \text{(15.1-r1)}$$

where `<...>` stands for expectation value and **delP** and **delQ** are the respective root-mean-square deviations from the mean (cf. Section 3.0). For example, if **P** is the momentum and **Q** the conjugate position coordinate, such that `[P,Q]== -I h`, then we recover the ubiquitous relation $\Delta Q \, \Delta P \geq h/2$.

The commutator of an observable **Q** with the hamiltonian **H** has an especially important role in the theory. One readily demonstrates with the Schrödinger wave equation that the expectation value of **Q** satisfies the equation of motion

$$I \; h \; D[<Q>,t] \; == \; <[Q,H]> \; + \; I \; h \; <D[Q,t]> \qquad \text{(15.1-r2)}$$

Thus, if **Q** is explicitly time independent, viz. `D[Q,t]= 0`, and commutes with **H**, the expectation value of **Q** is independent of time, and **Q** is said to be *conserved* and is termed a *constant of the motion*. In particular, we recover the fundamental result from Section 1.2 that if **H** is independent of time, i.e. *conservative*, it is also a constant of the motion (since any operator commutes with itself), and energy (= `<H>`) is conserved.

We turn now to examples of using the fundamental commutation relations to establish basic properties of a system. As we shall see, we can often learn a great deal from essentially formal considerations.

Problem 15.1-2

If **H** is a hamiltonian with potential **V[x]**, use relation (15.1-r1) to show that

$$delH \; delx \; >= \; h/(2m) \; Abs[<p>]$$

Investigate this relation for the free-particle gaussian wavepacket in Chapter 4 and for the coherent-state wavepacket in Chapter 8 (refer to Problem 8.2-1).

Problem 15.1-3

(a) Consider a *1D* harmonic oscillator defined by the hamiltonian `hHO = p**p/(2m)` `+ m w^2 x**x/2`. Use relation (15.1-r2) to set up and solve the equations of motion *directly* for the expectation values `<x(t)>` and `<p(t)>` subject to the initial conditions `<x(0)> = xPo`, `<p(0)> = h kPo`. (Refer to the classical solution from Exercise 8.0-1.) Thus, rederive the squeezed-state results we obtained in Section 8.3.

(b) Evaluate the commutators `[[hHO, x], x]` and `[[hHO, p], p]` and use the results to evaluate formally the sums (with paper and pencil if necessary)

```
Sum[(e[n] - e[k]) Abs[xme[k,n]]^2, {n,0,Infinity}]
Sum[(e[n] - e[k]) Abs[pme[k,n]]^2, {n,0,Infinity}]
```

where `e[n]` are the energy eigenvalues of `hHO` and `xme[k,n]` and `pme[k,n]` are the matrix elements of `x` and `p` between eigenstates of `hHO` (from Section 9.1). Which of these results easily generalizes to an arbitrary potential?

Two-Body Relative Coordinates   15.2

Consider the transformation from space-fixed coordinates to relative coordinates for a two-body system by requiring that the *new* coordinates and their conjugate momenta satisfy the fundamental commutation relations. Although we derive in this way the relative momenta, our approach to this familiar problem is useful in constructing a proper set of coordinates for a more general few-body system.

For simplicity, we may assume that the particles are constrained to a straight line, e.g. the space-fixed *x* axis shown in Figure 15.2-1, with particle `i` (`i = 1, 2`) located at `x[i]` with mass `m[i]` and momentum `p[i]`. We then introduce the center of mass position **CM** and the relative separation **R** of the two particles according to:

```
CM = m[1]/M x[1] + m[2]/M x[2];     R = x[1] - x[2];
```

where `M = m[1] + m[2]` is the system's total mass. To derive the momenta conjugate to **CM** and **R**, we form linear combinations of `p[1]` and `p[2]` with unknown coefficients `a[i]` and `b[i]`,

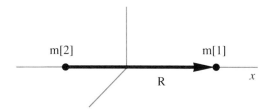

Figure 15.2-1.   Two particles constrained to a space-fixed axis.

```
pCM = a[1] p[1] + a[2] p[2];
pR  = b[1] p[1] + b[2] p[2];
```

and determine the coefficients by requiring that the new coordinates satisfy the appropriate commutation relations, `[R,pR] == I h`, etc. since the `x[i]` and `p[i]` do. We can do this by appending the rule `p[i_] ** x[i_] -> x[i] ** p[i] - I h` to `xpCommute`.

We readily verify, however, that a single rule is insufficient to simplify fully expressions generated by **Commutator**. In addition, we must arrange products involving `x[i]` and `p[j]` into a specific order, which we can do by noting that variables belonging to particle **1** commute with those belonging to particle **2**. Commutativity ensures from the uncertainty principle that a measurement on one particle doesn't affect the other if the two particles are noninteracting. We thus require four new rules:

```
xpCommute[expr_] :=
    Expand[ expr //.
        {p ** x :> x ** p - I h,
        p[i_] ** x[i_] :> x[i] ** p[i] - I h,
        p[i_] ** x[j_] :> x[j] ** p[i],
        x[i_] ** x[j_] :> x[j] ** x[i] /; i > j,
        p[i_] ** p[j_] :> p[j] ** p[i] /; i > j}
    ]
```

That is, we arrange **x** before **p** but otherwise **1** before **2**. An example demonstrates:

```
{p[1]**x[1], p[1]**x[2], x[2]**x[1]} //xpCommute

{-I h + x[1] ** p[1], x[2] ** p[1], x[1] ** x[2]}
```

Hence, we set up and solve a system of four equations in four unknowns `a[i]` and `b[i]` after making some necessary declarations in order to simplify `**`.

```
m /: NumberQ[m[_]] = True;    M /: NumberQ[M]    = True;
a /: NumberQ[a[_]] = True;    b /: NumberQ[b[_]] = True;

{Commutator[R,pR ] == I h, Commutator[CM,pCM] == I h,
 Commutator[R,pCM] == 0,   Commutator[CM,pR ] == 0} //
        xpCommute //ExpandAll
```

$$\{I\ h\ b[1]\ -\ I\ h\ b[2] == I\ h,\ \frac{I\ h\ a[1]\ m[1]}{M} + \frac{I\ h\ a[2]\ m[2]}{M} == I\ h,$$

$$I\ h\ a[1]\ -\ I\ h\ a[2] == 0,\ \frac{I\ h\ b[1]\ m[1]}{M} + \frac{I\ h\ b[2]\ m[2]}{M} == 0\}$$

```
ab = Simplify[Solve[%, {a[1],a[2],b[1],b[2]}] [[1]]] /.
       m[1]+m[2] -> M
```

$$\{a[1] \rightarrow 1,\ a[2] \rightarrow 1,\ b[1] \rightarrow \frac{m[2]}{M},\ b[2] \rightarrow -(\frac{m[1]}{M})\}$$

Thus, we have

```
{pCM, pR} /. ab
```

$$\{p[1] + p[2],\ \frac{m[2]\ p[1]}{M} - \frac{m[1]\ p[2]}{M}\}$$

and therefore that **pCM** is the total momentum of the system, as required. Moreover, we find that **pR** has the desired classical interpretation in that if we introduce classical velocities according to `p[i] -> m[i]v[i]` then

```
pR /.ab /. p[i_] -> m[i] v[i] //Factor
```

$$\frac{m[1]\ m[2]\ (v[1] - v[2])}{M}$$

That is, **pR** is the time derivative of **R** times a *reduced mass* `mR = m[1]m[2]/M`, since `v[i]` is the time derivative of `x[i]`.

The utility of these coordinates becomes apparent when the two-particle hamiltonian is transformed. Of particular interest are *central-force* systems in which the potential energy **V** is a function only of the separation **R** of the two particles (see Section 18.4 below). One finds (Exercise 15.2-1) that the kinetic energy transforms as

$$\frac{p[1]^2}{2\ m[1]} + \frac{p[2]^2}{2\ m[2]} == \frac{pCM^2}{2\ M} + \frac{pR^2}{2\ mR} \tag{15.2-r1}$$

The relative coordinates have the feature that no cross term, viz. **pCM · pR**, occurs in the kinetic energy, and therefore that the hamiltonian breaks up into two commuting parts. One is the kinetic energy of the center of mass motion, which can be ignored if the system is isolated. The other is the hamiltonian of a single "particle" with reduced mass **mR** moving in the central potential **V[R]**. This interpretation is reinforced by the transformation of the angular momentum (Exercise 16.0-3 below).

Although these results are likely familiar, the method readily generalizes to three or more particles where the choice of coordinates is less clear. Thus, starting with the center of mass position and the relative separation of any pair of particles, one can derive a complete set of relative coordinates and momenta for a system of *n* particles. We do this in Problem 15.2-1 below for a three particle system. We obtain in this way coordinates introduced last century by Jacobi to study planetary motion. Of course, Jacobi didn't have quantum commutation relations to rely on, he simply chose coordinates which didn't alter the form of the equations of motion. (We note in this regard that everything we've done here can also be formulated in terms of classical Poisson brackets analogous to quantum commutators.)

Exercise 15.2-1

(a) Solve for the space-fixed coordinates and momenta in terms of the relative coordinates derived above and show that

$$\{x[1] == CM + \frac{mR\ R}{m[1]},\ x[2] == CM - \frac{mR\ R}{m[2]},$$

$$p[1] == pR + \frac{pCM\ m[1]}{M},\ p[2] == -pR + \frac{pCM\ m[2]}{M}\}$$

Verify the transformation of the kinetic energy, relation (15.2-r1). Evaluate the limit m[2] >> m[1] appropriate to the Earth-Sun system or the hydrogen atom.

(b) Consider the transformation to relative coordinates in three dimensions and the relation of the relative- coordinate *vectors* to the space-fixed position vectors **rvec[i]** = {x[i], y[i], z[i]} and pvec[i] = {px[i], py[i], pz[i]} for i = 1, 2. Again, verify the transformation of the kinetic energy, relation (15.2-r1).

(c) Show that if **R** is held fixed at **R = Ro**, the transformed kinetic energy (15.2-r1) reduces to K = pCM^2/(2M) + lRvec • lRvec/(2 mR Ro^2), where lRvec = Rvec × pRvec is the relative angular momentum of particles 1 and 2 (cf. Exercise 16.0-3). (Hint: Introduce results from Exercise 2, Appendix V, as replacement rules.) This is just the hamiltonian of a *rigid rotor*, i.e. two point masses with reduced mass **mR** at the ends of a rigid massless rod of length **Ro** (see also Problem 19.1-1).

Problem 15.2-1

(a) Derive a set of Jacobi relative coordinates for a *three-body* system. Starting with the three-body center of mass position and momentum and the two-particle position R and momentum pR derived in the text, relate a third coordinate r and momentum pr to space-fixed variables (x[i] and p[i] in *1D*). Refer to Figure 15.2-2. Show that r = r[3] – cm[12] where cm[12] is the center of mass position of particles 1 and 2. Show that classically pr = mr D[r,t] where mr is the reduced mass of particle 3 relative to the center of mass of 1 and 2, i.e. mr = (m[1] + m[2])m[3]/M with M the system's total mass. This is just one of three sets of Jacobi coordinates that transform into each other by a cyclic permutation of the three particles.

(b) Solve for the space-fixed variables x[i] and p[i] in terms of the relative coordinates and *reduced masses* and show for example that

$$\{x[1] == CM + \frac{mR\ R}{m[1]} - \frac{mr\ r}{m[1] + m[2]},$$

$$x[2] == CM - \frac{mR\ R}{m[2]} - \frac{mr\ r}{m[1] + m[2]},\ x[3] == CM + \frac{mr\ r}{m[3]}\}$$

The three-particle center of mass lies along one of the relative vectors. Which one? Evaluate the three limits: m[1] = m[2] = m[3] (e.g. a carbon nucleus composed of three alpha particles); m[1] = m[2] >> m[3] (e.g. the hydrogen molecular ion with

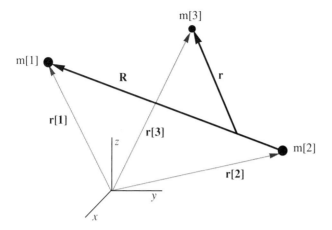

Figure 15.2-2.   Three-particle Jacobi relative coordinates.

two protons and one electron); and $\mathtt{m[1]=m[2]<<m[3]}$ (e.g. the two-electron helium atom). Sketch the relative vectors in each limit.

(c) Transform the system kinetic energy and show that

$$
\frac{\mathtt{p[1]}^2}{2\ \mathtt{m[1]}} + \frac{\mathtt{p[2]}^2}{2\ \mathtt{m[2]}} + \frac{\mathtt{p[3]}^2}{2\ \mathtt{m[3]}} == \frac{\mathtt{pCM}^2}{2\ \mathtt{M}} + \frac{\mathtt{pr}^2}{2\ \mathtt{mr}} + \frac{\mathtt{pR}^2}{2\ \mathtt{mR}}
$$

Again, we see that the center of mass motion of an isolated system can be ignored. Also, the Jacobi coordinates retain the feature that no cross terms occur in the kinetic energy.

Bra-Ket Formalism **15.3**

We can extend our formal considerations to include Dirac *bra-ket* algebra. Although it's difficult to do justice to a truly elegant and compact mathematical formalism, this generalization is a good example of the flexibility we have in defining our own rather abstract quantities on the computer.

   We thus introduce objects **ket[a]**, called *ket-vectors*, to represent wavefunctions and define them according to how they transform when acted upon by operators to produce other ket-vectors, specifically, according to rules such as **Q ** ket[a] -> ket[b]**. The complex conjugate of the wavefunction is represented by a *bra-vector*, **bra[a]**, connected to the ket-vector **ket[a]** by a scalar product which we shall define as **bra[a] ** ket[a]**. We shall thus express general matrix elements of **Q** as **bra[a] ** Q ** ket[b]**, which symbolize for example integrals over the position coordinates in the coordinate representation. The bra and ket vector spaces are said to be *dual* to one another.

A hermitian operator **Q** representing a system observable is characterized as having matrix elements which satisfy

$$bra[b] ** Q ** ket[a] == Conjugate[bra[a] ** Q ** ket[b]]$$
(15.3-r1)

It follows that the matrix corresponding to a hermitian operator equals the complex conjugate of its transpose, i.e. a hermitian matrix as defined in Section 9.2. Of course, this property ensures that diagonal matrix elements of **Q**, i.e. its *expectation values*, are real numbers, a crucial point if **Q** is to represent a physical quantity.

It is useful to introduce in this connection the more general (hermitian) *adjoint* of an arbitrary operator **Q** defined by

$$bra[b] ** Q ** ket[a] ==$$
$$Conjugate[bra[a] ** Adjoint[Q] ** ket[b]]$$
(15.3-r2)

It follows that **bra[a] ** Adjoint[Q]** is the bra vector dual to the ket vector **Q ** ket[a]**. A hermitian operator is thus one which is identical with its adjoint (or is *self-adjoint*). By definition, a *unitary* operator **U** is one whose adjoint equals its inverse, i.e. **Adjoint[U] ** U == 1**. Hence, from relation (15.3-r2) a unitary matrix is one whose inverse equals the complex conjugate of its transpose, as defined in Section 10.6. Taken together, these properties ensure that a unitary transformation preserves normalization, i.e. **U ** ket[a]** is normalized if **ket[a]** is.

Exercise 15.3-1

Prove with paper and pencil that a unitary operator preserves normalization. Also, prove the useful property **Adjoint[A ** B] == Adjoint[B] ** Adjoint[A]** for any two operators **A** and **B**.

Finally, show that **E^(I Q)** is unitary for any hermitian operator **Q**. It follows that the time-development operator **E^(-I H t/h)** is unitary as well as our approximate time-development operator **E^(-I K dt/h) E^(-I V dt/h)** from Section 12.6.

Much of the effort in quantum mechanics goes into solving the *eigenvalue problem* for hermitian operators: **Q ** ket[q] == q ket[q]**, where **q** is the (real) *eigenvalue* and **ket[q]** the *eigenket*. The most familiar example is the determination of a system's energy spectrum as eigenvalues **e[n]** of the hamiltonian **H**, i.e. as solutions of the Schrödinger equation.

This brief description of *bra-ket* vectors and the quantum operators which transform them gives a rather simplified account of the so-called *Hilbert space* of a physical system. Although we have enough ideas for what we shall do in the following, a good quantum mechanic is familiar with the details. An excellent starting point is still Dirac's book [16].

Harmonic Oscillator Spectrum   15.4

As an example of the bra-ket formalism, we determine in the following problem the energy spectrum of the harmonic oscillator algebraically. That formal considerations give such a complete solution in this case is in part due to the symmetric roles of **x** and **p** in the hamiltonian. We shall see shortly for the angular momentum and later on for the hydrogen atom that similar solutions can be obtained when certain dynamical variables display enough symmetry.

Problem 15.4-1

Using operator methods and the fundamental commutation relations, compute the energy spectrum of a *1D* harmonic oscillator defined by the hamiltonian **hHO == p**p/(2m)+m w^2 x**x/2**. Thus, work out the properties of the *energy representation* and calculate the matrix elements of **x** and **p**.

(a) Begin by introducing from Section 6.7 the *raising* and *lowering* operators **aR** and **aL** with a pair of replacement rules:

```
aRRep = aR -> Sqrt[m w/(2h)](x - I p/(m w));
aLRep = aL -> Sqrt[m w/(2h)](x + I p/(m w));
```

Given that **[x, p] == I h**, show that **[aL, aR] == 1**.

(b) Solve for **x** and **p** in terms of **aR** and **aL** and transform the hamiltonian as **hHO -> hNum == (Num + 1/2) h w**, where **Num == aR ** aL** is the "number" operator. (Use **[aL, aR] == 1**.)

(c) Define eigenkets of the number operator **Num** with (refer to Exercise 1.13, Appendix III)

```
ket/: aR ** aL ** ket[n_] := n ket[n]
bra/: bra[np_] ** ket[n_] := KD[np,n]
```

where **n** is the eigenvalue to be determined and **KD[np,n]** is the Kronecker delta function we introduced in Section 9.1 (see also Exercise 2.6, Appendix III). Declare **n /: NumberQ[n] = True** and verfiy that the expectation value of the hamiltonian

```
bra[n] ** hNum ** ket[n]  == h w (n+1/2)
```

At this point, the eigenvalue **n** is an arbitrary number. However, since **x** and **p** are hermitian, **aR** is the *adjoint* of **aL**, such that **bra[n] ** aR** is the bra-vector dual to the ket-vector **aL ** ket[n]**, and vice versa. Therefore, **Num** is hermitian and **n** is real, as required. Moreover, **Num** is positive definite since **bra[n] ** Num ** ket[n]** is the square of the norm of the ket-vector **aL ** ket[n]** (for any **ket[n]**). Hence, the eigenvalues are also positive definite **n $\geq$ 0** and the spectrum of **hNum** has as a lower bound **h w/2**.

(d) Evaluate the commutators **[Num, aR and aL]** using **[aL, aR] == 1**. Show that **aR ** ket[n]** and **aL ** ket[n]** are eigenkets of **Num** with eigenvalues **n+1** and **n-1**, respectively.

Thus, **aR ** ket [n] == cp ket [n+1]** and **aL ** ket [n] == cm ket [n-1]**, where **cp** and **cm** are constants, which may be taken to be real. Because **aR** and **aL** are adjoint to one another, we have that

```
bra[n]** aL**aR **ket[n]  ==  cp^2 bra[n+1]**ket[n+1]
bra[n]** aR**aL **ket[n]  ==  cm^2 bra[n-1]**ket[n-1]
```

Evaluate these equations and show that **cp = Sqrt[n+1]** and **cm = Sqrt[n]**. Hence, define the action of **aR** and **aL** on **ket[n]** with

```
ket /: aR ** ket[n_]  := Sqrt[n+1] ket[n+1]
ket /: aL ** ket[n_]  := Sqrt[n]   ket[n-1]
```

In words, **aR** *raises* the eigenvalue **n** by one and **aL** *lowers* it by one. Since **n $\geq$ 0**, we require that **aL ** ket [n < 1] == 0**. Moreover, we see that the only allowed value for **n < 1** is **n = 0**. Consequently, **n** is a nonnegative *integer*, and the energy spectrum **h w (n+1/2)** of the oscillator is determined.

(e) Show that the *nth* eigenstate **ket[n]** can be generated from the ground state according to **aR^n** ket[0]/Sqrt[n!]**. (You can do this with **aR^n** by introducing **PowertoNCM** from Problem 15.1-1.)

(f) Finally, calculate matrix elements of **x** and **p** and show that **xme[np,n] ==**

$$
\frac{\mathrm{Sqrt}\left[\dfrac{h}{m\ w}\right]\ (\mathrm{Sqrt}[n]\ \mathrm{KD}[np,-1 + n]\ +\ \mathrm{Sqrt}[1 + n]\ \mathrm{KD}[np,1 + n])}{\mathrm{Sqrt}[2]}
$$

and that **pme[np,n] ==**

$$
\frac{\mathrm{I}\ \mathrm{Sqrt}[h\ m\ w]\ (-(\mathrm{Sqrt}[n]\ \mathrm{KD}[np,-1 + n])\ +\ \mathrm{Sqrt}[1 + n]\ \mathrm{KD}[np,1 + n])}{\mathrm{Sqrt}[2]}
$$

and thus verify the expressions introduced in Section 9.1.

As we have already seen in Problem 6.7-1 and Exercise 11.9-2, the results of this problem lead directly to the wavefunctions of the harmonic oscillator in the coordinate or momentum representation.

Angular Momentum

The *orbital* angular momentum operator of a single particle moving about the origin **rvec = 0** is taken from classical mechanics to be the cross product of the particle's position vector with its momentum vector, **lvec = rvec × pvec**. (Our vector notation is discussed in Appendix V, especially Section 2.6.)

It is convenient to use the definition of the cross product developed in Appendix V, Section 1.0, in order to introduce noncommutative multiply ******. For the time being, we shall work exclusively with the cartesian representation of vectors and label their **x**, **y** and **z** components as **x[1]**, **x[2]** and **x[3]** for compactness. We thus define

```
lvec =
Table[
    Sum[ Signature[{i,j,k}] x[i]**p[j],{i,1,3}, {j,1,3}],
    {k,1,3}
]

  {x[2] ** p[3] - x[3] ** p[2], -x[1] ** p[3] + x[3] ** p[1],
    x[1] ** p[2] - x[2] ** p[1]}
```

This definition is straightforward because only commuting operators appear as products here. Hence, it's easy to show that **lvec** is hermitian, as required. Checking first that our rules for ****** are loaded, we thus extract the components of the angular momentum vector and calculate its square:

```
Needs["Quantum`NonCommutativeMultiply`"]

lx  = lvec[[1]];    ly = lvec[[2]];    lz = lvec[[3]];
```

```
lsq = lx ** lx + ly ** ly + lz ** lz

  x[1] ** p[2] ** x[1] ** p[2] - x[1] ** p[2] ** x[2] ** p[1] +

    x[1] ** p[3] ** x[1] ** p[3] - x[1] ** p[3] ** x[3] ** p[1] -

    x[2] ** p[1] ** x[1] ** p[2] + x[2] ** p[1] ** x[2] ** p[1] +

    x[2] ** p[3] ** x[2] ** p[3] - x[2] ** p[3] ** x[3] ** p[2] -

    x[3] ** p[1] ** x[1] ** p[3] + x[3] ** p[1] ** x[3] ** p[1] -

    x[3] ** p[2] ** x[2] ** p[3] + x[3] ** p[2] ** x[3] ** p[2]
```

Because the labelling of the coordinate axes is arbitrary (except that we stick to right-handed coordinate systems by convention), the components are symmetric under cyclic permutation of **x**, **y** and **z**, i.e. **1 -> 2 -> 3 -> 1**, etc. Hence,

```
{ (lx /.{2 -> 1, 3 -> 2}) == lz,
  (ly /.{1 -> 3, 3 -> 2}) == lx,
  (lz /.{1 -> 3, 2 -> 1}) == ly}
```

```
    {True, True, True}
```

As an example, we show that the commutator of **lsq** with a component of the angular momentum vanishes. Since we express the position and momentum components as **x[i]** and **p[i]**, we can order products and simplify results using our enhanced version of **xpCommute** from Section 15.2:

```
h /: NumberQ[h] = True;

xpCommute[expr_] :=
   Expand[ expr //.
       {p[i_] ** x[i_] :> x[i] ** p[i] - I h,
        p[i_] ** x[j_] :> x[j] ** p[i],
        x[i_] ** x[j_] :> x[j] ** x[i] /; i > j,
        p[i_] ** p[j_] :> p[j] ** p[i] /; i > j}
   ]
```

Thus, (this brute force approach takes about 14 minutes)

```
Commutator[lx,lsq] //xpCommute
```

```
   0
```

It follows from the permutation symmetry of the components that **lsq** commutes with **lvec**. The components themselves, however, don't commute with each other. In particular,

```
Commutator[lx,ly] == I h lz //xpCommute //ExpandAll
```

```
   True
```

and similarly for the cyclic permutations **[ly,lz]** and **[lz,lx]**. This of course is the basis of the fact that the wavefunction of a particle can be a simultaneous eigenfunction of **lsq** and only one of the components of **lvec**, usually chosen to be **lz**. (Recall from Section 15.1 that a common set of eigenfunctions can be constructed for two observables **A** and **B** if **A** and **B** commute.) Hence, unlike in classical mechanics, only **lsq** and, for example, **lz** can be simultaneously measured, and one refers to the z axis as the *axis of quantization*. We see that this result is a direct consequence of the fundamental commutation relations and thus related to the uncertainty principle. (See Problem 16.1-2 below.)

Moreover, we can use the angular momentum commutation relations to show that the spectrum of eigenvalues of **lsq** and say **lz** is discrete, which again is in sharp contrast to classical mechanics. We demonstrate this in Problem 16.1-1 below.

Exercise 16.0-1

Verify and explain that **[lx, lsq] == [lx, ly ** ly + lz ** lz]**, which evaluates faster. Show that **[lx, lsq]** vanishes by expanding it in terms of **[lx, lz]** and **[ly, lz]** using the properties of commutators from Section 15.1. (Hint: Define temporary quantities **lxtmp**, etc. with the appropriate properties.) Finally, evaluate **[lx, x[2]]** and **[lx, p[2]]**, etc. in cyclic order.

Exercise 16.0-2

Consider a *spherical* oscillator defined by the hamiltonian **hHO = Sum[p[i]^2/(2m) + k x[i]^2/2, {i,1,3}]**, and show that **hHO** commutes with each component of the angular momentum. Use this result (and not brute force!) to show that **lsq** commutes as well. (Hint: Define temporary quantities **lxtmp**, etc. with the appropriate properties.) Thus, each component of the angular momentum is a constant of the motion, a result which can be demonstrated for *any* hamiltonian with spherical symmetry. (See Section 18.4 below.)

It is useful in a variety of applications and essential in the formal determination of the angular momentum spectrum to introduce the *raising and lowering operators*, **lR** and **lL**, respectively. If we seek simultaneous eigenfunctions of **lsq** and **lz**, these operators are defined as

    lR = lx + I ly;   lL = lx - I ly;

Their usefulness is analogous to that of their counterparts in the algebraic solution of the harmonic oscillator problem. They thus help to establish the properties of the eigenkets **ket[l,m]** of **lsq** and **lz**.

In this regard, we note that **lR** and **lL** are adjoint to one another, since **lvec** is hermitian, such that **bra[l,m] ** lR** is the vector dual to **lL ** ket[l,m]** (cf. Problem 15.4-1). One also establishes the useful identities (see Problem 16.0-1)

```
lR ** lL == lsq - lz ** (lz - h)
lL ** lR == lsq - lz ** (lz + h)                        (16.0-r1)
```

Exercise 16.0-3

(a) Show that the total orbital angular momentum of two particles about a space-fixed origin transforms to the relative coordinates introduced in Section 15.2 as

```
lvec[1] + lvec[2] -> Cross[CMvec,pCMvec] + Cross[Rvec,pRvec]
```

(b) Show that the *three* -particle orbital angular momentum about a space-fixed origin transforms to the Jacobi relative coordinates introduced in Problem 15.2-1 as

```
lvec[1] + lvec[2] + lvec[3] ->
    Cross[CMvec,pCMvec] + Cross[Rvec,pRvec] + Cross[rvec,prvec]
```

Hint: Refer to Exercises 15.2-1 and 16.0-3 above. Use the definition of cross product **Cross** from the package **LinearAlgebra`CrossProduct`**. You will have to add, however, rules to **Cross** analogous to those we defined for **NonCommutativeMultiply** to handle multiplication and division by numbers.

Problem 16.0-1

(a) Verify that the *raising* and *lowering* operators satisfy the commutators **[lR,lL] == 2h lz** and **[lz,lR or lL] == h (lR or -lL)**.

(b) Verify the identities (16.0-r1). Obtain the same results without brute force using the commutation relation for **lx** and **ly**. (Hint: Define temporary quantities **lxtmp**, etc. with the appropriate properties.)

16.1 Angular Momentum Spectrum

The formal representation is appropriate for generalizing orbital angular momentum to a system with internal degrees of freedom which can be attributed to an intrinsic *spin angular momentum*. One can adopt the basic commutation rules among the orbital angular momentum components (Section 16.0) as the defining equations of an angular momentum vector **jvec** in the general case: **[jx, jy] = I h jz** and cyclic permutations. These algebraic relations can be solved to obtain the eigenvalue spectrum of **jsq** and **jz**, which is discrete and includes half-integer multiples of **h** characteristic of a particle with spin. This is the subject of the following problem.

Problem 16.1-1

Let the commutation rule **[jx, jy] = I h jz** and cyclic permutations define the components of a generalized angular momentum. Define raising and lowering operators and use operator methods to determine the eigenvalue spectrum of `jsq` and `jz` and the matrix elements of all the operators. Refer to Problems 15.4-1 and 16.0-1.

(a) Begin by defining normalized eigenkets `ket[j,m]` of `jsq` and `jz` according to

```
ket /: jsq ** ket[j_,m_] := h^2 j(j+1) ket[j,m]
ket /: jz  ** ket[j_,m_] := h m ket[j,m]
bra /: bra[jp_,mp_] ** ket[j_,m_] := KD[jp,j] KD[mp,m]
```

Without loss of generality, we assume that the eigenvalues of `jsq` and `jz` are (real) numbers of the form `h^2 j(j+1)` and `h m`, respectively. Compute the expectation value of the positive definite operator `jx^2 + jy^2` and show that `j(j+1)` $\geq$ `m^2`. Let `mMin` and `mMax` be the upper and lower bounds on `m` such that `{m, -mMin, mMax}`.

(b) Show that `jR ** ket[j,m]` and `jL ** ket[j,m]` are eigenkets of `jsq` with eigenvalues `h^2 j(j+1)` and of `jz` with eigenvalues `h(m+1)` and `h(m-1)`, respectively.

Thus, `jR` applied repeatedly to an eigenket `ket[j,m]` generates a sequence of eigenfunctions of `jz` with eigenvalues ascending a ladder as `m+1`, `m+2`, etc. Similarly, one descends the ladder with `jL`. Hence, `jR` operating on `ket[j,mMax]` has to be a null vector since otherwise it would be an eigenvector with `j(j+1) < m^2`, contrary to `j(j+1)` $\geq$ `m^2` from part **(a)**. Likewise, `jL ** ke t[j,mMin]` has to be a null vector.

(c) Compute the norms of `jR ** ket[j,m]` and `jL ** ket[j,m]` and show that `mMax = -mMin = j` or `mMax - mMin = 2j`. (Use relation (16.0-r1).) Thus, show that

```
ket /: jR ** ket[j_,m_] := h Sqrt[(j-m)(j+m+1)] ket[j,m+1]
ket /: jL ** ket[j_,m_] := h Sqrt[(j+m)(j-m+1)] ket[j,m-1]
ket /: ket[j_,j_+1] := 0
ket /: ket[j_,j_-1] := 0
```

Because `m` changes by one as the ladder is ascended or descended, `mMax - mMin` has to be either zero or a positive integer. Hence, `2j` has to be either zero or a positive integer, or `j = 0, 1/2, 1, 3/2, 2, ...` Moreover, for a given `j` value there are `2j+1` values of `m` in the range `{m, -j, j}`. This determines the eigenvalue spectrum.

(d) Calculate general matrix elements of `jR, jL, jx, jy, jz` and `jsq`. Show for example that `jxme[jp,mp,j,m]==`

$$\frac{h\ \mathrm{Sqrt}[(1 + j - m)\ (j + m)]\ KD[jp,j]\ KD[mp,-1 + m]}{2} +$$

$$\frac{h\ \mathrm{Sqrt}[(j - m)\ (1 + j + m)]\ KD[jp,j]\ KD[mp,1 + m]}{2}$$

and `jyme[jp,mp,j,m]==`

$$\frac{I}{2}\ h\ \mathrm{Sqrt}[(1 + j - m)(j + m)]\ KD[jp,j]\ KD[mp,-1 + m]\ -$$

$$\frac{I}{2}\ h\ \mathrm{Sqrt}[(j - m)(1 + j + m)]\ KD[jp,j]\ KD[mp,1 + m]$$

The matrices of `jx` and `jy` thus have only vanishing matrix elements linking kets with different `j` values. Hence, these matrices are block diagonal in `j` with each block having dimension 2j+1. Explain. As desired, the matrices of jz and jsq are diagonal, since ket[j,m] is an eigenket of both operators. Moreover, the diagonal elements are just the eigenvalues of jz and jsq.

Generate explicit matrices for j = 1/2, 1 and 2 using **Table** and display your results in **MatrixForm**. For j = 1/2, compare your results with the *Pauli spin matrices*

```
sx = {{0,  1},{1,  0}};
sy = {{0,-I},{I,  0}};
sz = {{1,  0},{0,-1}};
```

and for j = 1 with the example in the next section, and for j = 2 with the expressions given in Problem 16.4-1 below.

Problem 16.1-2

That the magnitude h Sqrt[j(j+1)] of the angular momentum vector is not equal to the magnitude of its maximum projection h j is a consequence of the uncertainty principle, which does not allow complete alignment of **jvec** along a given direction.

(a) Show that the root-mean-square uncertainties djx and djy satisfy the relation

```
djx^2 + djy^2 == h^2 (j(j+1) - m^2)
```

Hence, show that the uncertainty perpendicular to the z axis is minimal for m = ±j.

(b) You can interpret the eigenvalue h^2 j(j+1) with the following argument. Assume that the quantum number m lies in the range {m,-j,j}. An average over m is equivalent to an average over all spatial directions such that the average values <jx^2>, <jy^2> and <jz^2> are all equal. It follows that

```
<jsq> == 3<jz^2>/(2j+1) == 3 h^2 Sum[m^2,{m,-j,j}]/(2j+1)
```

Evaluate the sum symbolically by loading the package **Algebra`SymbolicSum`** and show that <jsq> = h^2 j(j+1). Since jsq is a scalar (with rotational invariance), it follows that it is equivalent to h^2 j(j+1).

For particles without spin described by three classical degrees of freedom, for example radial displacements and polar and azimuthal rotations, the requirement that the wavefunction be single-valued further restricts the (orbital) angular momentum spectrum to integer values only. We shall return to this point in Section 19.1 when we examine the *spherical harmonics*, i.e. the eigenfunctions of angular momentum in the coordinate representation.

16.2 Matrix Representation

The matrix representation of operators is common in quantum mechanics, where noncommutative multiplication (our ******) is ordinary matrix multiplication. Wavefunctions are represented as vectors or single-column matrices.

For the description of spin degrees of freedom, this is the most common representation. It is natural therefore to take an excursion here and investigate angular momentum via matrix algebra.

As we have seen in Problem 16.1-1, a representation in which the matrices of **jsq** and say **jz** are both diagonal and hence commute (z axis the axis of quantization) corresponds to one in which both have a common set of eigenfunctions or eigenkets **ket[j,m]**. The diagonal elements are just the eigenvalues of **jsq** and **jz**. Because none of the components of **jvec** has any nonvanishing matrix elements connecting states of different **j**, the eigenvector space of the system at hand breaks up into subspaces or *blocks* of dimension **2j+1** with elements labelled by **m** in the range **{m,-j,j}** in integer steps. For example, for **j = 1** we have the *two* submatrix blocks

```
{jsq = 2h^2 IdentityMatrix[3],
 jz = h DiagonalMatrix[{1,0,-1}]} //
       TableForm[#,TableDirections -> {Row,Row},
                   TableSpacing -> {10,3,1},
                   TableAlignments -> Center
       ]&
```

```
         2
    2 h       0       0                  h   0   0

                 2
     0      2 h       0                  0   0   0

                         2
     0       0      2 h                  0   0  -h
```

where we've used **TableForm** rather than **MatrixForm** to control the output format better.

Since these are diagonal matrices, the *m-kth* element of the normalized **j = 1** eigenvector in this representation is simply the Kronecker delta function **KD[m,k]**. Thus,

```
j = 1;
ez[m_] := Table[ KD[m,k], {k,-j,j}] //Reverse

{ez[1], ez[0], ez[-1]}

   {{1, 0, 0}, {0, 1, 0}, {0, 0, 1}}
```

where **Reverse** is needed because the iterator **k** starts with **-j**, just the opposite of the diagonal elements of **jz**. We easily verify that these are indeed eigenvectors with, for example,

```
{jz.ez[-j] == -h ez[-j], jsq.ez[-j] == 2h^2 ez[-j]}

   {True, True}
```

The raising and lowering operators are represented for **j = 1** by the two submatrices

```
{jR = h Sqrt[2] {{0,1,0}, {0,0,1}, {0,0,0}},
 jL = h Sqrt[2] {{0,0,0}, {1,0,0}, {0,1,0}}} //
     TableForm[#,TableDirections -> {Row,Row},
                TableSpacing -> {12,3,1},
                TableAlignments -> Center
     ]&
```

0	0	0	0	Sqrt[2] h	0
Sqrt[2] h	0	0	0	0	Sqrt[2] h
0	Sqrt[2] h	0	0	0	0

Their form ensures that a null vector is obtained if **Abs[m] > j**:

```
{jR.ez[j], jL.ez[-j]}
```

```
{{0, 0, 0}, {0, 0, 0}}
```

Moreover, we can generate all eigenvectors in the range **{m,-j,j}** by repeated application of **jR** and **jL** on just one of the vectors. For example, starting with the bottom rung **ez[-j]**, we easily climb the ladder **2j** rungs to the top rung:

```
ez[1] == MatrixPower[jR,2j].ez[-j]/(2h^2)
```

```
True
```

Here the factor **1/(2h^2)** cancels the accumulation of the factors **Sqrt[(j-m)(j+m+1)]** for **j = 1** from Problem 16.1-1, Part **(c)**, due to multiplying **jR** twice. Hence, the result is normalized, as is **ez[1]**.

Since **jvec** is a hermitian operator, the matrix representing any one of its components (and thus **jsq** as well) is hermitian and therefore equal to the transpose of its complex conjugate (see Section 15.3). This ensures, of course, that the diagonal matrix elements are real numbers, in particular the eigenvalues of **jz** and **jsq**. We can easily construct **jx** and **jy** from the raising and lowering operators and check their hermiticity. (We will need, however, our symbolic **Conjugate** rule from the package **Quantum`QuickReIm`**. See Exercise 2.4, Appendix III.) Thus,

```
Needs["Quantum`QuickReIm`"]

jx = (jR + jL)/2;     jy = (jR - jL)/(2I);

{jx == Transpose[jx], jy == Transpose[Conjugate[jy]]}
```

```
{True, True}
```

since **jx** is real.

Similarly, **jR** and **jL** are *adjoint to one another*, such that **jR** is equal to the transpose of the complex conjugate of **jL**, and vice versa from relation (15.3-r2) . Since these matrices are real for **j = 1**, we have simply

```
jR == Transpose[jL]
```

```
     True
```

If a measurement of **jz** is made, one of the eigenvalues **h m** will result and the system will be left in the corresponding eigenstate, represented in the formal representation by **ket[j,m]** and in the matrix representation by **ez[j,m]**. This state will describe the angular momentum of the system until another measurement is made or the system decays.

In general, wavefunctions are complex and their corresponding vectors in the matrix representation are also complex (cf. the end of Section 10.2 and also the end of Section 12.1). Hence, the scalar product of two vectors **e[a]** and **e[b]** must be computed in general as **Conjugate[e[a]].e[b]** and is symbolized by the product **bra[a]** ** **ket[b]**.

Exercise 16.2-1

Show that the matrices **jx**, **jy**, **jz** and **jR** and **jL** satisfy the necessary commutation relations from Section 16.0 and Problem 16.0-1 with respect to matrix multiplication. (See also Problem 16.4-1.)

Exercise 16.2-2

Show that the Pauli spin matrices defined in Problem 16.1-1, besides being hermitian, are also *unitary* such that their inverse is given by the transpose of their complex conjugate (cf. Section 10.6). Thus, show that **sx.sx == sy.sy == sz.sz == IdentityMatrix[2]**. Show also that **sx.sy == I sz** and cyclic permutations and that they *anticommute*, i.e. **sx.sy + sy.sx == 0**.

New Axis of Quantization   16.3

In many applications, it's useful to construct eigenfunctions of another component of **jvec** as a linear combination of eigenkets **ket[j,m]** of **jsq** and **jz** developed in Problem 16.1-1. For example, a measurement of **jy** can only turn up one of its eigenvalues **h my** and leave the system in a corresponding eigenstate, represented by the superposition. One rotates in effect to a new axis of quantization along the desired component of **jvec**. The linear combination is restricted, however, for a given value of **j** to a finite sum over **m**, since it must also remain a simultaneous eigenfunction of **jsq** (i.e. **jvec** and **jsq** commute).

Working with **jy**, consider for example the superposition

```
kety[j_,my_] := Sum[ ket[j,m] Dy[j,m,my], {m,-j,j}]
```

and determine the expansion coefficients **Dy[j,m,my]** by requiring that **kety[j,my]** be an eigenfunction of **jy** with eigenvalue **my**.

The most straightforward way to do this is to transform to the matrix representation (cf. Exercise 10.1-1) and for each value of **j** solve the *matrix* eigenvalue problem **jy.ey[j,my] == h my ey[j,my]** for each submatrix block of **jy** of dimension **2j+1**. (We shall connect with the coordinate representation in Section 19.2.) The *elements* of the resulting eigenvectors **ey[j,my]** are easily seen to be just the expansion coefficients **Dy[j,m,my]**. Formally, **Dy[j,m,my] == ket[j,m] ** kety[j,my]**. The linear combination **kety[j,my]** will be normalized if the eigenvectors are, since the **ket[j,m]** are normalized, and we thus refer to the unitary transformation of the **ket[j,m]** to **kety[j,my]** (cf. Section 15.3).

Now in order for the set of homogeneous equations defined by the matrix eigenvalue problem to have a nontrivial solution, we require in the usual way (cf. Section 10.2) that the determinant of the matrix of coefficients vanishes. The roots of this *secular equation* are the eigenvalues **my**. Thus, for **j = 1**

```
Solve[ Det[jy - h my IdentityMatrix[3]] == 0, my ] //Sort

    {{my -> -1}, {my -> 0}, {my -> 1}}
```

Alternatively, we can let the built-in function **Eigenvalues** set up and solve the secular equation. If we divide out **h**, we obtain numbers we can **Sort** into ascending order:

```
Eigenvalues[jy/h] //Sort

    {-1, 0, 1}
```

Note that these values are consistent with the general results of Problem 16.1-1 that the eigenvalues of a component of **jvec** have only the allowed values **{m,-j,j}** in integer steps. Finally, the eigenvalues and eigenvectors are conveniently computed with the built-in function **Eigensystem**:

```
{mys,vys} = Eigensystem[jy/h]

                     I             -I
    {{-1, 0, 1}, {{------, 1, ------}, {1, 0, 1},
                   Sqrt[2]      Sqrt[2]

       -I           I
     {------, 1, ------}}}
      Sqrt[2]     Sqrt[2]
```

Here the first list **mys** contains the eigenvalues, while the second list **vys** contains the corresponding eigenvectors. These symbolic eigenvectors are already sorted in the order we require, although they're not normalized (cf. the purely numerical computation in Section 10.2). Hence, normalized eigenvectors **ey** have to be calculated in an extra step:

```
Table[
    ey[my] =
        vys[[my+j+1]]/
```

```
          Sqrt[Conjugate[vys[[my+j+1]]].vys[[my+j+1]]],
      {my,-j,j}
  ]
```

$$\{\{-\frac{I}{2}, \frac{1}{Sqrt[2]}, \frac{-I}{2}\}, \{\frac{1}{Sqrt[2]}, 0, \frac{1}{Sqrt[2]}\}, \{\frac{-I}{2}, \frac{1}{Sqrt[2]}, \frac{I}{2}\}\}$$

Note that we shift the array subscripts as `[[my+j+1]]` to ensure that they are positive integers even though `{my,-j,j}`. We verify for example that

```
{jy.ey[1] == h ey[1], jsq.ey[1] == 2h^2 ey[1]}
```

```
{True, True}
```

Since the eigenvectors `ey[j,my]` of the hermitian submatrix `jy` are orthonormal, we construct in the usual way (cf. Section 10.6) a `2j+1` -dimensional *unitary* matrix `Dy[j]` of the eigenvectors (along the columns) which diagonalizes `jy` by a similarity transformation. Thus, for `j = 1`

```
Dy[j] = Transpose[{ey[1], ey[0], ey[-1]}];
```

```
Conjugate[Transpose[Dy[j]]].jy.Dy[j] //MatrixForm
```

```
  h    0    0
  0    0    0
  0    0   -h
```

Exercise 16.3-1

Verify the orthonormality of the eigenvectors `ey[my]`. Show in addition that the matrix `Dy[1]` is unitary, i.e. its inverse equals the complex conjugate of its transpose.

This matrix also determines the coefficients in the expansion of the eigen-kets `kety[j,my]` of `jy` according to

```
Dy[j_,m_,my_] := Dy[j][[-m+j+1,-my+j+1]]
```

(Here the array subscripts `-m + j` and `-my + j` ensure that we pick off the elements of `Dy[j]` in the order defined by the `ey[j,my]`.) We obtain for example in the formal representation for `j = 1, my = 1`

```
kety[j,1]
```

$$\frac{I}{2} ket[1, -1] + \frac{ket[1, 0]}{Sqrt[2]} - \frac{I}{2} ket[1, 1]$$

Exercise 16.3-2

Using the results of the formal representation from Problem 16.1-1, verify for `j = 1` that `kety[j,my]` is an eigenket of the formal operators `jy` and `jsq`. Show that it is not an eigenket of `jz`, as required by the commutation relations. Also, verify that `Dy[j,m,my] == ket[j,m] ** kety[j,my]`.

16.4 Quantum Rotation Matrix

The diagonalization of **jy** is closely connected to (rigid) rotations in quantum mechanics, an interpretation we shall investigate in Sections 19.2 and 19.3 when we construct the coordinate-space representation of **kety[j,my]**. In fact, the **2j+1**-dimensional matrix **Dy[j]** can be thought of as a unitary *rotation matrix* which rotates the eigenvectors **ez[j,m]** of **jz** into the eigenvectors **ey[j,my]** of **jy** in the sense that

```
Table[ey[m] == Dy[j].ez[m], {m,-j,j}]

   {True, True, True}
```

and back again

```
Table[ez[my] ==
    Conjugate[Transpose[Dy[j]]].ey[my], {my,-j,j}]

   {True, True, True}
```

We note that the normalizations are preserved, as required, since the transformation is unitary.

The following problems investigate these and other aspects of the matrix representation of angular momentum.

Exercise 16.4-1

Redo our calculations in Sections 16.3 and 16.4, but instead diagonalizing **jx** for **j = 1**. Show for a particular choice of eigenvector phases that the rotation matrix **Dx[1]** =

$$
\begin{bmatrix}
\dfrac{1}{2} & -\left(\dfrac{1}{\mathrm{Sqrt}[2]}\right) & \dfrac{1}{2} \\[2ex]
\dfrac{1}{\mathrm{Sqrt}[2]} & 0 & -\left(\dfrac{1}{\mathrm{Sqrt}[2]}\right) \\[2ex]
\dfrac{1}{2} & \dfrac{1}{\mathrm{Sqrt}[2]} & \dfrac{1}{2}
\end{bmatrix}
$$

We shall show in Section 19.3 that these phases agree with convention. Since this unitary matrix is also real, it's simply an *orthogonal* matrix such that its inverse is just its transpose.

Problem 16.4-1

We construct directly from the results of Problem 16.1-1 the angular momentum matrices for a particle with (orbital) angular momentum quantum number **l = 2** in a representation in which **lz** is diagonal according to

```
lsq = 6 h^2 IdentityMatrix[5]
lz = h DiagonalMatrix[{2,1,0,-1,-1}]
lR = h {{0,2,0,0,0},{0,0,Sqrt[6],0,0},{0,0,0,Sqrt[6],0},
        {0,0,0,0,2},{0,0,0,0,0}}
lL = h {{0,0,0,0,0},{2,0,0,0,0},{0,Sqrt[6],0,0,0},
        {0,0,Sqrt[6],0,0},{0,0,0,2,0}}
```

(a) Compute the matrices for `lx` and `ly` from `lR` and `lL` and show that `lsq = lx.lx + ly.ly + lz.lz`. Define `Commutator[A_,B_] := A.B - B.A` in terms of matrix multiplication, and check all the commutators in Exercise 15.2-1 and Problem 16.0-1 above.

(b) Consider the eigenvector `ez[2] = {1,0,0,0,0}` of `lz` corresponding to *maximum* angular momentum projection (eigenvalue) `+2h` along the z axis. Show that each application of `lL` on `ez[m]` generates a new eigenvector of `lz` with the projection quantum number lowered by `h`. Refer to Problem 16.1-1. Show that the null (or zero) vector is eventually obtained as required, since the minimum projection is `-2h`. Likewise, show that `lR.ez[2] = 0`.

Problem 16.4-2

(a) Consider the description of the system in Problem 16.4-1 *after* a measurement of `lx` has been made, so that the x axis is the axis of quantization. Thus, diagonalize `lx` for `l = 2` and show that its eigenvalues, i.e. the result of a measurement of `lx`, are `h mx` where `mx` is an integer in the range `{mx,-2h,2h}`. Determine the real, normalized eigenvectors `ex[mx]`. Check your results directly with `lx` and `lsq`.

(b) Define the real rotation matrix `Dx` whose columns are the eigenvectors `ex[mx]`, and show that `ex[mx] == Dx.ez[mx]`, where the `ez` are the (normalized) eigenvectors of `lz`. Show also that `Transpose[Dx].Dx == IdentityMatrix[5]`.

(c) Verify that `Dx` diagonalizes `lx` by the similarity transformation `Transpose[Dx].lx.Dx`. Also, show that `lsq` remains diagonal but that `lz` is undiagonalized, as required by the commutation relations.

Problem 16.4-3

In general, matrix diagonalization has to be done numerically, since the roots `r` of the secular equation `Det[M - r IdentityMatrix] == 0` can be obtained analytically for only the simplest matrices `M`. Redo Problem 16.4-2 for a system whose angular momentum has been measured along the more general axis defined by the unit vector `nvec = {1,1,1}/Sqrt[3]`.

   Express all quantities in appropriate units of `h` and resort to numerical approximation. Use `Chop` to shorten your output. Note that the eigenvectors `en` are complex and hence the rotation matrix `Dn` which diagonalizes the component `ln` of the angular momentum along `nvec` is unitary.

Problem 16.4-4

Consider the description of an electron's spin following a Stern-Gerlach experiment in which the magnetic field is oriented along an (arbitrary) axis of quantization defined by the unit vector `nvec = {nx,ny,nz}`. Redo Problem 16.4-2 for a spin-$1/2$ particle using `jx = h/2 sx`, `jy = h/2 sy`, and `jz = h/2 sz`, where `sx`, `sy`, `sz` are the Pauli spin matrices defined in Problem 16.1-1. The two-component complex eigenvectors in this problem are commonly referred to as *spinors*.

Introduce rules such as `c_. nx^2 + c_. ny^2 + c_. nz^2 -> c` to simplify your results. Check by setting, for example, `nvec = {0,1,0}` and then diagonalizing `jy` directly.

Angular Momentum Coupling

With tools at hand, it's natural that we investigate the quantum mechanical problem of adding two *independent* angular momentum vectors `jvec[1]` and `jvec[2]` to calculate a total angular momentum vector `Jvec = jvec[1] + jvec[2]`. For example, one might want to add the individual angular momenta of two particles to form a total two-particle angular momentum. Or one might wish to combine the intrinsic spin angular momentum `svec` of a single particle with its orbital angular momentum `lvec` in order to characterize the particle's total angular momentum `Jvec = svec + lvec`.

It shouldn't come as a surprise that we go about this problem in quantum mechanics quite differently than we would in classical mechanics, where it amounts to simple vector addition. The reason, of course, is due to the commutation relations satisfied by the operators `jvec[i]` and `Jvec` and thus ultimately to the uncertainty principle (see Problem 16.1-2).

In quantum mechanics, the idea is to construct eigenkets of the square `Jsq` and a component such as `J[z]` of the total angular momentum `Jvec`, starting with the eigenkets of `jsq[i]` and `j[i,z]` of the `jvec[i]`. One also refers to this problem depicted in Figure 17.0-1 as *angular momentum coupling*.

The commuting operators `jvec[1]` and `jvec[2]` can be thought of as belonging to two separate vector spaces which merge to form the *direct product space* of the coupled system. The simultaneous eigenvectors of the four operators `jsq[i]` and `j[i,z]` (with i = 1, 2) with quantum numbers `ji` and `mi` are given by the direct-product kets

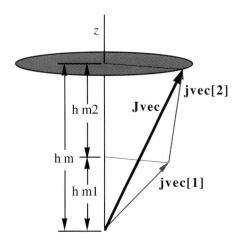

Figure 17.0-1. Addition of two angular momentum vectors `jvec[1]` and `jvec[2]` to produce a resultant vector **Jvec** with respect to a coordinate origin and z axis.

$$ket[j1,j2,\{m1,m2\}] == ket[j1,m1] \; ket[j2,m2] \qquad (17.0\text{-r1})$$

We define these eigenkets (using the results of Problem 16.1-1) by the eigenvalue equations

```
ket/: j[1,z] ** ket[j1_,j2_,{m1_,m2_}] :=
        h m1 ket[j1,j2,{m1,m2}]
ket/: j[2,z] ** ket[j1_,j2_,{m1_,m2_}] :=
        h m2 ket[j1,j2,{m1,m2}]
ket/: jsq[1] ** ket[j1_,j2_,{m1_,m2_}] :=
        h^2 j1(j1+1) ket[j1,j2,{m1,m2}]
ket/: jsq[2] ** ket[j1_,j2_,{m1_,m2_}] :=
        h^2 j2(j2+1) ket[j1,j2,{m1,m2}]

    h/: NumberQ[h] = True;
```

and by the properties of the raising and lowering operators

```
ket/: jR[1] ** ket[j1_,j2_,{m1_,m2_}] :=
        h Sqrt[(j1-m1)(j1+m1+1)] ket[j1,j2,{m1+1,m2}]
ket/: jL[1] ** ket[j1_,j2_,{m1_,m2_}] :=
        h Sqrt[(j1+m1)(j1-m1+1)] ket[j1,j2,{m1-1,m2}]
ket/: jR[2] ** ket[j1_,j2_,{m1_,m2_}] :=
        h Sqrt[(j2-m2)(j2+m2+1)] ket[j1,j2,{m1,m2+1}]
ket/: jL[2] ** ket[j1_,j2_,{m1_,m2_}] :=
        h Sqrt[(j2+m2)(j2-m2+1)] ket[j1,j2,{m1,m2-1}]
```

Also, we normalize these eigenkets according to

```
ket/: bra[j1_,j2_,{m1p_,m2p_}]**ket[j1_,j2_,{m1_,m2_}] :=
        KD[m1p,m1] KD[m2p,m2]
```

where `KD[jp,j]` is the Kronecker delta function we defined in Section 9.1. (See also Exercise 2.6, Appendix III.) Also, notice we've grouped the labels **m1** and **m2** in **ket** into a list, which will be convenient shortly when we construct the matrix of **Jsq**.

We refer to these eigenkets as the *direct-product basis set* or *direct-product representation*. In order to simplify our expressions involving **NonCommutativeMultiply** (******), we again check to make sure our rules from Section 15.1 (and Exercise 2.5, Appendix III) are loaded:

```
Needs["Quantum`NonCommutativeMultiply`"]
```

It is useful to have a set of rules for commuting and ordering the components of **jvec[1]** and **jvec[2]** in order to establish the properties of the *total* angular momentum **Jvec**, specifically, of the operators

```
J[z_] := j[1,z] + j[2,z];
Jsq := JR**JL + J[z]**J[z] - h J[z]
JR := jR[1] + jR[2];
JL := jL[1] + jL[2]
```

All that is necessary is to recognize that each component of **jvec[1]** commutes with each component of **jvec[2]** and that separately the components satisfy the basic angular momentum commutation relations. We thus define in analogy with **xpCommute** from Sections 15.2 and 16.0

```
j12Commute[expr_] :=
    Expand[expr //.
        {j[2,x_]**j[1,x_] :> j[1,x]**j[2,x],
         j[2,x_]**j[1,y_] :> j[1,y]**j[2,x],

         j[i_,y]**j[i_,x] :> j[i,x]**j[i,y] - I h j[i,z],
         j[i_,z]**j[i_,y] :> j[i,y]**j[i,z] - I h j[i,x],
         j[i_,x]**j[i_,z] :> j[i,z]**j[i,x] - I h j[i,y]}
    ]
```

Now we can easily verify, for example, that

```
Commutator[A_,B_] := A ** B - B ** A

{Commutator[j[1,z],j[2,z]], Commutator[j[1,x],j[2,y]],
 Commutator[j[1,x],j[1,y]], Commutator[j[2,x],j[2,y]]} //
    j12Commute

    {0, 0, I h j[1, z], I h j[2, z]}
```

We can also show that the components of **Jvec** satisfy the basic angular momentum commutation relations as well. For example,

```
Commutator[J[x],J[y]] == I h J[z] //
    j12Commute //ExpandAll
```

```
True
```

The operators **jsq[1]** and **jsq[2]** commute with each other and with every component of **jvec[1]** and **jvec[2]** and therefore with **Jvec** and **Jsq**. (But **j[1,z]** and **j[2,z]** don't commute with **Jsq**.) Thus, it is possible to construct a linear combination from the direct-product eigenkets which is simultaneously an eigenket of **Jsq** and **J[z]** as well as of **jsq[1]** and **jsq[2]**.

Exercise 17.0-1

Verify directly using **j12Commute** that **jsq[i]** commutes with **j[i,z]** and with **Jsq**. Show, however, that **j[i,z]** does not commute with **Jsq**. Hint: Evaluate **jsq[i]**, **jR[i]** and **jL[i]** with the replacement rules

```
jrep =
  {jsq[i_] :> j[i,x]**j[i,x] + j[i,y]**j[i,y] + j[i,z]**j[i,z],
   jR [i_] :> j[i,x] + I j[i,y],
   jL [i_] :> j[i,x] - I j[i,y]};
```

Denoting these new eigenkets as **ketp[j1,j2,{j,m}]**, where **j** and **m** are quantum numbers referring to **Jsq** and **J[z]**, we therefore have the expansion

```
ketp[j1_,j2_,{j_,m_}] :=
    Sum[
        c[j1,j2,{m1,m2},{j,m}] ket[j1,j2,{m1,m2}],
        {m1,-j1,j1}, {m2,-j2,j2}
    ]

brap[j1_,j2_,{j_,m_}] := ketp[j1,j2,{j,m}] /.
    ket -> bra
```

The sum is over all allowed values of **m1** and **m2** for specified (fixed) values of **j1** and **j2**. We refer to these eigenkets as the *coupled basis set* or *coupled representation*.

This expression reduces the problem of angular momentum addition to the determination of the expansion coefficients **c[j1,j2,{m1,m2},{j,m}]**. These elements of the transformation matrix connecting the direct-product and coupled representations are referred to variously as *vector-coupling, vector-addition, Wigner* or *Clebsch-Gordan coefficients*.

Formally, these coefficients are given by the overlap integrals **bra[j1, j2,{m1,m2}] ** ketp[j1,j2,{j,m}]** of the direct-product basis with the coupled basis, and establishing their general properties is relatively straightforward. For example, by enforcing the eigenvalue condition **J[z]**

** ketp[j1,j2,{j,m}] == h m ketp[j1,j2,{j,m}], we have that
(see also Exercise 17.3-1)

```
Sum[
    (m1+m2-m)  c[j1,j2,{m1,m2},{j,m}]  ket[j1,j2,{m1,m2}],
    {m1,-j1,j1},{m2,-j2,j2}
] == 0                                                       (17.0-r2)
```

Since the direct-product functions **ket[j1,j2,{m1,m2}]** are independent, we conclude that the Clebsch-Gordan coefficients vanish unless **m = m1 + m2**. We can also show (see Exercise 17.0-2) that the three angular momentum quantum numbers must satisfy the so-called *trianglar condition* **{j,Abs[j1-j2],j1+j2}**, which has a simple classical interpretation from Figure 17.0-1. Therefore, since **m = m1 + m2** ranges between **-j** and **j**, we have that **j** takes on only the values **j = j1+j2, j1+j2-1, ..., Abs[j1-j2+1]**, **Abs[j1-j2]**. Hence, if two of the three quantum numbers are half-integral, the third is an integer.

Exercise 17.0-2

Verify that **j** $\geq$ **Abs[j1-j2]** by the following argument. Clearly, **jmax = mmax = Max[m1+m2] = j1 + j2**. Moreover, a minimum **jmin** must exist since in any case **j** $\geq$ 0. The dimension of the direct-product subspace for fixed **j1** and **j2** is **(2j1 + 1)(2 j2 + 1)**. (Recall that **mi** takes on **2 ji + 1** values for each **ji**.) However, this subspace is also spanned by the coupled eigenkets **ketp[j1,j2,{j,m}]** so that its dimension is also given for fixed **j** by **Sum[2j+1, {j,jmin,j1+j2}]**. Solve for **jmin** by evaluating the sum symbolically with the package **Algebra`SymbolicSum`**.

Spin and Orbital Coupling   17.1

As a specific example, let's consider a single particle with orbital angular momentum **l = j1 = 1** and spin angular momentum **s = j2 = 1/2**. In this case, the allowed total angular momentum quantum numbers are **j = 1/2** and **3/2** from the triangular condition, and the **m1-m2** subspace (as well as the **j-m** subspace) has **(2 j1+1)(2 j2+1) = 6** dimensions.

Before we construct the matrices of **Jsq** in this subspace, it's helpful if we first collect all possible pairs of **m1** and **m2** and of **j** and **m**. For **m1** and **m2**, we thus form the table of pair labels

```
m1m2 = Table[{-m1,-m2}, {m1,-1,1},{m2,-1/2,1/2}] //
           Flatten[#,1]&
```

$$\{\{1, \tfrac{1}{2}\}, \{1, -(\tfrac{1}{2})\}, \{0, \tfrac{1}{2}\}, \{0, -(\tfrac{1}{2})\}, \{-1, \tfrac{1}{2}\}, \{-1, -(\tfrac{1}{2})\}\}$$

Here, **Flatten** unravels the nested lists into a single list of 6 pair-lists, while the syntax **{-m1,-m2}** arranges the elements in descending order of **m1 +**

m2 = m. For **j** and **m** it's convenient to put the pair labels in the corresponding order. In this particular case, we can use

```
jm =
    Table[
        {Abs[m1]+Abs[m2],-m1-m2},
        {m1,-1,1},{m2,-1/2,1/2}
    ] //Flatten[#,1]&
```

$$\{\{\tfrac{3}{2}, \tfrac{3}{2}\}, \{\tfrac{3}{2}, \tfrac{1}{2}\}, \{\tfrac{1}{2}, \tfrac{1}{2}\}, \{\tfrac{1}{2}, -(\tfrac{1}{2})\}, \{\tfrac{3}{2}, -(\tfrac{1}{2})\}, \{\tfrac{1}{2}, -(\tfrac{1}{2})\}, \{\tfrac{3}{2}, -(\tfrac{1}{2})\}\}$$

These orderings are useful for arranging the matrix of **Jsq** into *block diagonal form* (cf. the end of Problem 16.1-1 and the start of Section 16.2), although any ordering will give correct results. It is important, however, to have **m1** *before* **m2** in each **m1m2** pair, since this is the order we chose for **ket[j1,j2,{m1,m2}]**.

We can now calculate the matrix elements of **Jsq** in the direct-product subspace and set up the matrix as a table over the **m1m2**-list pairs (this takes about 3 minutes):

```
Jsqm1m2 =
    Table[
        bra[1,1/2,m1m2[[np]]] ** Jsq **
            ket[1,1/2,m1m2[[n]]],
        {np,1,Length[m1m2]},{n,1,Length[m1m2]}
    ];
Jsqm1m2 //TableForm[#,TableAlignments -> Center]&
```

$\dfrac{15\,h^2}{4}$	0	0	0	0	0
0	$\dfrac{7\,h^2}{4}$	$\text{Sqrt}[2]\,h^2$	0	0	0
0	$\text{Sqrt}[2]\,h^2$	$\dfrac{11\,h^2}{4}$	0	0	0
0	0	0	$\dfrac{11\,h^2}{4}$	$\text{Sqrt}[2]\,h^2$	0
0	0	0	$\text{Sqrt}[2]\,h^2$	$\dfrac{7\,h^2}{4}$	0
0	0	0	0	0	$\dfrac{15\,h^2}{4}$

This result, which is clearly in block-diagonal form, expresses **Jsq** in the *direct-product* or **m1m2** *representation*. The first block (the top left element) is a 1×1 submatrix for which **m = j1+j2** and **j = j1+j2**. The next block is a 2×2 submatrix for which **m = j1+j2-1** and **j = j1+j2** or **j = j1+j2-1**. The blocks evolve in this way until the last block (the lower right element) is reached, for which **m = -j1-j2** and **j = j1+j2**.

The corresponding matrix for **J[z]**, which of course is diagonal since the **j[i,z]** are, is also useful. The result, for example, makes the ordering of the **m1m2** pair labels clear.

```
Jzm1m2 =
    Table[
        bra[1,1/2,m1m2[[np]]] ** J[z] **
            ket[1,1/2,m1m2[[n]]],
        {np,1,Length[m1m2]},{n,1,Length[m1m2]}
    ];

Jzm1m2 //MatrixForm
```

$$
\begin{pmatrix}
\dfrac{3\ h}{2} & 0 & 0 & 0 & 0 & 0 \\
0 & \dfrac{h}{2} & 0 & 0 & 0 & 0 \\
0 & 0 & \dfrac{h}{2} & 0 & 0 & 0 \\
0 & 0 & 0 & \dfrac{-h}{2} & 0 & 0 \\
0 & 0 & 0 & 0 & \dfrac{-h}{2} & 0 \\
0 & 0 & 0 & 0 & 0 & \dfrac{-3\ h}{2}
\end{pmatrix}
$$

Exercise 17.1-1

Verify the matrix **Jsqm1m2** by construction of the matrices **J[x]** and **J[y]** and explicit matrix multiplication. Also, derive the identity

```
Jsq == jsq[1] + jsq[2] +
        jR[1]**jL[2] + jL[1]**jR[2] + 2 j[1,z]**j[2,z]
```

from the vector identity **Jsq == jsq[1]+ jsq[2]+ 2 jvec[1] jvec[2]** and verify it by matrix multiplication for **Jsqm1m2**.

17.2 Total Angular Momentum Spectrum

Continuing with our spin-orbital example, we obtain the Clebsch-Gordan coefficients by diagonalizing the matrix **Jsqm1m2**. First, we note that it has the correct eigenvalues

```
Eigenvalues[Jsqm1m2] //Sort
```

$$\left\{\frac{3\ h^2}{4},\ \frac{3\ h^2}{4},\ \frac{15\ h^2}{4},\ \frac{15\ h^2}{4},\ \frac{15\ h^2}{4},\ \frac{15\ h^2}{4}\right\}$$

corresponding to **h^2 j(j+1)** with **j = 1/2** and **3/2**. The degeneracy, or multiplicity, of the eigenvalues is explained by the **2j+1** allowed values of **Abs[m]** $\le$ **j** for a given **j** value. Likewise, the eigenvectors **ep[{j,m}]** of **Jsqm1m2** are easily obtained without numerical approximation because the matrix is small and *sparse*, i.e. most of its elements are zero. Although we choose the dimension of the matrix, sparseness is a feature characteristic of the Clebsch-Gordan transformation. Thus, we obtain

```
{ms,vs} = Eigensystem[Jsqm1m2/h^2]
```

$$\left\{\left\{\frac{3}{4},\ \frac{3}{4},\ \frac{15}{4},\ \frac{15}{4},\ \frac{15}{4},\ \frac{15}{4}\right\},\ \{\{0,\ 0,\ 0,\ 1,\ -\text{Sqrt}[2],\ 0\},\right.$$

$$\left\{0,\ 1,\ -\left(\frac{1}{\text{Sqrt}[2]}\right),\ 0,\ 0,\ 0\right\},\ \{0,\ 0,\ 0,\ 0,\ 0,\ 1\},$$

$$\left\{0,\ 0,\ 0,\ 1,\ \frac{1}{\text{Sqrt}[2]},\ 0\right\},\ \{0,\ 1,\ \text{Sqrt}[2],\ 0,\ 0,\ 0\},$$

$$\{1,\ 0,\ 0,\ 0,\ 0,\ 0\}\}\}$$

As in Section 16.3, these symbolic vectors have to be normalized separately. However, they are evidently ordered. We can determine just what the ordering is by computing the diagonal matrix elements of **Jzm1m2** with the new vectors, viz.

```
Table[
    vs[[m]].Jzm1m2.vs[[m]]/vs[[m]].vs[[m]],
    {m,Length[ms]}
]
```

$$\left\{\frac{-h}{2},\ \frac{h}{2},\ \frac{-3\ h}{2},\ \frac{-h}{2},\ \frac{h}{2},\ \frac{3\ h}{2}\right\}$$

This is just a rearrangement of the label list **jm** we set up in the previous section. Hence, consider the alternative ordering

```
jmp =
    Table[
        {Abs[m1]+Abs[m2],-m1+m2},
        {m1,-1,1},{m2,-1/2,1/2}
    ] //Flatten[#,1]& //RotateLeft[#,2]&
```

$$\{\{-\tfrac{1}{2}, -(-\tfrac{1}{2})\}, \{-\tfrac{1}{2}, -\tfrac{1}{2}\}, \{-\tfrac{3}{2}, -(-\tfrac{3}{2})\}, \{-\tfrac{3}{2}, -(-\tfrac{1}{2})\}, \{-\tfrac{3}{2}, -\tfrac{1}{2}\}, \{-\tfrac{3}{2}, -\tfrac{3}{2}\}\}$$

to pullout and label the eigenvectors from **vs**. We thus enter

```
Do[
    ep[jmp[[n]]] = vs[[n]]/Sqrt[vs[[n]].vs[[n]]],
    {n,Length[jmp]}
]
```

to set up the normalized eigenvectors. We can verify these results directly, as
for example,

```
{Jsqm1m2.ep[{1/2,-1/2}] ==  3h^2/4 ep[{1/2,-1/2}],
   Jzm1m2.ep[{3/2,-3/2}] == -3h/2   ep[{3/2,-3/2}]} //
      ExpandAll
```

```
{True, True}
```

Since **Jsqm1m2** is real and symmetric, the eigenvectors **ep[{j,m}]** are
also real and therefore unique to within a sign. As usual, they form the
columns of a unitary transformation matrix **c[j1,j2]** that diagonalizes
Jsqm1m2 by a similarity transformation. The elements of **c[j1,j2]** are the
Clebsch-Gordan coefficients in the **j1-j2** subspace. Ordering the eigenvectors **ep[{j,m}]** with our original label list **jm** to correspond to the matrices
Jsqm1m2 and **Jzm1m2**, we thus obtain

```
c[1,1/2] = Transpose[ Table[ ep[jm[[n]]],{n,Length[jm]} ] ];
c[1,1/2] //TableForm[#,TableAlignments -> Center]&
```

1	0	0	0	0	0
0	$\dfrac{1}{\text{Sqrt}[3]}$	$\text{Sqrt}\left[\dfrac{2}{3}\right]$	0	0	0
0	$\text{Sqrt}\left[\dfrac{2}{3}\right]$	$-\left(\dfrac{1}{\text{Sqrt}[3]}\right)$	0	0	0
0	0	0	$\dfrac{1}{\text{Sqrt}[3]}$	$\text{Sqrt}\left[\dfrac{2}{3}\right]$	0
0	0	0	$-\text{Sqrt}\left[\dfrac{2}{3}\right]$	$\dfrac{1}{\text{Sqrt}[3]}$	0
0	0	0	0	0	1

Since this unitary matrix is real, it is simply orthogonal, and clearly has the same block-diagonal form as **Jsqm1m2**. We easily check that it performs the desired task, the diagonalization of **Jsqm1m2** (this takes about 80 seconds):

```
Transpose[c[1,1/2]].Jsqm1m2.c[1,1/2] //
         ExpandAll //MatrixForm
```

$$
\begin{pmatrix}
\dfrac{15\,h^2}{4} & 0 & 0 & 0 & 0 & 0 \\[2mm]
0 & \dfrac{15\,h^2}{4} & 0 & 0 & 0 & 0 \\[2mm]
0 & 0 & \dfrac{3\,h^2}{4} & 0 & 0 & 0 \\[2mm]
0 & 0 & 0 & \dfrac{3\,h^2}{4} & 0 & 0 \\[2mm]
0 & 0 & 0 & 0 & \dfrac{15\,h^2}{4} & 0 \\[2mm]
0 & 0 & 0 & 0 & 0 & \dfrac{15\,h^2}{4}
\end{pmatrix}
$$

This result expresses **Jsq** in the *coupled* or *jm representation*.

Exercise 17.2-1

Referring to our spin-orbital example with $j1 = 1$, $j2 = 1/2$:

(a) Check that the **ep[{j,m}]** are eigenvectors of **Jzm1m2** and **Jsqm1m2** with the appropriate eigenvalues and that they form an orthonormal set. Generate the **ep[{j,m}]** using the raising and lowering matrices **JRm1m2** and **JLm1m2**.

(b) Verify that the matrix **c[j1,j2]** is orthogonal.

(c) Construct the direct-product eigenvectors **e[{m1,m2}]** of **Jzm1m2** and show that **c[j1,j2]** transforms them into the **ep[{j,m}]** according to **ep[j,m] == c[j1,j2].e[{m1,m2}]**. Also, verify the *inverse* transformation **e[{m1,m2}] == Transpose[c[j1,j2]].ep[j,m]**. (Compare with the rotation transformation **Dy** from Section 16.4.)

Problem 17.2-1

Diagonalize **Jsq** in the **j1-j2** subspace for **j1 = j2 = 1**. This might represent, for example, two spinless particles with orbital angular momenta **l1 = l2 = 1**. Construct the Clebsch-Gordan matrix **c[j1,j2]** of eigenvectors. (You can check it below in Section 17.3 against **Clebsch**.) Redo Exercises 17.1-1 and 17.2-1.

While matrix diagonalization may be the most straightforward way to calculate the Clebsch-Gordan coefficients, it isn't generally the most efficient. Instead, the Clebsch-Gordan coefficients satisfy two rather simple *recursion relations* which can be solved iteratively and quickly. These relations are obtained by application of the raising and lowering operators **JR** and **JL** to **ketp[j1,j2,{j,m}]**.

Moreover, the recursion relations can be evaluated in closed form and expressed in terms of a finite sum, which is relatively easy to program. This general formula is due to Racah and is the basis of the function **Clebsch[{j1, m1},{j2,m2},{j,m}]** defined for nonsymbolic (number) arguments in the package **Quantum`Clebsch`**. The *built-in* function **ClebschGordan** works similarly and has the same syntax. It evaluates by relating Racah's sum to terminating hypergeometric series and simplifies with symbolic arguments in some cases. (Refer to the package **ClebschGordan.m** located in the **StartUp** or **Preload** directory and references therein.) We shall examine some of the properties of these functions in the next section. In any case, you should consult the brief but excellent book by Brink and Satchler [12] for a compendium of properties of the Clebsch-Gordan coefficients and many other aspects of quantum angular momentum. Racah's formula is given on p. 34. For a treatise of techniques, formulas and references on all aspects of angular momentum, refer to Biedenharn and Louck [7].

The construction of the recursion relations and their use is outlined in Problems 17.4-1 and 17.4-2. Problem 17.4-3 shows that the Racah formula reduces to simple algebraic expressions if one argument of the Clebsch-Gordan coefficient is assigned a definite value, as does the built-in function **ClebschGordan**.

Clebsch Gordanary   17.3

Rather than pursue further the evaluation of the Clebsch-Gordan coefficients, we simply examine here some of their properties using the function **Clebsch**. We load the package and thus define generally the expansion coefficients which determine the eigenfunctions **ketp[j1,j2,{j,m}]** of **Jsq**, **J[z]**, **jsq[1]** and **jsq[2]** from Section 17.0:

```
Needs["Quantum`Clebsch`"]

c[j1_,j2_,{m1_,m2_},{j_,m_}] :=
    Clebsch[{j1,m1},{j2,m2},{j,m}]
```

Note how **m1** and **m2** and **j** and **m** have been regrouped on the left. This allows us to check easily our spin-orbital coupling matrix **c[1,1/2]** as a table of **Clebsch** functions over our label lists **m1m2** and **jm** from Section 17.1. That is,

```
c[1,1/2] == Table[
            c[1,1/2,m1m2[[np]],jm[[n]]],
            {np,1,Length[m1m2]},{n,1,Length[jm]}
        ]
```

```
True
```

However, the Clebsch-Gordan coefficients are also just the overlap integrals **bra[j1,j2,{m1,m2}] ** ketp[j1,j2,{j,m}]** of the direct-product basis with the coupled basis, so that alternatively

```
c[1,1/2] ==
    Table[
        bra[1,1/2,m1m2[[np]]] ** ketp[1,1/2,jm[[n]]],
        {np,1,Length[m1m2]},{n,1,Length[jm]}
    ]
```

```
True
```

Of course, the most important property of the Clebsch-Gordan coefficients is that they solve the angular-momentum coupling problem and allow us to construct directly eigenfunctions of **Jsq**, **J[z]**, **jsq[1]** and **jsq[2]**. For example,

```
{Jsq    **ketp[1,1/2,{3/2,1/2}] ==
            15h^2/4 ketp[1,1/2,{3/2,1/2}],
 J[z]   **ketp[1,1/2,{3/2,1/2}] ==
            h/2     ketp[1,1/2,{3/2,1/2}],
 jsq[1]**ketp[1,1/2,{3/2,1/2}] ==
            2h^2    ketp[1,1/2,{3/2,1/2}],
 jsq[2]**ketp[1,1/2,{3/2,1/2}] ==
            3h^2/4  ketp[1,1/2,{3/2,1/2}]} //
                ExpandAll
```

```
{True, True, True, True}
```

Unitarity of the Clebsch-Gordan transformation means that the coupled eigenkets **ketp[j1,j2,{j,m}]** are orthonormal if the direct-product eigenkets **ket[j1,j2,{m1,m2}]** are. Thus,

```
Table[
    brap[1,1/2,jm[[np]]] ** ketp[1,1/2,jm[[n]]],
    {np,1,Length[jm]},{n,1,Length[jm]}
] ==
    IdentityMatrix[Length[jm]]

True
```

Unitarity also means that the Clebsch-Gordan coefficients determine the inverse transformation which expresses the **ket[j1,j2,{m1,m2}]** as a linear combination of the **ketp[j1,j2,{j,m}]** summed over **j** and **m**. Thus, for example

```
ket[1,1/2,{0,-1/2}] ==
    Sum[ c[1,1/2,{0,-1/2},{j,m}] ketp[1,1/2,{j,m}],
        {j,1/2,3/2},{m,-j,j}
    ] //ExpandAll

True
```

Hence, the Clebsch-Gordan coefficients satisfy a pair of *orthogonality relations*, or *normalization conditions*, which are the subject of Exercise 17.3-3.

Exercise 17.3-1

(a) Show with a few explicit examples that the Clebsch-Gordan coefficients vanish unless **j1**, **j2** and **j** satisfy the triangular condition and **m1 + m2 = m**.

(b) Check with a few explicit examples the identity (17.0-r2), which we used to establish that **m1 + m2 = m**.

Exercise 17.3-2

Construct the *spin* eigenkets **ketp[1/2,1/2,{s,ms}]** of two spin-$1/2$ particles such that **s1 = s2 = j1 = j2 = 1/2**. Define the total-spin operators **Ssq == Jsq** and **S[z] == J[z]** with quantum numbers **s = j** and **ms = m**. Do this using **Clebsch**. Show that the **ketp** are orthonormal and eigenfunctions of **Ssq**, **S[z]**, **ssq[1]** and **ssq[2]**.

Check your results by redoing Problem 17.2-1 with **j1 = j2 = 1/2**. As we noted at the beginning of Section 16.2, the matrix representation of spin is the most common.

From spectroscopy, the **s = 0** state with spins *antiparallel* is said to form a *singlet*, and the three **s = 1** states with spins *parallel* are said to form a *triplet*. These states are important in the description of any system with a pair of identical particles, and, for example, account for *ortho* and *para* helium.

Exercise 17.3-3

(a) Show that the normalization of **ketp** requires that the Clebsch-Gordan coefficients satisfy the *orthogonality relation* (sum over **m1** and **m2**)

```
Sum[ c[j1,j2,{m1,m2},{jp,mp}] c[j1,j2,{m1,m2},{j,m}],
    {m1,-j1,j1},{m2,-j2,j2}
] == KD[jp,j] KD[mp,m]
```

and check it explicitly for **j1**, **j2** $\leq$ **2**.

(b) The expression in Part **(a)** is the matrix-element equivalent of the matrix relation **Transpose[c[j1, j2]].c[j1,j2]==1** (see Exercise 17.2-1). Thus, verify that the Clebsch-Gordan coefficients also satisfy the following (inverse) *orthogonality relation* (sum over **j** and **m**)

```
Sum[ c[j1,j2,{m1p,m2p},{j,m}] c[j1,j2,{m1,m2},{j,m}],
     {j,Abs[j1-j2],j1+j2},{m,-j,j}
] == KD[m1p,m1] KD[m2p,m2]
```

and check it explicitly for **j1, j2** $\leq$ **2**.

(c) Use the relation in Part **(b)** (and paper and pencil if necessary) to write **ket[j1, j2,{m1,m2}]** as the linear combination

```
ket[j1,j2,{m1,m2}] ==
    Sum[ c[j1,j2,{m1,m2},{j,m}] ketp[j1,j2,{j,m}],
         {j,Abs[j1-j2],j1+j2},{m,-j,j}
    ]
```

and check it explicitly for **j1, j2** $\leq$ **2**.

As we find in Problem 17.4-1, the recursion relations in conjunction with a normalization condition determine all of the Clebsch-Gordan coefficients to within a phase. This phase is chosen by convention by demanding that the "top-rung" coefficient **c[j1,j2,{j1,j-j1},{j,j}]** be real and positive. Consequently, all of the Clebsch-Gordan coefficients are real and therefore, as we have already seen, the unitary Clebsch-Gordan matrix **c[j1,j2]** is orthogonal.

If **j = 0**, the Clebsch-Gordan coefficients must vanish unless **j1 = j2** and **m2 = -m1**, as required by the triangular condition, in which case one finds that (see Problems 17.4-2 and 17.4-3)

```
c[j1,j1,{m1,-m1},{0,0}]
```

$$\frac{(-1)^{j1-m1}}{\text{Sqrt}[1 + 2\ j1]}$$

Angular momentum algebra or "Clebsch Gordanary" has many applications in quantum physics and in realizations of the so-called *rotation group*, for which one might even define mathematical objects which are not angular momenta but which nevertheless satisfy the angular-momentum commutation relations. For example, we will show later on that such *pseudo angular momenta* with half-integral spin can be employed to determine the energy spectrum of the hydrogen atom.

Wigner 3j Symbols 17.4

Clebsch-Gordan coefficients display a number of symmetries which are so useful in practice one usually finds it convenient to work with another related function that emphasizes them. One thus defines the *Wigner 3-j symbol* according to

```
Threej[{j1_,m1_},{j2_,m2_},{j3_,m3_}] :=
    (-1)^(j1-j2-m3)/Sqrt[2j3+1] *
        Clebsch[{j1,m1},{j2,m2},{j3,-m3}]
```

This function is also defined in the package **Quantum`Clebsch`** and given by the built-in function **ThreeJSymbol**. Thus, the *3-j* symbol vanishes unless **j1**, **j2** and **j3** satisfy the triangular condition or **m1 + m2 + m3 = 0**, since **−m3** appears in the defining **Clebsch** function (see Exercise 17.3-1). The **mi** are therefore now symmetrically related.

The **Threej** are *invariant* to cyclic (even) permutations of their argument list, and they are multiplied by **(-1)^(j1+j2+j3)** for non-cyclic (odd) permutations. Thus, for example

```
Threej[{2,2},{3,3},{4,4}] ==
    Threej[{4,4},{2,2},{3,3}] ==
        (-1)^(2+3+4) Threej[{3,3},{2,2},{4,4}]
```

```
    True
```

Finally, they are multiplied by **(-1)^(j1+j2+j3)** if the signs of **m1, m2** and **m3** are all reversed:

```
Threej[{2,2},{3,3},{4,4}] ==
    (-1)^(2+3+4) Threej[{2,-2},{3,-3},{4,-4}]
```

```
    True
```

The origin of these symmetries can be traced to the geometric nature of the Clebsch-Gordan coefficients (see Biedenharn and Louck [7]).

Exercise 17.4-1

(a) Verify the symmetries of the *3-j* symbol with some examples. What are the corresponding symmetries of the Clebsch-Gordan coefficients?

(b) Show that the triple scalar product of three vectors **uvec** • (**vvec** × **wvec**) can be calculated as (cf. Exercise 1, Appendix V)

```
-I Sqrt[6] Sum[
                u[i] v[j] w[k] Threej[{1,i},{1,j},{1,k}],
                {i,-1,1},{j,-1,1},{k,-1,1}
            ]
```

where **u[i]**, **v[i]** and **w[i]** are the *spherical tensor components* of the vectors **uvec**, **vvec** and **wvec**. These are defined as **u[1]= -(ux + I uy)/Sqrt[2]**, **u[0]=uz** and `u[-1]= (ux - I uy) / Sqrt [2]`, and similarly for the `v[i]` and `w[i]`, and have many useful applications (see Exercise 19.3-5).

Problem 17.4-1

Apply the raising and lowering operators `JR` and `JL` to `ketp` and derive (with paper and pencil if necessary) the following two *recursion relations* among the Clebsch-Gordan coefficients:

```
Sqrt[(j+m)(j-m+1)] c[j1,j2,{m1,m2},{j,m-1}] ==
    Sqrt[(j1-m1)(j1+m1+1)] c[j1,j2,{m1+1,m2},{j,m}] +
    Sqrt[(j2-m2)(j2+m2+1)] c[j1,j2,{m1,m2+1},{j,m}]

Sqrt[(j-m)(j+m+1)] c[j1,j2,{m1,m2},{j,m+1}] ==
    Sqrt[(j1+m1)(j1-m1+1)] c[j1,j2,{m1-1,m2},{j,m}] +
    Sqrt[(j2+m2)(j2-m2+1)] c[j1,j2,{m1,m2-1},{j,m}]
```

Implement these rules as a *dynamic program* (refer to Exercise 3.1, Appendix III) which, for specific values of `j1`, `j2` and `j`, computes all of the Clebsch-Gordan coefficients in terms of just one, `c[j1,j2,{j1,j-j1},{j,j}]`. Assume, following convention, that this top-rung coefficient is a real, positive constant, and determine the constant from the normalization condition

```
Sum[c[j1,j2,{m1,m2},{j,j}]^2, {m1,-j1,j1},{m2,-j2,j2}] == 1
```

Demonstrate that your program works for all `j1`, `j2` $\leq$ **2** by comparing your results with `ClebschGordan`. Hint: Convert the recursion relations to a set of definitions for a function `cg[j1_,j2_,{m1_,m2_},{j_,m_}]`. You will need separate definitions for the cases `m` $\neq$ `j` and `m = j`. Be sure to include the boundary conditions for `Abs[m]> j`, etc. While debugging, you might find it convenient to use your definitions as a set of *replacement* rules which can be applied by hand to a single coefficient.

Problem 17.4-2

Using the recursion relations and the normalization condition in the previous problem, show that `Clebsch[{A,a},{A,-a},{0,0}]== (-1)^(A-a)/Sqrt[2A+1]` for *symbolic* arguments `A` and `a`. Hint: Write a rule to deduce that `Clebsch[{A,a},{A,-a},{0,0}]` and the top rung `Clebsch[{A,A},{A,-A},{0,0}]` are equal to within a phase. Determine the phase and then write a rule connecting `Clebsch[{A,a},{A,-a},{0,0}]` with `Clebsch[{A,A},{A,-A},{0,0}]`. (Take care to introduce boundary conditions to prevent your rules from chasing their tails and hence *infinite recursion*. Refer to Exercise 3.1, Appendix III.) Compute `Clebsch[{A,A},{A,-A},{0,0}]` by evaluating the normalization condition symbolically with the package `Algebra`SymbolicSum`.

Problem 17.4-3

By assigning definite nonsymbolic values to one argument of the Clebsch-Gordan coefficient, show that Racah's general formula (used in the package `Quantum-`Clebsch`) reduces to simple algebraic expressions. In particular, show that Racah's sum can be evaluated in closed form.

First, load the package **Algebra`SymbolicSum`** so that **Sum** evaluates symbolically. Then, from the package **Quantum`Clebsch`**, redefine the function **f1[C, A, B, a, b]** as a new function **Symbolicf1[C, A, B, a, b][kmin, kmax]** that performs the sum over **k** symbolically. You will have to determine the limits **kmin** and **kmax** on the sum by hand or invent something more elaborate in order to ensure that arguments of the factorial functions are nonnegative.

Thus, define two new functions **SymbolicClebsch[{A,a},{B,b},{C,c}] [{kmin,kmax}]** and **SymbolicThreej[{A,a},{B,b},{C,c}][{kmin,kmax}]** and evaluate them for definite values of one argument, e.g. **C = 0, 1,** ... Simplify your results with **PowerContract** from the package **Quantum`PowerTools`** and with the rules (refer to Problem 6.5-4)

```
FactorialSimplify[expr_] := expr //.
    {Factorial[a_] :> Gamma[a+1],
     Gamma[k_Integer  + a_] :> Gamma[a] Pochhammer[a,k],
     Gamma[k_Rational + a_] :>
                 Gamma[a+1/2] Pochhammer[a+1/2,k-1/2]}
```

Try also the built-in function **SimplifyGamma**. Compare your results with the built-in functions **ClebschGordan** and **ThreeJSymbol** and with Brink and Satchler [12], Table 3, p.36.

Recoupling Coefficients 17.5

The next step in this already protracted story is to construct coupled eigenkets out of the direct-product representation of a three-particle angular momentum **Jsq == Abs[jvec[1]+ jvec[2]+ jvec[3]]^2** and **J[z] == j[1,z]+ j[2,z]+ j[3,z]**. Even though this is relatively straightforward if we build on what we've already done, we see that the labels **j** and **m** are not enough to specify uniquely the coupled eigenkets.

There are instead three possibilities corresponding to three possible pair combinations of the **jvec[i]**. For example, we can couple **jvec[1]** with **jvec[2]** to form **jvec[1,2]** and then add **jvec[3]** to get **Jvec**. Likewise, we can form **jvec[2,3]** and **jvec[1,3]**. This procedure is followed in Problem 17.5-2.

We can then relate the eigenkets corresponding to **jvec[1,2]** and **jvec[2,3]**, for example, and define a new transformation referred to as angular momentum *recoupling*. The elements of the corresponding transformation matrix define *recoupling coefficients* or *6-j symbols*. These functions are also built-in as **SixJSymbol[{j1,j2,j12},{j3,j,j23}]** and can be expressed as sums over *3-j* symbols. The recoupling coefficients were first defined by Racah and thus are also referred to as *Racah W-functions* **W[j1, j2, j, j3, {j12, j23}]**, which differ from the built-in **SixJSymbol** by just the phase **(-1)^(j1+j2+j3+j)**. (The labelling here and in Problem 17.5-2 follows Brink and Satchler [12], Section 3.3.)

Of course, the process need not end with three angular momenta. One can define in principle *3n-j* symbols although this rapidly complicates with increasing n. In practice, one usually manages with combinations of *3-j, 6-j* and perhaps *9-j* symbols.

Exercise 17.5-1

Use the built-in function **SixJSymbol** to verify that the Racah *W-function* **W[a, a+1/2, b, b+1/2, {1/2,c}]==**

$$
\frac{(-1)^{a+b-c} \; \text{Sqrt}\left[\dfrac{(1+a+b-c)\,(2+a+b+c)}{(1+a)\,(1+2\,a)\,(1+b)\,(1+2\,b)}\right]}{2}
$$

in agreement with Brink and Satchler [12], Table 4, p. 43. Hint: Simplify with the function **PowerContract** from the package **Quantum`PowerTools`**.

Problem 17.5-1

Define a direct-product basis **ket[j1,j2,j3,{m1,m2,m3}]** which provides simultaneous eigenkets of **jsq[i]** and **j[i,z]** for **i = 1, 2, 3**. Show with some specific examples that

```
ketp[j1_,j2_,j3_,{0,0}]  :=
    Sum[
        Threej[{j1,m1},{j2,m2},{j3,m3}] *
            ket[j1,j2,j3,{m1,m2,m3}],
        {m1,-j1,j1},{m2,-j2,j2},{m3,-j3,j3}
    ]
```

is a *coupled-representation* eigenket of the total angular momentum **Jsq** with eigenvalue zero.

Problem 17.5-2

Determine a spin-coupled basis set of total **Jsq** and **J[z]** of a three-electron system (each electron has spin-$1/2$) by forming linear combinations of the direct-product eigenkets **ket[j1,j2,j3,{m1,m2,m3}]** defined in Problem 17.5-1. Build on what we have already done using Clebsch-Gordan coefficients. Calculate from your results the general recoupling coefficients connecting eigenkets of **jsq[1,2]** and **jsq[2,3]** and demonstrate that they can be expressed in terms of *6-j* symbols.

(a) First, construct generally eigenkets **ketp[j1_,j2_,{j12_,m12_},{j3_, m3_}]** of **jsq[1,2]** and **jsq[3]** as a sum over the **ket[j1,j2,j3,{m1,m2,m3}]** from Problem 17.5-1.

Then, construct generally eigenkets **ket12[{{j1_,j2_},j12_,j3_,{j_, m_}}]** of **jsq[1,2]** and **Jsq** as a sum over the **ketp[j1,j2,{j12,m12},{j3, m3}]** and show that they form an orthonormal set. The ordering of the arguments of **ket12** is conventional (cf. Brink and Satchler [12], Section 3.3). Refer to this as the **j12jm** representation.

Specialize to three spin-$1/2$ electrons and compute the allowed values of **j12** and **j**. Show that the matrices of **jsq[1,2]** and **Jsq** are diagonal in the **j12jm** representation. Hint: Define an array **j12jm** (analogous to the array **jm** defined in Section 17.1) of all possible combinations of **{{j1,j2},j12,j3,{j,m}}**. You can check your ordering of the elements of `j12jm` by computing the matrix of `J[z]`. Note that the arguments of `ket12` have been wrapped in a list for convenience in using the `j12jm` array elements to set up the matrices.

(b) Now, construct generally eigenkets `ketq[{j1_,m1_},j2_,j3_,{j23_, m23_}]` of `jsq[1]` and `jsq[2,3]` as a sum over the `ket[j1,j2,j3,{m1,m2,m3}]` from Problem 17.5-1.

Then, construct generally eigenkets `ket23[{j1_,{j2_,j3_},j23_,{j_, m_}}]` of `jsq[2,3]` and `Jsq` as a sum over the `ketq[{j1,m1},j2,j3,{j23, m23}]` and show that they form an orthonormal set. Again, the ordering of the arguments of `ket23` is conventional. Refer to this as the `j23jm` representation.

Compute the allowed values of `j23` and `j`. Show that the matrices of `jsq[2,3]` and `Jsq` are diagonal in the `j23jm` representation.

(c) Because `ket12` and `ket23` both span the direct-product subspace, we can express one as a linear combination of the other. For example, (cf. Brink and Satchler [12], eq. 3.8)

```
ket12[{{j1,j2},j12,j3,{j,m}}] ==
    Sum[
        w[{j12,j23}] ket23[{j1,{j2,j3},j23,{j,m}}],
        {j23,Abs[j2-j3],j2+j3}
    ]
```

Since `ket12` and `ket23` both are simultaneous eigenkets of `Jsq` and `J[z]`, the sum is only over `j23`. The expansion coefficients are determined by the orthonormality of the `ket23`, that is

```
w[{j12,j23}] ==
    ket23[{j1,{j2,j3},j23,{j,m}}] **
    ket12[{{j1,j2},j12,j3,{j,m}}]
```

These overlaps define recoupling coefficients *proportional* to Racah *W* functions and therefore provide an expression for computing *6-j* symbols. The latter are defined in this connection as (refer to Brink and Satchler [12], Section 3.3)

```
Sixj[{j1_,j2_,j12_},{j3_,j_,j23_}] :=
    ket23[{j1,{j2,j3},j23,{j,m}}] **
    ket12[{{j1,j2},j12,j3,{j,m}}] *
    (-1)^(j1+j2+j3+j)/(Sqrt[2j12+1] Sqrt[2j23+1])
```

where the normalization `Sqrt[2 j12+1] Sqrt[2 j23+1]` is introduced to simplify the symmetry properties of the `Sixj`. Note that this result defines and can be used to compute *6-j* symbols generally if `ket12` and `ket23` have been defined generally for arbitrary arguments.

Evaluate `Sixj` for some physically acceptable arguments and compare with the built-in function `SixJSymbol`. Note that `m` doesn't appear in the argument of `SixJSymbol`. The reason is *6-j* symbols are scalars independent of the orientation of the *z* axis and therefore of projection quantum numbers. Thus, demonstrate that `Sixj` is also independent of `m`.

18
Coordinate and Momentum Representations

We will now consider the more conventional representation of the components of the linear momentum **pvec** as derivatives with respect to their *conjugate position coordinates*. We thus take **pvec** to be a gradient with respect to the position vector **rvec**, viz. we set **pvec -> -I h grad** such that **px -> -I h Dt[..., x]**, etc. We ensure in this way that the fundamental commutation relations are satisfied in the *coordinate representation* when the system wavefunction is a function of **x**, **y** and **z**. On the other hand, we can take **rvec** to be a gradient with respect to the momemtum vector **pvec** and the system wavefunction a function of the momenta **px**, **py**, and **pz** in the *momentum representation*. The two representations are mathematically equivalent; their wavefunctions are Fourier-transform pairs (see Chapter 11 and also Problem 20.2-1).

Note that Appendix V on vector calculus provides many of the mathematical tools we will need.

Position and Momentum Operators   18.1

We shall use the function **Dt** to define derivative operators, as discussed in Section 2.1, Appendix V, because of the compact expressions it affords (see also Problem 18.2-4 below). We therefore need to declare constants and specify that the position and momentum coordinates are independent. Thus, introducing our function from Section 2.1, Appendix V, we enter

```
IndependentVariables[x_,y_,z_] :=
    Module[{},
            x/: Dt[x,y]  := 0;    x/: Dt[x,z]  := 0;
            y/: Dt[y,x]  := 0;    y/: Dt[y,z]  := 0;
            z/: Dt[z,x]  := 0;    z/: Dt[z,y]  := 0
    ]

IndependentVariables[ x, y, z];
IndependentVariables[px,py,pz];
SetAttributes[{h,m}, Constant]
```

Here the constant **m** is the particle's mass and, as usual, **h** is Planck's constant divided by **2Pi**. Thus, we define the momentum operator **pvec** in the coordinate representation in terms of the gradient from Section 2.2, Appendix V, and extract its cartesian components. Likewise, we define the position operator **rvec** in the momentum representation and extract its components. Thus,

```
grad[f_]  := {Dt[f, x],Dt[f, y],Dt[f, z]}
pvec @ psi_  := -I h grad[ psi];
px @ psi_  := pvec[psi][[1]];
py @ psi_  := pvec[psi][[2]];
pz @ psi_  := pvec[psi][[3]];

gradp[f_]  := {Dt[f,px],Dt[f,py],Dt[f,pz]}
rvec @ phi_  :=  I h gradp[phi]
x @ phi_  := rvec[phi][[1]]
y @ phi_  := rvec[phi][[2]]
z @ phi_  := rvec[phi][[3]]
```

The *prefix* form **Q @ psi** (**== Q[psi]**) of operator application (cf. Exercise 1.8, Appendix III) is used to mimick **∗∗** from the formal representation.

It is useful here to define the kinetic energy operator with **pvec • pvec/ (2m)** and therefore with the *laplacian* from Section 2.5, Appendix V. Thus,

```
laplacian[f_]  = Dt[f,{x,2}] + Dt[f,{y,2}] + Dt[f,{z,2}];

K @ psi_  = -h^2 laplacian[psi]/(2m)
```

$$-\frac{(h^2 \; (Dt[psi, \{x, 2\}] + Dt[psi, \{y, 2\}] + Dt[psi, \{z, 2\}]))}{2\; m}$$

Exercise 18.1-1

Show that the *free-particle planewave* `psiPW = E^(I kvec.rvecC)` with `rvecC = {x,y,z}` and *wavevector* `kvec = {kx,ky,kz}` is an eigenfunction of momentum `pvec` and kinetic energy `K` with eigenvalues `kvec` and `(h k)^2/(2m)`, respectively. Here `k = Sqrt[kvec.kvec]` is the *wavenumber* such that `2Pi/k` is the particle's *deBroglie wavelength*. (Refer to Exercise 1.1-1.)

Commutation Relations   18.2

We need to extend our definition of **Commutator** from Chapter 15 to handle derivative operators. We only have to represent the basic definition, **[A, B] == A @ B − B @ A**, since differentiation already has all the properties we need built in, in particular, those we had to define for noncommutative multiply ****** in the formal representation. Indeed, this equivalence justifies the derivative forms of the momentum and position operators in the coordinate and momentum representations, respectively.

Since we are introducing derivatives, we need to make sure we have something to differentiate. Hence, it's appropriate to have **Commutator** operate explicitly on a wavefunction **psi**. We thus enter the following *single rule*:

```
Commutator[A_, B_] @ psi_ := A @ B @ psi - B @ A @ psi //
    Expand
```

Now in order for this rule to work properly, both **A** and **B** have to be *Mathematica* functions which operate on the wavefunction, such as **px @ psi**. We can ensure this compactly and elegantly by entering quantities which are not already operators (expressions involving the position coordinates **x**, **y**, and **z** in the coordinate representation and the momentum coordinates **px**, **py**, and **pz** in the momentum representation) as *pure functions*.

For example, if *in the coordinate representation* **A** is just the coordinate **x**, then the operation is ordinary multiplication and we use **x # &** in the commutator with a *slot* **#** for **psi** and an **&** to signal the end of the pure function (refer to Exercise 1.9, Appendix III). Thus, we verify the fundamental commutation relations in the coordinate representation according to

```
{Commutator[x # &, px] @ psi,
 Commutator[py, y # &] @ psi}/psi
```

```
{I h, -I h}
```

We divide out **psi** in the end to emphasize that these are operator relations which are essentially independent of the function being operated on.

In the momentum representation, on the other hand, we use simply **x** in the commutator but enter **px** for example as **px # &**. Hence,

```
{Commutator[x, px # &] @ phi,
 Commutator[py # &, y] @ phi}/phi
```

```
{I h, -I h}
```

Without pure functions we would have to invent extraneous names and rules, such as **xTimes[f_] : = x f**, or attach more rules to **Commutator**.

Our single rule also works for more complicated commutators. Consider for example the angular momentum commutator **[lx,y]== I h z** from Exercise 16.0-1 (see also Exercise 18.3-1 below). In the coordinate representation, we can enter this in one of two ways:

```
{Commutator[y pz @ # - z py @ # &, y # &] @ psi ==
    I h z psi,
 Commutator[y pz[#] - z py[#] &,   y # &] @ psi ==
    I h z psi}
```

```
{True, True}
```

where in the second version we simply compacted **pz @ #** with **pz[#]**, etc.

In the momentum representation, we can use either of the following:

```
{Commutator[y @ # pz - z @ # py &, y @ # &] @ phi ==
    I h z @ phi,
 Commutator[y[#] pz - z[#] py &,   y[#] &] @ phi ==
    I h z @ phi}
```

```
{True, True}
```

Note that in all four expressions only a single **&** is needed at the end of each argument of **Commutator** in order to declare each argument a pure function. (Apply **//FullForm** to one of the arguments.)

We enter the slightly more complicated expression **px x** in the coordinate representation as **px[# x]&** or as **px @ (# x)&**. Compare, for example, evaluation of **[x,px^2 x] - [x, x px^2]** in the coordinate representation

```
(Commutator[x # &, px @ px @ (x #) &] @ psi -
 Commutator[x # &, x px @ px[#] &]      @ psi)/psi
```

$$2 \; h^2$$

and in the momentum representation

```
(Commutator[x, px^2 x @ # &]      @ phi -
 Commutator[x, x @ (px^2 #) & ] @ phi)/phi
```

$$2 \; h^2$$

We thus find our single commutator rule very efficient and, when applicable, allows us to redo quickly much of what we proved in the formal representation and in some instances to generalize previous results (compare for example Problem 15.1-1 and Exercise 18.2-2).

When working in the coordinate representation, it's a good idea to label the position vector as something other than **rvec**, for example **rvecC =** **{x,y,z}**, in order to avoid collision with the position operator **rvec** should you decide to switch to the momentum representation. Likewise, you should label the momentum vector differently, for example **pvecC = {px,py,pz}** (see also Section 2.6, Appendix V). Keep in mind that expectation values **<...>** and in general matrix elements in the coordinate or momentum representations are integrals over **rvecC** or **pvecC** respectively.

Exercise 18.2-1

(a) Show as well that **[px, x^2 px] - [px, px x^2] == 2 h^2**.

(b) Show that **pvec** commutes with the kinetic energy. Note that a common set of eigenfunctions were constructed in Exercise 18.1-1.

Exercise 18.2-2

Show that the commutator of **px/(-I h)** with a function **v[x]** equals **v'[x]** (cf. Problem 15.1-1). What is the generalization of this result to three dimensions?

Exercise 18.2-3

(a) Consider again the *1D* harmonic oscillator defined by the hamiltonian **hHO =** **px^2/(2m) + m w^2 x^2/2**. Evaluate the commutators **[[hHO,x],x]** and **[[hHO,** **px],px]** in the coordinate and momentum representations and compare with your results from Problem 15.1-3.

(b) Define the raising and lowering operators from Problem 15.4-1 and verify that **[aL,aR]== 1** and evaluate **[hHO, aR and aL]** in both the coordinate and momentum representations.

Problem 18.2-1

(a) Evaluate the commutator of **rvec • pvec** with the hamiltonian **H = K + V** of a particle in a conservative field, and show generally that *d***<rvec • pvec>**/*dt* **==** **<2K> - <rvec • grad[V]>**, where **<...>** is an expectation value. Thus, prove for a *stationary state* that **<2K> == <rvec • grad[V]>**, which is known as the *virial theorem*. Hint: Refer to relation (15.1-r2) and work in the coordinate representation.

(b) Verify the theorem explicitly with a Coulomb potential and a hydrogenic *2s* excited state. Thus set **V = -Z/r** and use the wavefunction **R2s = norm (2 - Z r) E^ (-Z r/2)**, where **r** is the spherical-polar magnitude of **rvec**, **Z** is the nuclear charge, and **norm** is a normalization constant. (Here all quantities are expressed in *atomic units* with **h = m = e = 1**. See Exercise 19.0-3 and also Section 20.1 below.) Hint: Load the package **Quantum`integExp`** we introduced in Section 13.4 to evaluate the necessary integrals.

Problem 18.2-2

Given the hamiltonian $H = K + V$ of a particle in a conservative field, show with relation (15.1-r2) that the time derivatives of the expectation values $< \ldots >$ of the position and momentum vectors are given by $d<rvec>/dt == <pvec>/m$ and $d<pvec>/dt == -<grad[V]>$. This result is known as *Ehrenfest's theorem*: the laws of classical mechanics (*Hamilton's equations*) hold for the expectation values. (See Section 2.5 on the two-state box wavepacket, Section 4.2 on the free-particle wavepacket, Sections 8.3 and 11.11 on squeezed states, and Problem 15.1-3.)

Problem 18.2-3

Recall that a classical particle with charge **e** moving in the presence of electric **Evec** and magnetic **Bvec** fields with velocity $uvec = drvec/dt$ experiences a *Lorentz force* $m\,duvec/dt = e(Evec + uvec \times Bvec/c)$ (in gaussian or cgs units with **c** the speed of light). Verify the generalization of Ehrenfest's theorem (Problem 18.2-2) given the hamiltonian $HEB = (pvec - e/c\ Avec)\hat{}2/(2m) + V$, where **Avec** is the vector potential such that $Evec = -grad[V] - 1/c\ \partial Avec/\partial t$ and $Bvec = curl[Avec]$.

Thus, define a velocity operator $uvec = (pvec - e/c\ Avec)/m$ and show that $d<rvec>/dt == <uvec>$ and that $d<uvec>/dt == <e\ Evec + e/(2c)(uvec \times Bvec - Bvec \times uvec)>$. Note that the two terms $uvec \times Bvec$ and $-Bvec \times uvec$ are identical classically but differ in quantum mechanics since **uvec** doesn't commute with **Bvec**. Also, their symmetric average is hermitian, although the separate terms are not.

Hint: Work in the coordinate representation. You will find it convenient to define a function `OperatorCross[uvec,bvec] @ psi` to compute the cross product of a vector (gradient) operator **uvec** with a vector field **bvec**. It should work like an ordinary cross product except that it involves function application with @ (refer to Section 1.0, Appendix V). Use `OperatorCross` to compute $Bvec = curl[Avec]$ and then $uvec \times Bvec$.

Problem 18.2-4

We use the derivative **Dt** here so that we can operate on arbitrary functions without having to specify their dependence on coordinates explicitly. It is instructive, however, to take a look at what we've done using the derivative operator **D**. Thus, redefine the components of **pvec** and **rvec** using **D**. Do this first by writing all functions with an explicit coordinate dependence, as in `f[x,y,z]`. Then do the same using the **NonConstants** option for **D** and passing this option along to the operators **pvec**, **rvec** and **Commutator** with triple blanks (see Exercise 1.10, Appendix III).

18.3 Angular Momentum in Cartesian Coordinates

We transform the *orbital* angular momentum $lvec = rvec \times pvec$ to a derivative operator in the *coordinate* representation by substituting $pvec \to -I\,h\,grad$. Here, we can simply define the cross product by introducing the function **Cross** from the package **LinearAlgebra`CrossProduct`**. Thus,

```
Needs["LinearAlgebra`CrossProduct`"]

rvecC = {x,y,z};
lvec @ psi_ = Cross[rvecC, pvec[psi]];
```

For definiteness, we work in the coordinate representation, although we could just as well do everything in the momentum representation. In analogy with the formal representation from Section 16.0, we extract the components and calculate the raising and lowering operators and the square of the angular momentum:

```
lx @ psi_ = lvec[psi][[1]];
ly @ psi_ = lvec[psi][[2]];
lz @ psi_ = lvec[psi][[3]];

lR @ psi_ = lx @ psi + I ly @ psi;
lL @ psi_ = lx @ psi - I ly @ psi;

lsq @ psi_ =
    lx @ lx @ psi + ly @ ly @ psi + lz @ lz @ psi //
        Expand;
```

With these definitions, we easily verify that the components satisfy the basic commutation relations, $[lx, ly] == I h lz$, etc. in cyclic order. For example,

```
Commutator[lx,ly] @ psi == I h lz @ psi //ExpandAll

    True
```

Exercise 18.3-1

(a) Examine the form of **lsq** and of the components of **lvec** and compare with our definitions from the formal representation. Verify that **lvec** commutes with **lsq**. Also, evaluate $[lx, y]$ and $[lx, py]$, etc. in cyclic order and compare your results with Exercise 16.0-1. Finally, show that $[lR, lL] == 2h\,lz$ and $[lz, lR$ or $lL] == I h\,(lR$ or $-lL)$ and compare your results with Problem 16.0-1.

(b) Show that **lvec** commutes with the kinetic energy **K**. Hence, establish that angular momentum is a constant of the motion for a *free* particle (cf. Exercise 18.2-1 and Problem 18.2-2). We shall construct a common set of eigenfunctions in Exercise 19.1-3.

Rotational Symmetry   **18.4**

As an example, consider an otherwise arbitrary potential energy **V[r]** which is a function only of the magnitude of **rvecC**, i.e. the radial distance $r =$ **Sqrt[x^2 + y^2 + z^2]** from the origin. As we have seen in Exercise

5, Appendix V, the gradient of this function and thus the classical force are proportional to **rvecC** and hence act along a line through the origin. Such forces are spherically symmetric and are designated *central forces*; the potentials from which they derive are called *central potentials*.

We can demonstrate that the *orbital* angular momentum **lvec** commutes with any central-force hamiltonian of the general form

```
H @ psi_ = K @ psi + V[Sqrt[x^2 + y^2 + z^2]] psi;
```

We thus find that

```
{Commutator[lvec,H] @ psi, Commutator[lsq,H] @ psi}

    {{0, 0, 0}, 0}
```

Since the commutator of an operator with **H** determines from relation (15.1-r2) the time rate of change of the expectation value of the operator, we have proven that the orbital angular momentum is a constant of the motion for central forces. This of course is the quantum counterpart of *Kepler's second law* in classical mechanics. We also have that the eigenfunctions of **H** can be chosen to be simultaneous eigenfunctions of any one of the components of **lvec** (since the components don't commute with each other) and hence of **lsq**. We associate spherical symmetry with these constants of the motion: in the absence of external torques, the particle experiences no preferred direction when held a fixed distance from the origin.

Problem 18.4-1

Consider a central-force system whose spherical symmetry is broken by a uniform external electric field. Thus, consider a dipole interaction of the form **Eo nvec • rvecC**, where **Eo** is the field strength and **nvec** is a unit vector which defines the field direction. Show that in the surviving cylindrical symmetry only the projection of the angular momentum along the direction of the electric field is conserved. Do this assuming first the field points along the *z* axis and then along an arbitrary axis **nvec**.

The above discussion hints at the connection between angular momentum and rigid rotations, which is in fact extensive and has many useful applications. To be specific, consider the change in a function **f[x,y,z]** due to an *infinitesimal rotation* by an angle **dp** about an axis **nvec** through the origin. The geometry is depicted in Figure 18.4-1. We can easily show by Taylor expansion (see Problem 18.4-2) that **d[f] == -I/h dp nvec • lvec @ f**, where the operator **lvec/h** is said to be the *generator of infinitesimal rotations*. We then note that the change can be re-expressed as **d[f] == -I/h dp [nvec • lvec, f]**, which you can verify in Exercise 18.4-1. This expression is the rotational analogue of the change in a function due to an infinitesimal translation (see Problem 18.4-3).

The change **d[f]** can thus be used to give physical meaning to the commutator of angular momentum with a central force hamiltonian. By definition,

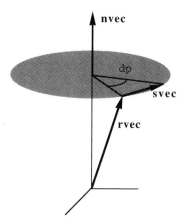

Figure 18.4-1. Geometry of an infinitesimal rotation about an axis **nvec** through the origin.

a function **f** is rotationally symmetric or *invariant* about an axis **nvec** if **d[f] = 0**. It follows that such a function commutes with the generator of infinitesimal *rotations* about **nvec**: **[nvec • lvec, f] = 0**. We conclude that rotational invariance of the potential about an axis implies conservation of angular momentum about that axis, and vice versa, since the angular momentum commutes with the kinetic energy alone (see Exercise 18.3-1 and also Section 19.0 below). Of course, this is just another example of the powerful notion that symmetries give rise to conservation laws, and vice versa.

Exercise 18.4-1

Show that **[lvec, f] == lvec @ f** for an arbitrary function **f** of **x**, **y** and **z**. Also, transform to the momentum representation and show that **[lvec, g] == lvec @ g** for an arbitrary function **g** of **px, py** and **pz**.

Problem 18.4-2

(a) Derive **d[f] == -I/h dp nvec • lvec @ f** by expanding **f[rvec – svec]** for a *displacement* **svec** and then substituting **svec -> dp nvec × rvec** for an infinitesimal rotation, retaining terms to only first order in **dp**. Refer to Figure 18.4-1 and to Section 2.3, Appendix V. (We consider here that the *value* of the function **f[rvec]** is displaced to the new point **rvec + svec** such that **f[rvec – svec]** gives the new value of **f** at the point **rvec**.) Hint: You will need to change the *head* of the argument **rvec – svec** of f from **List** to **f** using **f @@ (rvec – svec)** [i.e. **Apply**] in order for **Series** to work properly.

(b) Check directly that **d[f]** vanishes in the case of (i) spherical symmetry and rotation about any axis and (ii) cylindrical symmetry and rotation about the z axis.

(c) Check that rotation of an arbitrary function followed by an identical *inverse* rotation leaves the function unchanged to first order in **dp**.

Problem 18.4-3

Show that **pvec/h** is the *generator of infinitesimal translations* and that translational invariance implies conservation of *linear* momentum, and vice versa. Thus, show that **d[f] == +I/h dn [nvec • pvec, f]** is the change in a function **f** due to an infinitesimal linear translation **dn** along **nvec** (cf. Exercise 18.2-2). This is the linear analogue of the rotational expression derived in the previous problem.

Problem 18.4-4

Redo Problem 18.4-1 for a particle with charge **e** moving in a uniform external magnetic field. Refer to Problem 18.2-3 and use **Avec == Bo/2 nvec × rvec** for the vector potential, where **Bo** is the field strength (see also Exercise 9, Appendix V).

This problem is often done by ignoring contributions to the hamiltonian quadratic in the field, i.e. the (small) term proportional to **(Bo/c)^2**. Show that if this is the case the magnetic field can be accounted for in first approximation by a magnetic energy term **−Bo nvec • Mvec**, where **Mvec == e / (2m c) lvec** is the effective *magnetic moment* of the system. This approximate interaction clearly commutes with the generator of infinitesimal rotations about **nvec**.

18.5 Dynamical Symmetry

The recognition a system's symmetries at the outset can be a great help in constructing a complete set of commuting observables for the system. If we can calculate the eigenvalue spectrum of an observable, we obtain another label (or labels) which can be used to specify more fully and uniquely the state of the system. This can include of course the energy and hence the solution of the system Schrödinger equation. In addition, we associate commuting observables with separation of variables in the differential equations that define the observables.

For example, we have seen that if the potential is also spherically symmetric, then the orbital angular momentum commutes with the hamiltonian and thus provides an additional set of labels. As we shall see later on, central potentials permit the reduction of the two-particle Schrödinger equation from six independent variables to just one, the the radial separation of the particles. Such *geometrical symmetries* are usually easy to identify: they correspond to invariant displacements and rotations of the system in space and time and include space reflections and inversions (*parity*) and time-reversal. (See Chapter 5 and Exercise 5.0-6 and refer to Schiff [59]).

Of special interest are systems whose force law permits a solution of the Schrödinger equation (and usually of the classical equations of motion as well) in terms of a finite number of simple functions, i.e. in *closed* form. Their wave equations often separate in more than one system of coordinates or in a single system of coordinates oriented in different ways. The most prominent examples are the harmonic oscillator and the hydrogen atom. In such cases, the question naturally comes to mind: Are additional hidden or *dynamical symmetries* at work other than the obvious geometrical ones?

In fact, Pauli's [51] seminal contribution to the foundations of quantum mechanics was a derivation of the energy spectrum of the hydrogen atom based on conservation of a second vector besides the angular momentum. The additional conserved quantity is now widely known as the *Runge-Lenz vector* (see Heintz [32]), although its origins can be traced to Laplace and his work on the law of gravitation, which is also an inverse-square central-force (see Goldstein [25,26]). We shall investigate its properties in the next section.

An additional conserved quantity can also be constructed in the case of a *two-dimensional isotropic* harmonic oscillator. Like the classical Kepler problem, the isotropic oscillator also has elliptical orbits as solutions. Unlike the Kepler problem, however, where the center of attraction is a focus of the ellipse, the center of the ellipse in the oscillator is the center of attraction. This ostensibly higher symmetry in the oscillator means that the additional constant of the motion is more complicated than a vector (see Problem 18.7-2). Nevertheless, the isotropic oscillator and Kepler problems are closely related, and one can even be transformed into the other (see Problem 20.5-4).

Thus, the hydrogen atom, the harmonic oscillator, and a few other problems in quantum mechanics can be solved fully in closed form. Although only Kepler-type `1/r^2` forces are involved, it's striking that such solutions have never been found for systems seemingly as simple as the helium atom or the Earth, Sun and Moon. Despite masterful efforts for some two hundred years, this situation remains one of the outstanding gaps in the story of modern physics.

For a modern discussion of these issues and the deeper connections of quantum and classical mechanics, refer to the book by Gutzwiller [30]. The reader interested in the few-body Coulomb problem might begin with the recent and accessible review by Rau [55] and references therein.

Remark on Group Theory

A systematic study of the subject of symmetry can be undertaken using an elaborate mathematical formalism known as *group theory*. While its detailed methods are mostly not required in physical applications, the basic ideas and jargon arise and are useful to refer to. A *group* is a set of mathematical operations associated with a physical transformation of a system which leaves the system unchanged. A group is characterized by the commutator algebra or *group algebra* satisfied by certain constants of the motion, specifically, the infinitesimal generators of transformation. For example, the commutation relations (Problem 16.1-1) among the components of the angular momentum, i.e. the generators of infinitesimal rotations, define the rotation group $SO(3)$ in three dimensions. Not surprisingly, one can show there is a direct link (or *homeomorphism*) with the algebra of the rotation matrix **RotationMatrix3D** defined in Section 4.3, Appendix V (see also Sections 19.2 and 19.3 and Exercise 19.3-5 below). Hence, the label O designates real orthogonal matrices in three dimensions while the label S indicates *special* or proper matrices with determinant equal to **+1** (see Exercise 21, Appendix V).

We shall find that the six components of the angular momentum and Runge-Lenz vectors in the Coulomb problem satisfy 15 commutation relations which define the algebra of the *SO(4)* group of proper rotations in four dimensions. This space is quite distinct from *3D* physical space, and *SO(4)* is said to describe the hidden or dynamical symmetry of the hydrogen atom. (See also Goldstein [25] and Schiff [59].)

18.6 Runge-Lenz Vector

We will now specialize our central-force hamiltonian to the Coulomb potential of a one-electron or *hydrogenic* atom with nuclear charge **Z e** at the coordinate origin **r = 0**:

```
H @ psi_ = K @ psi - Z e^2/Sqrt[x^2 + y^2 + z^2] psi;

SetAttributes[{Z,e,m,h}, Constant];
```

The quantity **Z** is called the *atomic number*. The mass **m** in the kinetic energy **K** is the electron-nucleus reduced mass, although if the nuclear mass is assumed to be infinite, **m** is just the electron mass (see Section 15.2 and Exercise 15.2-1).

It is straightforward to show classically using Newton's second law that the Runge-Lenz vector **Avec == (pvec × lvec)/m − Z e^2 rvecC/r** is a constant of the motion, where **r = Sqrt[rvecC.rvecC]** (see for example Goldstein [25] and Heintz [32]). In translating this quantity into quantum mechanics, we must recognize that the components of **pvec** and **lvec** don't commute (see Exercise 18.3-1) so that **pvec × lvec** is not a hermitian operator. We can replace it, however, by its symmetric average (**pvec × lvec − lvec × pvec)/2**, which is. (A similar situation was encountered in Problem 18.2-3.)

We thus construct the quantum Runge-Lenz vector, its cartesian components, and its square. We can compute the cross products of the vector operators using the definition from Section 1.0, Appendix V. (This takes about 90 sec. We suppress the output here since it's long and uninteresting for our present needs.)

```
Avec @ psi_ =
    Table[
        Sum[
            Signature[{i,j,k}] *
            (pvec[#][[i]]& @ lvec[#][[j]]& @ psi -
             lvec[#][[i]]& @ pvec[#][[j]]& @ psi),
                {i,1,3}, {j,1,3}
        ],
```

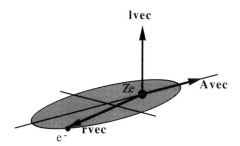

Figure 18.6-1.   An electron in a Kepler orbit with position **rvec**, angular momentum **lvec**, and Runge-Lenz vector **Avec**.

```
        {k,1,3}
    ]/(2m) - Z e^2 rvecC/Sqrt[rvecC.rvecC] psi //
        Expand;
 Ax @ psi_ = Avec[psi][[1]];
 Ay @ psi_ = Avec[psi][[2]];
 Az @ psi_ = Avec[psi][[3]];

 Asq @ psi_ =
     Ax @ Ax @ psi + Ay @ Ay @ psi + Az @ Az @ psi //
         Expand;
```

We now establish three important results. First, we show that **Avec** commutes with the hamiltonian (this takes about 80 sec):

```
Commutator[ Avec, H ] @ psi //Map[Factor,#]&
```

```
    {0, 0, 0}
```

We then note that **Avec** lies along the major axis of the *classical* orbit and is therefore perpendicular to the angular momentum vector, **lvec · Avec = 0**, as depicted in Figure 18.6-1. Quantum mechanically, we therefore have that

```
{lx @ Ax @ # + ly @ Ay @ # + lz @ Az @ # & @ psi,
  Ax @ lx @ # + Ay @ ly @ # + Az @ lz @ # & @ psi} //
     Expand
```

```
    {0, 0}
```

Finally, we relate **Asq** to the hamiltonian in analogy with the classical expression according to

```
2/m H @ (lsq @ # + h^2 #) + Z^2 e^4 # & @ psi //Expand;
```

The quickest way to check that this is equivalent to **Asq @ psi** is to verify that the difference of the two expressions vanishes. Thus,

```
Asq @ psi - % //Together
```

```
    0
```

as desired.

Although we've established these three results will little effort, examination of each expression shows that considerable algebra is involved, and their verification by hand is at best tedious (cf. Goldstein [25] and also Schiff [59]).

Exercise 18.6-1

Derive an alternative expression for **Avec** in terms of the commutator of **pvec** and **lsq**. Such a result is useful for evaluating matrix elements involving **Avec**.

18.7 Hydrogen Atom Spectrum

Remarkably, the energy spectrum **e[n]** of the hydrogen atom can be constructed at this point without much further effort. We have only to recognize an analogy with angular momentum. In the case of *negative* energies, which of course includes the ground state, we need to show that if **Avec** is replaced by the operator

$$Mvec == Sqrt[m/(-2 \ e[n])] \ Avec \tag{18.7-r1}$$

and both **lvec** and **Mvec** by the pair

$$Jlvec == (lvec + Mvec)/2$$
$$J2vec == (lvec - Mvec)/2 \tag{18.7-r2}$$

then **J1vec** and **J2vec** commute and their components satisfy the angular-momentum commutation relations. Hence, their eigenvalue spectrums are known (from Problem 16.1-1) and can be used to compute the spectrum of **Asq** and therefore of **H**. This elegant trick was introduced by Pauli. Its details are left to the next problem. Note that the **Jivec** are mathematical constructions and not the angular momenta of physical particles. They are thus sometimes referred to as *pseudo* angular momenta.

As we shall see, we can even construct wavefunctions of the hydrogen atom from these operators in analogy with Problem 6.7-1 (see also Exercise 11.9-2). It is appropriate, however, to wait until we've introduced eigenfunctions of **lsq** and **lz**, namely, the spherical harmonics.

Problem 18.7-1

Derive the negative energy spectrum of a one-electron atom with nuclear charge **Z e** from the fact that the angular momentum and Runge-Lenz vectors commute with the hamiltonian.

(a) First, establish the commutation relations among the components of **lvec** and **Avec**. Thus, show that

```
Commutator[Ax,lx]  ==   0
Commutator[Ax,ly]  ==   I h Az
Commutator[Ax,lz]  ==  -I h Ay
Commutator[Ax,Ay]  ==  -2I h/m H @ lz
```

and all cyclic permutations. The resulting 12 commutators along with the 3 from **lvec** define the algebra of *SO(4)*.

Because **[Ax,Ay]** involves **H**, it's useful to work only in the subspace of energy eigenfunctions corresponding to a particular eigenenergy. Then, **H** can be replaced by its eigenvalue **e[n]**. However, in order to enforce the above commutation relations, it's convenient to switch to the formal representation and introduce ****** from Chapter 15.

(b) If the operator **Avec** is replaced by **Mvec == Sqrt[m/(-2 e[n])] Avec** in the formal representation, the above 15 commutators become cyclic permutations of **[Mx,ly] == I h Mz**. Define a function **MlCommute** analogous to **j12Commute** from Section 17.0 which will implement these commutators with respect to ******.

Use **MlCommute** to show that the pseudo angular momentum operators **J1vec == (lvec + Mvec)/2** and **J2vec == (lvec - Mvec)/2** commute and that their components satisfy the basic angular momentum commutation relations. Hence, their squares **Jisq** have eigenvalues $h^2 ji(ji+1)$ with $ji = 0, 1/2, 1, 3/2, \ldots$

Show that **J1sq + J2sq == (lsq + Msq)/2** and that **J1sq - J2sq == lvec • Mvec**. Thus, **J1sq == J2sq**, since **lvec • Mvec = 0**, and therefore **j1 = j2**.

(c) Use the relation between **Asq** and **H** to show that the eigenenergies are given by the *Balmer-Bohr* formula $e[n] = -Z^2 Ry/n^2$, with $n = 2 j1+1$ and $Ry = e^2/(2ao) = 13.6\ eV$ the Rydberg constant and $ao = h^2/(m e^2) = 0.529$ the Bohr radius.

Hence, the *principal quantum number* has the allowed values $n = 1, 2, 3, \ldots$, since $j1 = 0, 1/2, 1, \ldots$. Thus, the ground-state energy of the hydrogen atom (**Z = 1**) is $-13.6\ eV$.

(d) That the energy spectrum is determined entirely by the quantum number **j1** or **n** is a circumstance peculiar to the Coulomb force. Here the Runge-Lenz vector is a constant of the motion. Any deviation from a pure $1/r^2$ force causes it to precess about the origin as the electron orbits. Hence, the energy eigenstates can be chosen to be simultaneous eigenstates of **J1z** and **J2z**. Then, since **J1z** and **J2z** each have **2 j1+1 = n** independent eigenvalues for a given **n**, there are n^2 *degenerate* states for each energy level (n^2 states with the same energy).

It is usual, however, to choose energy eigenstates which are simultaneous eigenstates of **lsq** and **lz**. It follows that the quantum numbers of **lsq** for **lvec == J1vec + J2vec** lie in the range $0 = Abs[j1-j2] \leq l \leq j1 + j2 = n-1$ from the triangular condition for the addition of angular momenta (Section 17.0). Thus, show again that the number of degenerate states is n^2 given that there are **2 l + 1** eigenvalues of **lz** for each value of **l**. (Evaluate the required sum symbolically with the package **Algebra`SymbolicSum`**.)

Problem 18.7-2

Consider the *2D isotropic* harmonic oscillator defined by the hamiltonian **hHO = Sum[(p[i]^2 + m^2 w^2 x[i]^2)/(2m), {i,1,2}]** in the *x-y* plane with **x[1] = x** and **x[2] = y**. Clearly, **lz** is one of the constants of the motion, since the *z* axis is an axis of symmetry for the system. See also Problems 20.5-3 and 20.5-4.

(a) Show that the *2D* symmetric matrix or *tensor* operator **A** with components `A[i,j]` = `(p[i]p[j]+m^2 w^2 x[i]x[j])/(2m)` is also a constant of the motion. Explain on the basis of a physical argument why the diagonal elements are conserved.

(b) Show that the three operators

```
S1 = (A[1,2][#] + A[2,1][#])/(2 w) &
S2 = (A[2,2][#] - A[1,1][#])/(2 w) &
S3 = lz[#]/2 &
```

satisfy the basic angular momentum commutation relations and commute with the sum of their squares `Ssq=S1 @ S1[#]+ S2 @ S2[#]+ S3 @ S3[#]&`. These operators are therefore the generators of infinitesimal rotations in a nonphysical *3D* space. (It turns out, however, that *SO(3)* is not quite the correct symmetry group for this system. The correct one is the closely related (*homeomorphic*) group *SU(2)* of all special unitary 2×2 matrices. See Goldstein [25].)

(c) Find the relationship between `Ssq` and `hHO` and show that the energy spectrum of the oscillator is given by `eHO[s] = h w (2s+1)`, where `s = 0, 1/2, 1, ...` is the quantum number of `Ssq`. For a given value of `s` there are `2s+1` degenerate energy levels corresponding to eigenfunctions of one of the components `Si`. If the energy eigenfunctions are chosen to be also eigenfunctions of `S3`, what are the allowed eigenvalues of `lz`? Explain the degeneracy by writing down the energy spectrum (from Problem 15.4-1) in terms of the independent **x** and **y** motions.

19
Angular Momentum in Spherical Coordinates

As an important example of introducing curvilinear coordinates, we shall now transform the *orbital* angular momentum, we shall now transform the orbital angular momentum and calculate **lvec = -I h rvec × grad** in spherical-polar coordinates **r**, **t** and **p**. These coordinates are discussed in Section 4.0, Appendix V, and depicted there in Figure 4.0-1. The polar angle **t** and the azimuthal angle **p** for **{t,0,Pi}** and **{p,0,2Pi}** map the unit sphere with radius **r = 1**.

We easily accomplish our task with the position vector and the gradient expressed in spherical coordinates. In terms of the spherical-polar unit vectors **er**, **et** and **ep**, we have from Section 4.4, Appendix V, that

```
grad[f_] =
    er Dt[f,r] + et/r Dt[f,t] + ep/(r Sin[t]) Dt[f,p];
```

In order to be able to refer the angular momentum to either spherical or Cartesian axes, we also need from Section 4.1, Appendix V, the replacement rules **eSrep** and **eCrep** for the unit vectors referred to spherical or Cartesian axes:

```
eSrep = {er -> {1,0,0}, et -> {0,1,0}, ep -> {0,0,1}};
```

```
eCrep =
    {er -> {Cos[p] Sin[t], Sin[p] Sin[t],  Cos[t]},
     et -> {Cos[p] Cos[t], Cos[t] Sin[p], -Sin[t]},
     ep -> {-Sin[p], Cos[p], 0}};
```

Following Section 18.3 and taking **rvec = r er**, we thus calculate the angular momentum *referred to the spherical axes* as

```
Needs["LinearAlgebra`CrossProduct`"]

lvecS @ psi_ = -I h Cross[r er, grad[psi]] /. eSrep

   {0, I h Csc[t] Dt[psi, p], -I h Dt[psi, t]}
```

In effect, we have substituted the gradient **gradS** referred to the spherical axes from Section 4.4, Appendix V. From this simple result, it's apparent that **lvec** commutes with any function of **r**, in particular with any central potential. Moreover, the absence of derivatives with respect to **r** is required by the fact that **lvecS/h** is the generator of infinitesimal rotations.

In quantum mechanics, we require the Cartesian components **lx**, **ly** and **lz**, even when working in spherical coordinates. We obtain them by substituting the gradient **gradC** referred to the Cartesian axes (from Section 4.4, Appendix V). Also, since our calculations here are trig intensive, it helps if we load our enhanced version of **TrigReduce** from Exercise 2.2, Appendix III. Thus,

```
Needs["Quantum`Trigonometry`"]

lvecC @ psi_ = -I h Cross[r er, grad[psi]] /. eCrep //
        TrigReduce //Expand;

{lx @ psi_ = lvecC[psi][[1]],
 ly @ psi_ = lvecC[psi][[2]],
 lz @ psi_ = lvecC[psi][[3]]}

  {I h Cos[p] Cot[t] Dt[psi, p] + I h Dt[psi, t] Sin[p],

   -I h Cos[p] Dt[psi, t] + I h Cot[t] Dt[psi, p] Sin[p],

   -I h Dt[psi, p]}
```

As a check, we can again calculate one of the basic commutation relations. Since we are using **Dt**, we need to declare the spherical coordinates to be independent as we did with the Cartesian coordinates in Section 18.1:

```
IndependentVariables[x_,y_,z_] :=
    Module[{},
        x/: Dt[x,y] := 0;    x/: Dt[x,z] := 0;
        y/: Dt[y,x] := 0;    y/: Dt[y,z] := 0;
        z/: Dt[z,x] := 0;    z/: Dt[z,y] := 0
    ]

IndependentVariables[r,t,p];        SetAttributes[h, Constant]
```

```
Commutator[A_,B_] @ psi_ := A @ B @ psi - B @ A @ psi //
    Expand

Commutator[lx,ly] @ psi == I h lz @ psi //TrigReduce
```

```
    True
```

The Cartesian components **lx**, **ly** and **lz** are required in order to calculate **lsq** correctly. The spherical unit vectors are themselves functions of the coordinates (actually angles) **t** and **p**, so we have to take care that we properly differentiate them when we calculate **lsq**, as we do in calculating **div** and **curl** in Section 4.5, Appendix V. In particular, we would not obtain the correct result if we used **lvecS**. We thus calculate **lsq** as the sum of the squares of the components and simplify:

```
lsq @ psi_ =
    lx @ lx @ psi + ly @ ly @ psi + lz @ lz @ psi //
        Expand //TrigReduce

       2                        2       2
  -(h   Cot[t] Dt[psi, t]) - h   Csc[t]   Dt[psi, {p, 2}] -

    2
   h   Dt[psi, {t, 2}]
```

It's apparent from this result that **lsq** also commutes with any function of **r**, as required since **lvec** does. As further proof of the appropriateness of spherical coordinates, you might compare how much simpler this result is than its Cartesian counterpart (cf. Exercise 18.3-1).

Finally, we construct the raising and lowering operators and put them into a conventional form involving complex exponentials **E^ (±I p)** with the function **TrigToComplex**, which is also defined in the package **Quantum`Trigonometry`**:

```
{lR @ psi_ = lx[psi] + I ly[psi] //.
                {Cos[p] :> TrigToComplex[Cos[p]],
                 Sin[p] :> TrigToComplex[Sin[p]]} //
                Expand //Factor,

 lL @ psi_ = lx[psi] - I ly[psi] //.
                {Cos[p] :> TrigToComplex[Cos[p]],
                 Sin[p] :> TrigToComplex[Sin[p]]} //
                Expand //Factor}

     I p
  {E     h (I Cot[t] Dt[psi, p] + Dt[psi, t]),

      -I p
  -(E     h (-I Cot[t] Dt[psi, p] + Dt[psi, t])))}
```

It's informative at this point to transform the kinetic energy operator **K** to spherical coordinates and relate it to **lsq**. We find in Exercise 19.0-2 that

```
K = -h^2 Dt[r^2 Dt[#,r], r]/(2m r^2) + lsq[#]/(2m r^2) &
```

$$-\left(\frac{h^2 \; Dt[r^2 \; Dt[\#1, \; r], \; r]}{2 \; m \; r^2}\right) + \frac{lsq[\#1]}{2 \; m \; r^2} \; \&$$

a result anticipated from the classical expression (see Exercise 15.2-1). Here we have defined a pure function without reference to a wavefunction to prevent for compactness the two contributions from evaluating (enter **K @ psi** and compare). The first term is referred to as the *radial* kinetic energy, while the second term involving **lsq** is referred to as the *centrifugal potential*, since its negative derivative with respect to **r** gives a centrifugal force which pushes the particle away from the origin for **lsq** $\neq$ **0**. One also refers to a *centrifugal barrier* about the origin. (See Exercise 19.1-3 and also Section 20.1 below).

In this form, **K** clearly commutes with **lsq**, since **lsq** is independent of **r** and hence commutes with derivatives with respect to **r**. Moreover, it's again apparent (see Exercise 18.3-1) that **K** commutes with **lvec** since **lvec** is also independent of **r** and commutes with **lsq**. We therefore conclude that both **lsq** and **lvec** commute with any central-force hamiltonian, as we did in Section 18.4.

Exercise 19.0-1

Redo Exercise 18.3-1, Part **(a)**. Also, compare the **t** derivatives in **lsq** with the more conventional and compact form **-h^2 Dt[Sin[t] Dt[psi,t],t]/Sin[t]**, which *Mathematica* automatically expands.

Exercise 19.0-2

Using the laplacian from Exercise 27, Appendix V, transform the kinetic energy operator **K** for a particle with mass **m** defined in Section 18.1 from Cartesian to spherical coordinates and verify the expression given in the text.

Exercise 19.0-3

Check the *2s* wavefunction we introduced in Problem 18.2-1 by showing that it is a solution of the Schrödinger equation with the *Balmer-Bohr* energy **-Z^2/(2 n^2)** for **n = 2**. (All quantities are expressed in *atomic units* with **h = m = e = 1** such that 1 a.u. of energy equals 27.2 eV and the ground state energy of hydrogen equals − 13.6 eV.) Refer to Problem 18.7-1 and to Sections 20.1 and 20.2 below.

Clearly, this spherically symmetric wavefunction has angular momentum quantum number **l = 0**. Hence, the label *s* from the conventional scheme **l = 0, 1, 2, ... => s, p, d, ...**

Problem 19.0-1

Transform the angular momentum to *parabolic coordinates* as defined in Problem 8, Appendix V. Derive and simplify its *Cartesian* components with **PowerContract** from the package **Quantum `PowerTools`** and show that **lvecCuvp @ psi ==**

```
                (-u + v) Cos[p] Dt[psi, p]
{-I h (---------------------------------- +
                    2 Sqrt[u v]


        Sqrt[u v] Dt[psi, u] Sin[p] -

     Sqrt[u v] Dt[psi, v] Sin[p]),

        -I h (-(Sqrt[u v] Cos[p] Dt[psi, u]) +

     Sqrt[u v] Cos[p] Dt[psi, v] +

        (-u + v) Dt[psi, p] Sin[p]
        -----------------------------), -I h Dt[psi, p]}
              2 Sqrt[u v]
```

(The notation **lvecCuvp** is defined in Section 2.6, Appendix V.) Calculate the square **lsq**, and the raising and lowering operators. Verify the basic angular momentum commutation relations and redo Exercise 18.3-1, Part **(a)**.

Problem 19.0-2

Transform the angular momentum to *elliptical coordinates* as defined in Problem 9, Appendix V, and redo the previous problem. Show that **lvecClmp @ psi ==**

```
              l m Cos[p] Dt[psi, p]
{-I h (-(---------------------------) -
                    2       2
          Sqrt[(-1 + l ) (1 - m )]


              2          2
    m Sqrt[(-1 + l ) (1 - m )] Dt[psi, l] Sin[p]
    ----------------------------------------------- +
                      2     2
                     l  - m

          2          2
  l Sqrt[(-1 + l ) (1 - m )] Dt[psi, m] Sin[p]
  ----------------------------------------------),
                    2     2
                   l  - m


                  2          2
        m Sqrt[(-1 + l ) (1 - m )] Cos[p] Dt[psi, l]
    -I h (----------------------------------------------- -
                            2     2
                           l  - m

          2          2
  l Sqrt[(-1 + l ) (1 - m )] Cos[p] Dt[psi, m]
  ------------------------------------------------ -
                    2     2
                   l  - m

    l m Dt[psi, p] Sin[p]
    ----------------------------), -I h Dt[psi, p]}
            2       2
      Sqrt[(-1 + l ) (1 - m )]
```

19.1 Spherical Harmonics

The built-in functions **SphericalHarmonicY** provide a common set of normalized eigenfunctions of **lsq** and the component **lz**. We conveniently introduce these eigenfunctions in terms of the angular coordinates **t** and **p** by entering

```
Y[l_,m_,t_,p_] := SphericalHarmonicY[l,m,t,p]
```

where the quantum numbers **l** and **m** determine the eigenvalues **h^2 l (l+1)** and **h m** of **lsq** and **lz**, respectively, with **Abs[m]** $\leq$ **l**. These functions are just the coordinate-space representation for *positive integer* **l** of the normalized angular-momentum eigenkets **ket[l,m]** from the formal representation we developed in Problem 16.1-1. (In effect, **Y[l,m,t,p] == bra[t,p] ** ket[l,m]**.) That spherical harmonics are defined for integer **l** only is a consequence of the requirement that the wavefunction be single-valued (see Problem 19.1-2 and also Exercise 20.2-2). However, Pauli also showed that **lvec** fails for half-integral **l** to be a hermitian operator (refer to Biedenharn and Louck [7]).

We relegate the derivation of these functions to the problems below and presently examine their properties instead.

For example, for **l = m = 1** we verify the eigenvalue equations

```
{lsq @ Y[1,1,t,p] == 2h^2 Y[1,1,t,p],
 lz  @ Y[1,1,t,p] ==  h   Y[1,1,t,p]} //TrigReduce
```

```
{True, True}
```

We easily demonstrate with the raising and lowering operators that the spherical harmonics vanish if **Abs[m]** $\geq$ **1**. For **l = 1**, we obtain for example

```
{1R @ Y[1,1,t,p], 1L @ Y[1,-1,t,p]}
```

```
{0, 0}
```

in analogy with our example from Section 16.2 in the matrix representation. Similarly, we can generate all of the spherical harmonics in the range **{m,-1,1}** by repeated application of **1R** and **1L**. Thus, starting with **Y[1,-1,t,p]**

```
Y[1,1,t,p] == Nest[1R, Y[1,-1,t,p], 2]/(2h^2)
```

```
True
```

For some applications it's useful to note that spherical harmonics for **m** negative are related to their complex conjugates for **m** positive. The connection is **Conjugate[Y[l,m,t,p]]== (-1)^m Y[l,-m,t,p]**. For example, for **l = 3** (using our symbolic **Conjugate** rule from Exercise 2.4, Appendix III, and the package **Quantum`QuickReIm`**)

```
Needs["Quantum`QuickReIm`"]

Y[3,-3,t,p] == (-1)^3 Conjugate[Y[3,3,t,p]]
```

    True

We demonstrate in Exercise 19.1-1 and Problem 19.1-2 that spherical harmonics are orthonormal with respect to the *weight function* `Sin[t]` for `{t,0,Pi}` and `{p,0,2Pi}`.

Spherical harmonics, when multiplied by certain functions of **r**, also define simultaneous eigenfunctions of the kinetic energy **K** and hence any central-force hamiltonian. The introduction of a product wavefunction of the form `f[r]Y[l,m,t,p]` leads therefore to a separation of variables. For example, with the kinetic energy we generate an equation for the *radial* wavefunction `f[r]` for `l = 2` with

```
(K @ (f[r] Y[2,1,t,p]))/Y[2,1,t,p] //Expand //TrigReduce

    2            2            2
 3 h  f[r]      h  f'[r]     h  f''[r]
 ──────────  -  ────────  -  ─────────
     2            m r          2 m
   m r
```

Clearly, the first term derives from the centrifugal potential `h^2 l(l+1)/(2m r^2)` for `l = 2`. We show in Exercise 19.1-3 that the free-particle radial wavefunction is a *spherical Bessel function* of order `l`.

If we want to plot spherical harmonics, we can use the function **SphericalPlot3D** from the package **Graphics`ParametricPlot3D`**. (Refer to Section 4.2, Appendix V.) We obtain interesting *3D* graphics with rich structure if `l` and the difference `l - m` are large. In Figure 19.1-1 we thus make a "cut-away" plot of `Abs[Y[5,1,t,p]]^2` (this takes about 5 minutes).

Further properties of the spherical harmonics and their relation to the angular momentum operators are developed in the following exercises and problems.

Exercise 19.1-1

Verify the orthonormality of the spherical harmonics by integrating explicitly with respect to the weight function `Sin[t]`. Verify similarly the matrices of `lsq`, `lz`, `lx` and `ly` for `l = 1, 2` derived in Problem 16.1-1 (see also Section 16.2 and Problem 16.4-1).

Exercise 19.1-2

In some applications it is useful to transform the spherical harmonics to Cartesian coordinates **x**, **y**, **z**. Set up replacement rules to do this for arbitrary `l` and `m` and generate the so-called *solid harmonics* `Y[l,m,x,y,z]` for which the point `rvec = {x,y,z}` is no longer restricted to the unit sphere. Show for example that `{Y[1,1,x,y,z], Y[1,0,x,y,z], Y[1,-1,x,y,z]} ==`

$$\left\{-\left(\frac{\mathrm{Sqrt}[\frac{3}{8\,\mathrm{Pi}}]\,(x + I\,y)}{r}\right),\ \frac{\mathrm{Sqrt}[\frac{3}{\mathrm{Pi}}]\,z}{2\,r},\ \frac{\mathrm{Sqrt}[\frac{3}{8\,\mathrm{Pi}}]\,(x - I\,y)}{r}\right\}$$

```
Needs["Graphics`ParametricPlot3D`"]
Y51sq = Conjugate[Y[5,1,t,p]] Y[5,1,t,p] //N ;

SphericalPlot3D[
    Evaluate[Y51sq],
    {t,0,Pi,Pi/60}, {p,0,3Pi/2,2Pi/15},
    Axes -> None, Boxed -> False, BoxRatios->{1,1,1},
    ViewPoint->{1.623, -2.730, 1.168}
];
```

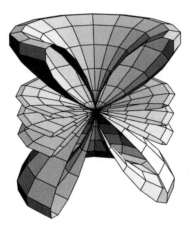

Figure 19.1-1. A cut-away spherical polar plot for {**p**, **0**, **3Pi/2**} of the absolute square of a spherical harmonic with **l = 5** and **m = 1**.

Verify for **l = 1** and **5** that these functions are eigenfunctions of **lsq** and **lz** expressed in Cartesian coordinates from Section 18.3. Finally, verify the properties of the raising and lowering operators expressed in Cartesian coordinates.

Exercise 19.1-3

Demonstrate for selected values of **l** that **j[l,k r]Y[l,m,t,p]** is an eigenfunction of the kinetic energy **K** with eigenenergy **(h k)^2/(2m)**, where **k** is the wavenumber (cf. Exercise 18.1-1) and **j[l,k r]** is a *spherical Bessel function* of order **l** defined by the built-in Bessel function **BesselJ** according to

```
j[l_,z_] := Sqrt[Pi/(2z)] BesselJ[l+1/2,z]
```

Simplify the resulting functions with **PowerExpand** and with **PowerContract** from the package **Quantum`PowerTools`** and show for example that {**j[0,z]**, **j[1,z],j[2,z]**} ==

$$
\left\{ \frac{\text{Sin}[z]}{z},\ -\left(\frac{\text{Cos}[z]}{z}\right) + \frac{\text{Sin}[z]}{z^2},\ \frac{-3\,\text{Cos}[z]}{z^2} + \left(\frac{3}{z^3} - \frac{1}{z}\right)\text{Sin}[z] \right\}
$$

One can define in fact an entire family of spherical Bessel functions in direct correspondence with the full family of ordinary Bessel functions. The function `j[l,z]` is the one member which is regular (nonsingular) at the origin `z = 0`. Thus, examine the series expansion `j[l,z]+O[z]^2` for a few values of `l`.

The function `j[l,k r]` describes the motion of a free-particle along the radial coordinate `r` for a particular value of the angular momentum quantum number. We have shown in Exercise 18.1-1 that the free-particle planewave `E^(I kvec.rvecC)` is also an eigenfunction of `K` with eigenenergy `(h k)^2/(2m)` and `k = Sqrt[kvec.kvec]`. The planewave, however, is not an eigenfunction of angular momentum, but rather can be shown to be a linear superposition of the functions `j[l,k r]Y[l,m,t,p]` (see e.g. Brink and Satchler [12], Section 4.10).

Take `h = m = 1` and make plots of `j[l,k r]` for selected values of `l` along with the centrifugal potential `l(l+1)/(2 r^2)`. Observe in particular how this function is pushed further and further away from the origin by the centrifugal barrier as `l` increases and how the wavelength varies with kinetic energy (cf. the discussion of the kinetic energy operator `K` at the end of the previous section). Shift each function by the corresponding energy `k^2/2` and compare the position of each function's inflection point closest to the origin with the potential barrier.

Problem 19.1-1

(a) Show that the built-in functions `SphericalHarmonicY[l,m,t,p]` for l = 2, 3 and {m, -1, 1} are eigenfunctions of `lsq` and `lz` with eigenvalues `h^2 l(l+1)` and `h m`, respectively. That `l` is an integer is the subject of Problem 19.1-2.

(b) Generate `SphericalHarmonicY[l,m,t,p]` explicitly for l = 2, 3 and {m, -1, 1} by repeated application of the raising and lowering operators `lR and lL`.

(c) For fixed `r = ro`, `K` is the hamiltonian of a *rigid rotator* (see Exercise 15.2-1), which gives a pretty good approximation of a rotating *diatomic molecule*. Use it to estimate the first three energy levels of CO given `ro = 1.13 Å`. Give your answers in eV. Compute the wavelengths of light emitted in a l = **2**-to-**1** transition and a l = **1**-to-**0** transition.

Problem 19.1-2

Construct the spherical harmonics by solving the eigenvalue problems of `lsq` and `lz` as differential equations in the coordinate representation. Prove that the orbital angular momentum quantum numbers `l` and `m` are integers. Refer to the contruction of the HO and Morse eigenfunctions in Chapters 6 and 13, and also the solution of the hydrogen-atom Schrödinger equation in Chapter 20 below.

(a) Begin by integrating the *1st-order* differential equation `lz @ f[p]==h m f[p]` for the eigenfunction `f[p]`. Show that `f[p] = E^(I m p)`. In keeping with the fundamental postulates of quantum mechanics, demand that the wavefunction be a *single-valued* function of the particles position. Show that the periodicity condition `f[p + 2Pi]== f[p]` restricts m to be an integer, m = 0, ±1, ±2, …

We thus conclude that the *orbital* angular momentum quantum number `l` is also an integer, since `Abs[m] ≤ l` from the general results of Problem 16.1-1. We reach the same conclusion, however, in Part **(b)** by requiring that the wavefunction be everywhere finite.

(b) Show that the eigenvalue problem `lsq @ y[t,p] == h^2 l(l+1) y[t,p]` can be transformed by the substitution `y[t,p] = P[Cos[t]] f[p]` into the *generalized Legendre* differential equation

```
D[(1-z^2) P'[z],z] - m^2/(1-z^2) P[z] == -l(l+1) P[z]
```

in the variable $z = \text{Cos}[t]$. Hence, we conclude that `P[z]` is a linear combination of the two independent solutions `LegendreP[l,m,z]` and `LegendreQ[l,m,z]` known as *associated Legendre functions*, which are built-in *Mathematica* functions.

Note that `{z,-1,1}` when `{t,0,Pi}`. Since the above version of the Legendre equation is already in Sturm-Liouville form, it follows from Problem 6.4-1 that solutions `P[z]` belonging to different eigenvalues `l` are orthogonal on the interval `{z,-1,1}` with respect to the weight function `w = 1`. Since `Dt[z] == -Sin[t]Dt[t]`, it follows that the functions `P[Cos[t]]` on the interval `{t,0,Pi}` are orthogonal with respect to `Sin[t]`. This is the origin of the weight function `Sin[t]` for the spherical harmonics.

Use **Series** to show that `LegendreQ` has in general singularities at $z = \pm 1$ ($t = 0$ or `Pi`) and must therefore be rejected if `y[t,p]` is to represent a probability amplitude. Show that `LegendreP` has singularities at $z = \pm 1$ unless `l` is an integer, in which case `LegendreP` is an associated Legendre polynomial. These functions actually involve powers of `Sqrt[1-z^2]` times polynomials, although for `m = 0` they're just the Legendre polynomials `P[l,z]`.

(c) Normalize the functions `y[l,m,t,p] = norm[l,m] LegendreP[l,m,Cos[t]] E^(I m p)`, with `norm[l,m]` the normalization constant. Do this for arbitrary `l` and `m` using the general result

```
Integrate[LegendreP[l,m,z]^2, {z,-1,1}]  ==
     (l+m)!/((l-m)!(l+1/2))
```

Compare your results with `SphericalHarmonicY[l,m,t,p]`. Check that `y[l,m]` satisfies `y[l,-m] == (-1)^m Conjugate[y[l,m]]`. What do you conclude about the normalization and phase of `LegendreP[l,m]` built into *Mathematica*?

Problem 19.1-3

Calculate the spherical harmonics from the general properties of the eigenkets `ket[l,m]` of `lsq` and `lz` developed in Problem 16.1-1.

Thus, solve `lR ** ket[l,l] == 0` as a *1st-order* differential equation in the coordinate representation. Determine in this way the "top rung in the ladder" `y[l,l,t,p]` for `m = l`. Use the eigenfunction of `lz`, `E^(I m p)`, from the previous problem and refer to Problem 6.7-1 and Exercise 11.9-2. Normalize your result and show for *arbitrary l* that `y[l,l,t,p] ==`

$$\frac{(-1)^l\ 2^{-1-l}\ E^{I\,l\,p}\ \text{Sqrt}\left[\dfrac{(1 + 2\,l)!}{\text{Pi}}\right]\ \text{Sin}[t]^l}{l!}$$

Here, we have included the phase `(-1)^l` in order to obtain agreement with the built-in functions `SphericalHarmonicY` (see the end of the previous problem). This analytical result for the top rung is often useful.

As we have seen, we can generate all the spherical harmonics in the range `{m,-l,l}` by applying the lowering operator `lL` to `y[l,l,t,p]` repeatedly.

New Axis of Quantization   19.2

We can construct eigenfunctions of **lsq** and another component of **lvec**, say **ly**, in analogy with our construction of the eigenkets **kety[l,my]** in the matrix representation in Section 16.3. We simply form linear combinations of spherical harmonics **Y[l,m,t,p]** with expansion coefficients given by the components of an eigenvector **ey[l,my]** of **ly** in the matrix representation and therefore by elements of the rotation matrix **Dy[l]**. The result, of course, is a new representation in which **lsq** and **ly** are diagonal, instead of **lsq** and **lz**.

We set up in this way the coordinate-space representation **Yy[l,my,t,p]** of the eigenket **kety[l,my]** we constructed in the matrix representation. (Formally, **Yy[l,my,t,p]==bra[t,p]**kety[l,my]** whereas **Y[l, m,t,p]==bra[t,p]**ket[l,m]**.)

For variety, let's introduce the results of Exercise 16.4-1 and examine how this works for a rotation which diagonalizes **lx**. We thus consider **l = 1** and enter the matrix

```
Dx[1] = {{1/2,        -1/Sqrt[2],   1/2      },
         {1/Sqrt[2],   0,          -1/Sqrt[2]},
         {1/2,         1/Sqrt[2],   1/2      }};
```

In analogy with **Dy[l,m,my]** and **kety[l,my]** from Section 16.3, we define expansion coefficients and form the linear combination **Yx[l,mx, t,p]** according to

```
Dx[l_,m_,mx_]      := Dx[l][[-m+l+1,-mx+l+1]]
Yx[l_,mx_,t_,p_] := Sum[ Y[l,m,t,p] Dx[l,m,mx], {m,-1,1}]
```

It is relatively straightforward to check that this superposition is an eigenfunction of **lx** (this takes about 90 sec):

```
Table[
    lx @ Yx[1,mx,t,p]  == h mx Yx[1,mx,t,p],
    {mx,-1,1}
] //. {Cos[p] -> TrigToComplex[Cos[p]],
       Sin[p] -> TrigToComplex[Sin[p]]} //ExpandAll

   {True, True, True}
```

Exercise 19.2-1

Verify that **Yx[1,m,t,p]** is normalized and also an eigenfunction of **lsq**. Show, however, that it's no longer an eigenfunction of **lz**, as required by the commutation relations. (Compare with Exercise 16.3-2.) Finally, construct raising and lowering operators appropriate to the new axis of quantization and show that they have the desired properties when acting on **Yx[1,m,t,p]**.

In effect, we have rotated the axis of quantization. We can demonstrate this explicitly by showing that **Yx** will transform into a *single* spherical harmonic, albeit as a function of new spherical coordinates defined relative to a new z' axis along the original x axis. The transformation to new axes $x'y'z'$ is conveniently generated by a rotation through Euler angles. It is appropriate here to use the y convention from Section 4.3, Appendix V, such that the rotation from the z to the x axis is given by the Euler angles **ay -> 0, by -> Pi/2** and **gy -> 0**. Recalling the rotation matrix in the y convention, we thus have

```
Needs["Geometry`Rotations`"]

Ry[ay_,by_,gy_] =
    RotationMatrix3D[ay+Pi/2,by,gy-Pi/2] //TrigReduce;
Ry[0,Pi/2,0] //MatrixForm
```

```
    0    0   -1

    0    1    0

    1    0    0
```

We see that this matrix has the desired form and rotates for example the Cartesian unit vector along the original x axis to the new z' axis:

```
{0,0,1} == Ry[0,Pi/2,0].{1,0,0}
```

```
   True
```

The choice of the other two Euler angles **ay -> 0** and **gy -> 0** is not unique in this regard, rather it determines the phase of the rotated wavefunction **Yx**.

Let **b** and **a** be the new polar and azimuthal angles such that the unit position vector **er** referred to the new primed axes has components (see Figure 19.2-1)

```
erPba = {Sin[b] Cos[a], Sin[b] Sin[a], Cos[b]};
```

We then set up an effective set of transformation rules for the coordinates **{t,p} -> {b,a}** by equating, via a rotation, components of **er** referred to the old and the new axes (which can also be read off the figure). Thus, introducing our replacements **eCrep** for the unit vectors referred to the Cartesian axes from Section 19.0, we have that

```
erPba == Ry[0,Pi/2,0].(er /. eCrep)
```

```
   {Cos[a] Sin[b], Sin[a] Sin[b], Cos[b]} ==
      {-Cos[t], Sin[p] Sin[t], Cos[p] Sin[t]}
```

We obtain the transformation rules by solving for **{Cos[t], Sin[p], Cos[p]}**:

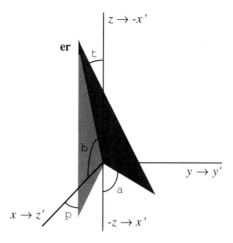

Figure 19.2-1. Polar and azimuthal angles of the unit position vector relative to the old (unprimed) and the new (primed) axes.

```
CtoP = Solve[%, {Cos[t],Sin[p],Cos[p]}] [[1]]

  {Cos[t] -> -(Cos[a] Sin[b]), Sin[p] -> Csc[t] Sin[a] Sin[b],

   Cos[p] -> Cos[b] Csc[t]}
```

Making these substitutions in **Yx[1,mx,t,p]** and simplifying, we obtain

```
ComplexToTrig[
    {Yx[1,-1,t,p], Yx[1,0,t,p], Yx[1,1,t,p]}
] //. CtoP //.
   {Cos[a] -> TrigToComplex[Cos[a]],
    Sin[a] -> TrigToComplex[Sin[a]]} //Expand
```

$$
\left\{ E^{-Ia}\ \mathrm{Sqrt}\!\left[\frac{3}{8\ Pi}\right]\ \mathrm{Sin}[b],\quad \frac{\mathrm{Sqrt}\!\left[\dfrac{3}{Pi}\right]\ \mathrm{Cos}[b]}{2},\right.
$$

$$
\left. -\!\left(E^{Ia}\ \mathrm{Sqrt}\!\left[\frac{3}{8\ Pi}\right]\ \mathrm{Sin}[b]\right)\right\}
$$

which as desired are just spherical harmonics *in the rotated frame*:

```
{Y[1,-1,b,a], Y[1,0,b,a], Y[1,1,b,a]}
```

$$
\left\{ E^{-Ia}\ \mathrm{Sqrt}\!\left[\frac{3}{8\ Pi}\right]\ \mathrm{Sin}[b],\quad \frac{\mathrm{Sqrt}\!\left[\dfrac{3}{Pi}\right]\ \mathrm{Cos}[b]}{2},\right.
$$

$$
\left. -\!\left(E^{Ia}\ \mathrm{Sqrt}\!\left[\frac{3}{8\ Pi}\right]\ \mathrm{Sin}[b]\right)\right\}
$$

This result shows that **Abs[Yx]** has cylindrical symmetry about the z' axis just as **Abs[Y]** has cylindrical symmetry about the original z axis. Of course, expressed in terms of the new coordinates **b** and **a**, the operators **lx** and **lsq** must have the same form as **lz** and **lsq** in the original frame but with **t** and **p** replaced by **b** and **a**, e.g. **lx -> -I h Dt[#,a]&**.

19.3 Quantum Rotation Matrix

For completeness, but by no means thoroughness, we point out that eigen-functions of **lsq** and a component of **lvec** along *any* direction can be constructed directly from a quantum analogue of the geometric rotation matrix **RotationMatrix3D**. According to the general principles of quantum mechanics, there exists a *unitary rotation operator* **E^(-I p nvec.jvec/h)** which relates states of a system under a finite rotation about the axis **nvec** by the angle **p**. In the case of orbital angular momentum, this operator can be obtained (see Schiff [59], p. 200) by integration of the change **d[f] == -I dp nvec.lvec/h f** in a function **f** under an infinitesimal rotation, which we derived in Section 18.4.

Among their many useful properties, the matrix elements of the rotation operator in the angular momentum representation give the coefficients of a linear superposition of eigenkets **ket[j,m]** of **jz** which is an eigenket of the component of **jvec** along **nvec**. Our package **Quantum`QuantumRotations`** defines the function **QuantumRotationMatrix** using a closed-form sum due to Wigner to compute the rotation matrix elements parametrized in terms of Euler angles **ay**, **by** and **gy** in the y convention. (Refer to Brink and Satchler [12], p. 22, and also to the notes and references in the package.)

For a given **j**, we thus define the **2j+1**-dimensional quantum counterpart of **Ry[a,b,g]** (formally, **DP[j,m,n,a,b,g] == bra[j,m] ** E^(-I p nvec.jvec/h) ** ket[j,n]**):

```
Needs["Quantum`QuantumRotations`"]

DP[j_,m_,n_,a_,b_,g_] :=
    QuantumRotationMatrix[j,m,n,a,b,g]
DP[j_,a_,b_,g_] := DP[j,a,b,g] =
    Table[ DP[j,-m,-n,a,b,g], {m,-j,j}, {n,-j,j} ];
```

The matrix **Dy[j]** from Section 16.3 and the matrix **Dx[j]** used in the previous section are just special cases of **DP[j]** with **nvec** referred to the y and the x axes, respectively. For example, with the particular choice of Euler angles **ay -> 0**, **by -> Pi/2** and **gy -> 0** from the previous section, we recover **Dx[j]**. That is, for **j = 1**

```
Dx[1]  == DP[1,0,Pi/2,0]
```

```
    True
```

However, `DP[1,0,Pi/2,0]` also allows us to extend our definition of the eigenfunctions `Yx` of `lsq` and `lx` in the previous section to *arbitrary* (integer) `l`. We simply set

```
Dx[l_,m_,mx_]      := DP[l,m,mx,0,Pi/2,0]
Yx[l_,mx_,t_,p_] := Sum[Y[l,m,t,p] Dx[l,m,mx], {m,-1,1}]
```

In addition, **QuantumRotationMatrix** is defined for *half-integer* **j** (spin angular momentum) and can be used to rotate for example *spinors* (defined in Problem 16.4-4).

Exercise 19.3-1

Determine the Euler angles for which `DP[j]` is equivalent to `Dy[j]`.

As with `Dy[j]` and `Dx[j]`, `DP[j]` is a unitary matrix. For example, for `j = 3/2` we have the *unitarity relation*

```
Needs["Quantum`QuickReIm`"]

Transpose[Conjugate[DP[3/2,a,b,g]]].DP[3/2,a,b,g] //
        Expand //TrigReduce //Expand //MatrixForm
```

```
1   0   0   0
0   1   0   0
0   0   1   0
0   0   0   1
```

This property ensures that the rotated wavefunctions are normalized when the initial wavefunctions are. It also means that the `Y[l,m,t,p]` can be expressed as linear combinations of the `Yx[l,mx,t,p]`. As with **Ry** (see Exercise 21, Appendix V), the inverse of **DP** is also obtained by reversing the angles of rotation and their order of application:

```
DP[3/2,-g,-b,-a].DP[3/2,a,b,g] //
        Expand //TrigReduce //Expand //MatrixForm
```

```
1   0   0   0
0   1   0   0
0   0   1   0
0   0   0   1
```

Comparing these last two results, we may conclude that

$$Conjugate[DP[j,n,m,a,b,g]] == DP[j,m,n,-g,-b,-a] \qquad (19.3\text{-r1})$$

In particular, note that the order of the indices **m** and **n** matters in general, and you have to pay attention to which index you sum over when constructing wavefunctions.

The rotation matrix elements themselves can serve as wavefunctions in certain applications. We note in this connection that when say **n = 0**, they're independent of **g** and for **j** equal to an integer **l** reduce to spherical harmonics according to

$$DP[l,m,0,a,b,g] == Sqrt[4Pi/(2l+1)] \ Conjugate[Y[l,m,b,a]]$$

(19.3-r2)

which you can easily check.

Finally, we remark that the rotation matrices for *half-integer* **j** have the curious property that for a rotation of **2Pi** about any axis, we do not obtain the identity matrix. (That is, one might think that a rotation by **2Pi** should be the same as no rotation at all.) Consider, for example, **j = 3/2** and a **2Pi** rotation about the *y* axis:

```
DP[3/2,0,2Pi,0] //MatrixForm
  -1    0    0    0
   0   -1    0    0
   0    0   -1    0
   0    0    0   -1
```

If we perform, however, a rotation by **4Pi**, we restore the identity matrix:

```
DP[3/2,0,4Pi,0] //MatrixForm
   1    0    0    0
   0    1    0    0
   0    0    1    0
   0    0    0    1
```

Note that for integer **j** we obtain the identity matrix with a **2Pi** rotation. For example, for **j = 2**

```
DP[2,0,2Pi,0] //MatrixForm
   1    0    0    0    0
   0    1    0    0    0
   0    0    1    0    0
   0    0    0    1    0
   0    0    0    0    1
```

We can visualize these results by plotting an element of **DP[3/2]** as a function of **b**, as in Figure 19.3-1. We thus see that **DP** is periodic in **b** but with period **4Pi**, which is a general result when **j** is half integer. If we think of **DP** as a function of the vector which specifies the magnitude and direction of rotation (refer to Figure 19.3-1), we conclude that it's a *double-valued* function.

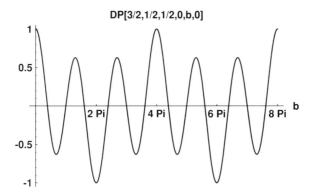

Figure 19.3-1.   An element of **DP[3/2]** with Euler angles **a = g = 0**. In order to obtain an accurate plot, the option **PlotPoints -> 30** was used.

Hence, eigenvectors or eigenkets representing half-integer spin systems change sign when the system is rotated by **2Pi**. We don't need to dwell on this result, however, since eigenkets aren't usually measurable (see, however, Aharonov and Susskind [2]). A quantity that is measurable, like an expectation value, doesn't change sign since it depends on the product of two quantities which do: a *ket* and its dual *bra*, i.e. the square of the wavefunction. Nevertheless, this is more than a mathematical curiosity and is related to the special properties of the rotation group *SO(3)* and its connection to the group *SU(2)* which involves quantum mechanics.

Analogous effects can also be observed on a macroscopic classical level. Hold for example your favorite coffee mug in the palm of your hand with your arm extended elbow down in front of you. Keeping the mug upright, rotate it towards you and under your arm and back out away from you to its original position. (You should be able to do this without spilling any coffee, but you might stay clear of your computer!) You will notice by watching the handle that the mug has turned through 360º (about the vertical). Your arm, however, is twisted palm and elbow up. To untwist it, continue rotating the mug upright *in the same direction* through another 360º. Thus, after 720º the cup and your arm return to their original positions. Hence, although a **2Pi** rotation restores an object's orientation, it doesn't necessarily disentangle it from its surroundings. Only after a **4Pi** rotation are both the orientation and *entanglement* of the object restored. This peculiar geometrical effect can be represented by the transformation of two-component spinors.

The interested reader is referred to the discussions and references in Misner, Thorne, and Wheeler [46], p. 1148, and to *Dirac's construction* in Biedenharn and Louck [7], p. 10.

Exercise 19.3-2

Using **QuantumRotationMatrix**, generate matrices **Dx** and **Dy** for **j = 1/2** which define a new z' axis along the x and y axes, respectively. Show that they diagonalize the Pauli spin matrices **sx** and **sy**, respectively, defined in Problem 16.1-1. Show in each case that the unitary transformation preserves the basic commutation relations, i.e. that the rotated angular-momentum matrices satisfy **[jxP, jyP]== I h jzP** and cyclic permutations.

Exercise 19.3-3

Verify the unitarity relation by summing explicitly over the components of **DP[3/2]**. Also, express the spherical harmonics **Y[l,m,t,p]** as a linear combination of the **Yx[l,mx,t,p]** and verify your results for **l = 1, 2**.

Exercise 19.3-4

(a) Use our expression in the text for **Yx** for arbitrary **l** to derive the *addition theorem* for the spherical harmonics:

```
LegendreP[l,Cos[b]] == 4Pi/(2l+1) *
     Sum[Y[l,m,t1,p1] Conjugate[Y[l,m,t2,p2]], {m,-l,l}]
```

(Hint: Use the relation between the spherical harmonics and the Legendre polynomials for **m = 0** (refer to Problem 19.1-2).) Here we interpret **t1, p1** and **t2, p2** as the spherical polar angles of two unit vectors **n1** and **n2**, respectively, (previously **er** and the *z'* axis) relative to the *xyz* axes such that **b** is the angle between **n1** and **n2**.

(b) Use this theorem to derive the familiar identity

```
Cos[b] == Cos[t1] Cos[t2] + Sin[t1] Sin[t2] Cos[p1-p2]
```

Exercise 19.3-5

We can explicitly relate **QuantumRotationMatrix** and **RotationMatrix3D** for **l = 1**. However, whereas **RotationMatrix3D** transforms the Cartesian components **{ax, ay, az}** of a vector, **QuantumRotationMatrix** for **l = 1** transforms its *spherical tensor* components defined by (see Exercise 17.4-1) **{a[1], a[0], a[-1]}** ==

$$\left\{ -\left(\frac{ax + I\ ay}{Sqrt[2]} \right),\ az,\ \frac{ax - I\ ay}{Sqrt[2]} \right\}$$

A vector operator in this form is referred to as an *irreducible tensor operator of rank 1*.

(a) Thus, show that the spherical tensor components **a[m]** have the same form as the solid harmonics **Y[l=1,m,x,y,z]** from Exercise 19.1-2.

(b) Show that the unitary matrix **A ==**

$$\begin{pmatrix} -\left(\dfrac{1}{Sqrt[2]} \right) & \dfrac{-I}{Sqrt[2]} & 0 \\ 0 & 0 & 1 \\ \dfrac{1}{Sqrt[2]} & \dfrac{-I}{Sqrt[2]} & 0 \end{pmatrix}$$

will transform the Cartesian components **{ax, ay, az}** of an arbitrary vector into its spherical tensor components and back again. Use **A** to generate **{x, y, z}** from **Y[l=1,m,x,y,z]**.

(c) Finally, show that **A** transforms **Transpose[QuantumRotationMatrix[l=1]]** into **RotationMatrix3D** in the *y* convention for an arbitrary set of Euler angles.

Problem 19.3-1

(a) Construct for l = 2 eigenfunctions **Yx** of **lsq** and **lx** and then eigenfunctions **Yy** of **lsq** and **ly** as linear combinations of spherical harmonics. Do this by diagonalizing the matrices **lx** and **ly** and then using **QuantumRotationMatrix**.

(b) Generate plots of **Abs[Y[l,m,t,p]]**, **Re[Y[l,m,t,p]]** and **Im[Y[l,m,t,p]]** for l = 2 and 3 using **SphericalPlot3D** from the package **Graphics `Paramet-ricPlot3D`**. Make similar plots of **Yx** and **Yy** and notice the effects of the change in the axis of quantization. In particular, compare the rotational symmetries of **Abs[Y]**, **Abs[Yx]** and **Abs[Yy]**. Explain.

Problem 19.3-2

Construct coordinate-space wavefunctions of **lsq** and the component of **lvec** along the axis defined in Problem 16.4-3 for l = 2. Use the eigenvectors from Problem 16.4-3 and then **QuantumRotationMatrix**.

20
Hydrogen Atom Schrödinger Equation

We turn now to the calculation of one-electron Coulomb wavefunctions. Although we shall do this shortly using the Runge-Lenz algebra, we examine first briefly the conventional approach and construct solutions of the Schrödinger equation as a differential equation in spherical polar coordinates. We construct the solutions in *parabolic* coordinates in Problem 20.5-1. The spherical and parabolic solutions are naturally connected via the Runge-Lenz algebra. In fact, both are needed to complete the story of the underlying dynamical symmetry. To that end, we shall also establish the connection with eigenfunctions of the *2D* harmonic oscillator in Problem 20.5-4.

States of the hydrogen atom are generally classified according to the solutions in spherical coordinates because they provide the simplest scheme for describing *radiative transitions*, i.e. the emission and absorbtion of photons. Of course, the *spectroscopy* of the hydrogen atom was historically one of the guiding lights in the development of quantum mechanics. The solution in parabolic coordinates provides an appropriate starting point for handling processes in which the atom's spherical symmetry is broken by an external disturbance which defines a *preferred direction* in space (e.g. the z axis). Familiar examples are the response of the atom to a uniform external electric field (*Stark effect*), the scattering of photons (*Compton effect*) and of electrons and *photoionization*.

You might note that all aspects of this problem (except the Runge-Lenz algebra) are discussed in the classic text of Bethe and Salpeter [6].

20.1 Separation in Spherical Coordinates

We locate the electron with the spherical polar coordinates **r**, **t** and **p** we introduced in Section 19.0 relative to an origin located at the nucleus **Z e**. We seek solutions of the Schrödinger equation in these coordinates which are simultaneous eigenfunctions of **lsq** and **lz** and therefore begin with wavefunctions proportional to spherical harmonics

```
psi[r_,t_,p_] := R[r] Y[l,mz,t,p]
Y[l_,mz_,t_,p_] := SphericalHarmonicY[l,mz,t,p]
```

where the radial function **R[r]** is to be determined. (We use **mz** for the projection quantum number to distinguish it from the mass **m**). This *ansatz* allows us to replace the operator **lsq** in the centrifugal potential defined at the end of Section 19.0 by its eigenvalue **h^2 l(l+1)** and thus define a *radial hamiltonian* as

```
H = -h^2/(2m) D[r^2 D[#,r],r]/r^2 + h^2 l(l+1)/(2m r^2) # -
       Z e^2/r # &
```

$$-\left(\frac{h^2\; D[r^2\; D[\#1,\; r],\; r]}{2\; m\; r^2}\right) + \frac{h^2\; l\;(l+1)\;\#1}{2\; m\; r^2} - \frac{Z\; e^2\;\#1}{r}\; \&$$

(Here we express **H** using *partial* derivatives, since we won't need the generality of **Dt**.) As in Section 18.6, the mass **m** is the electron-nucleus reduced mass and thus simply the electron mass if the nuclear mass is assumed infinite. In any case, the nucleus is at the coordinate origin **r = 0**.

In effect, we have separated variables and reduced the Schrödinger equation to a *1D* equation in the radial wavefunction **R[r]**. The angular momentum eigenvalue **h^2 l(l+1)** can be identified with the *separation constant*. Thus, operating on **psi** with **H - Energy**, we obtain

```
(H @ psi[r,t,p] - Energy psi[r,t,p])/Y[l,mz,t,p] == 0 //
    ExpandAll[#,r]&
```

$$-(Energy\; R[r]) + \frac{h^2\; l\;(l+1)\; R[r]}{2\; m\; r^2} - \frac{e^2\; Z\; R[r]}{r} -$$

$$\frac{h^2\; R'[r]}{m\; r} - \frac{h^2\; R''[r]}{2\; m} == 0$$

(Here **ExpandAll[#,r]&** expands terms which depend on **r** and avoids factors independent of **r**.) Because of the spherical symmetry, this result is independent of the projection quantum number **mz** and therefore of the orientation of the z axis.

The centrifugal and Coulomb potentials together define an *effective potential*

```
Veff[l_,r_] :=  -Z e^2/r + h^2 l(l+1)/(2m r^2)
```

In Figure 20.1-1, this function is plotted for $l = 1$ in dimensionless form as a function of `r` and as a function of the Cartesian coordinates `x = r Sin[t]` and `z = r Cos[t]`. It is evident that the centrifugal potential, which is everywhere positive, dominates **Veff** for small `r` but quickly falls off for large `r` relative to the Coulomb tail, which is everywhere negative. Thus, since $l > 0$, **Veff** has a repulsive *centrifugal barrier* for small `r` and hence a finite minimum for intermediate `r`. This minimum remains negative and, as we shall see, supports bound states for all `l`.

These figures help us to visualize how classically the effective potential supports bounded orbits (elliptical in the orbital plane) for **Energy < 0** by providing inner and outer turning points in `r` defined by **Energy == Veff**. For **Energy > 0**, we see that the orbits are unbounded (hyperbolic in the orbital plane) with only one inner turning point at the centrifugal barrier. For **Energy = 0**, the orbits are also unbounded (but parabolic in the orbital plane). It is interesting to note that classically the angular momentum is nonvanishing and therefore `r` as well, whereas quantum mechanically `l` can equal zero and the electron can pass through the origin.

For convenience and appropriate to hydrogenic atoms, we express distances in units of **ao/Z** and energies in units of **Z^2 Ry**. Here **ao = h^2/(m e^2)** = 0.529 Å is the *Bohr radius* and **Ry = e^2/(2ao)** = 13.6 eV the *Rydberg constant*, quantities we introduced in Problem 18.7-1. We can define these constants with a set of replacement rules

```
aoRy := {ao :> h^2/(m e^2), Ry :> e^2/(2ao)};
```

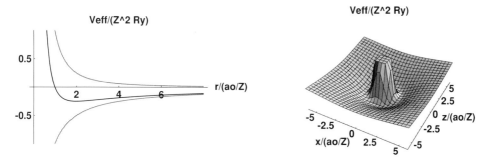

Figure 20.1-1.   Shown on the left is the effective radial potential **Veff** for $l = 1$ as a function of `r` along with the centrifugal potential (positive gray curve) and the Coulomb potential (negative gray curve). Shown on the right is the same potential as a function of the Cartesian coordinates `x = r Sin[t]` and `z = r Cos[t]`. Here **Ry** is the *Rydberg* and **ao** the *Bohr radius* (refer to the discussion on atomic units in the text).

When used to scale coordinates and energies, as in Figure 20.1-1, these quantities are closely related to *atomic units* in which **ao** is the unit of length and **e^2/ao = 2 Ry = 27.2 eV** is the unit of energy. Thus, to convert to atomic units we simply set **h = m = e = 1** such that **ao == 1** and **Ry == 1/2**.

Exercise 20.1-1

Generate the plots in Figure 20.1-1 by scaling length in units of **ao/Z** and **Veff** in units of **Z^2 Ry**. Make a family of plots of **Veff** as a function of **r** for different values of **l**.

20.2 Bound-State Wavefunctions

We will first construct the negative-energy **Energy < 0** wavefunctions in analogy with our solutions of the harmonic oscillator and Morse problems in Chapters 6 and 13. It is easy to verify that the **1/r^2** centrifugal terms are eliminated from the radial equation if we take **R[r]** proportional to **r^l**. This in fact determines the wavefunction for small **r**, since the **1/r^2** terms dominate near the origin. We thus see that the radial wavefunctions vanish at the origin if **l > 0**, i.e. under the centrifugal barrier. Moreover, the solution of the *asymptotic* radial equation for large **r** is easily seen to be **R[r] -> E^(-Z r/(ao n))** if we express the energy as **e[n] = -Z^2 Ry/n^2**, where **n** is a number to be determined (refer especially to Sections 6.2 and 13.1). These facts suggest we seek solutions of the form (see Exercise 20.2-1)

```
R[r_] := r^l E^(-Z r/(ao n)) w[2 Z r/(ao n)]
```

and transform the independent variable to **y = 2 Z r/(ao n)**. Thus, operating with **H - e[n]**

```
H @ R[r] + Z^2 Ry/n^2 R[r];
```

and changing variables and multiplying by certain factors to simplify, we obtain a *second-order* differential operator for the function **w[y]**:

```
(%/R[r] /.r -> n ao/Z y/2) y w[y]/(-4 Z^2 Ry/n^2) //.aoRy //
    Expand //Collect[#,{w[y],w'[y]}]&

(-1 - l + n) w[y] + (2 + 2 l - y) w'[y] + y w''[y]
```

When equated to zero, we recognize this to be *Kummer's* differential equation from Section 6.5 with solution the built-in confluent hypergeometric function **Hypergeometric1F1[a, c, y]**. Since this function equals unity at the origin and is therefore regular (nonsingular) there, it would appear to be a suitable candidate for the wavefunction, and we tentatively equate it to **w[y]**:

```
w1F1[n_,l_,y_] := Hypergeometric1F1[1+l-n,2l+2,y]
```

In particular, we reject the linearly independent second solution from Section 6.5 **HypergeometricU[1+l-n,2l+2,y]** which behaves as **y^(-1-2l)** for small **y** and is therefore too singular to represent the wavefunction for all **r** and **l** $\geq$ 0. That is, **R[r]** would behave as **r^(-1-l)** and therefore be too singular near the origin (see Exercise 20.2-1).

It follows, however, from Problems 6.5-2 and 6.5-4 that **w1F1[n,l,y]** diverges as **E^y** for large **y** and thus **R[r]** as **E^(+Z r/(n ao))** for large **r** unless **1+l-n** is a negative integer or zero. Hence, we require **n** to be an integer since **l** is, and **w1F1[n,l,y]** truncates to a polynomial of order **n-l-1**. Moreover, **l** is bounded for a given **n** by **l** $\leq$ **n-1**. We therefore conclude that **n** $\geq$ 1 since **l** $\geq$ 0. Thus, we recover the Balmer-Bohr energy spectrum and confirm the results of Problem 18.7-1. The integer **n** is known as the *principal quantum number*.

That the energy is independent of **mz** is due to space *isotropy:* a reorientation of the *z* axis does not affect the energy. This degeneracy holds for any system in the absence of external fields and is called *directional degeneracy*. That the energy is independent of **l** is a property of the Coulomb potential and is sometimes referred to as *accidental degeneracy*. As we have seen in Section 18.7, however, it's due to the special symmetry of the Coulomb interaction which supports another constant of the motion, the Runge-Lenz operator. This vector precesses about the origin if the symmetry is broken by an external disturbance with the consequence that the hydrogen atom is particularly influenced by external fields. (See, however, the discussion of the Stark effect in Section 21.5 below.)

For **a = -q**, a negative integer, **Hypergeometric1F1[-q,c,y]** is a polynomial of order **q**. In fact, for the **q** and **c** values defined here, these functions are proportional to the built-in *associated Laguerre polynomials* **LaguerreL[q,c-1,y]**, which we introduced in Exercise 13.4-2 in connection with the Morse potential. Therefore, the radial functions **w[y]** can also be expressed (to within a constant) as

```
wL[n_,l_,y_] := LaguerreL[n-l-1,2l+1,y]
```

It is easy to verify that **n-l-1** gives the number of nodes in **wL** and hence in **R[r]** for **r > 0**.

Collecting results, we thus define the bound-state radial wavefunction as

```
R[n_,l_,r_] := norm[n,l] *
    r^l E^(-Z r/(ao n)) wL[n,l, 2 Z r/(ao n)]
```

where **norm[n,l]** is a normalization constant. We show in Exercise 20.2-1 that

```
norm[n_,l_] =
    2^(1 + l) ao^(-3/2 - l) n^(-2 - l) Z^(3/2 + l) *
        ((-1 - l + n)!/(l + n)!)^(1/2)
```

$$2^{1+l}\ ao^{-(3/2)-l}\ n^{-2-l}\ Z^{3/2+l}\ \mathrm{Sqrt}\left[\frac{(-1-l+n)!}{(l+n)!}\right]$$

(Differences in this normalization and others found elsewhere are due to the definition of the associated Laguerre polynomials in *Mathematica*. See Exercise 13.4-2.) Since $l \le n-1$, we calculate for example the ground-state **n = 1** and the **n = 2** radial wavefunctions (and simplify with **PowerContract**) upon entering

```
Needs["Quantum`PowerTools`"]

{R[1,0,r], R[2,0,r], R[2,1,r]} //PowerContract
```

$$\left\{\frac{2\left(\frac{Z}{ao}\right)^{3/2}}{E^{(r\,Z)/ao}},\ \frac{\left(\frac{Z}{ao}\right)^{3/2}\left(2-\frac{r\,Z}{ao}\right)}{2\ E^{(r\,Z)/(2\,ao)}},\ \frac{r\left(\frac{Z}{ao}\right)^{5/2}}{\mathrm{Sqrt}[24]\ E^{(r\,Z)/(2\,ao)}}\right\}$$

We can easily check the normalization of these functions with our rules **integExp**, which we introduced in Section 13.4 for integrating combinations of exponentials and powers. Loading the package and integrating all three functions at once, we thus have

```
Needs["Quantum`integExp`"]

integExp[r^2 %%^2, {r,0,Infinity}]

    {1, 1, 1}
```

as desired. Finally, we define the complete, normalized bound-state wavefunctions, which are also eigenfunctions of **lsq** and **lz**, according to

```
psi[{n_,l_,mz_},r_,t_,p_] := R[n,l,r] Y[l,mz,t,p]
```

For historical reasons, the sequence of quantum numbers $l = 0, 1, 2, 3,$ ... is referred to as $l = s, p, d, f, ...$ Thus, the ground state for **n = 1** is an *s* state with angular momentum equal to zero, since $l \le n-1$. For **n = 2**, there are **n^2 = 4** degenerate states, *2s*, *2p* **(mz=0)**, and *2p* **(mz=1)**. We introduced the *2s* state in Problem 18.2-1 (see also Exercise 19.0-3 and Problem 18.7-1).

Exercise 20.2-1

(a) Show that **R[r]==r^l g[r]** or **r^(-l-1) g[r]** eliminates the centrifugal **1/r^2** terms from the radial equation. We reject the second possibility, however, since the wavefunction would be too singular near the origin. Refer to Section 1.3 and note that Park [50], p. 186, writes **R[r]** in the neighborhood of the origin as a power series proportional to **r^nu** and shows that hermiticity of the hamiltonian requires that **nu > -1/2**.

Show that **E^(-Z r/(ao n))** is a solution of the asymptotic radial equation in the limit **r -> Infinity**.

(b) Normalize the radial wavefunction **R[n, l, r]** and derive **norm[n, l]** given in the text for arbitrary **n** and **l** with

```
Integrate[
    E^-y y^(2l+2) LaguerreL[n-l-1,2l+1,y]^2,
    {y,0,Infinity}
] == 2n (n+l)!/((n-l-1)!)
```

and the procedure we used in Sections 6.6 and 13.5. Be sure to include the weight factor **r^2** and the differential **dr** in the change of variables. Check your results by explicit integration for the first few values of **n**. See also the more general formula in Problem 21.5-2 below.

Plot the radial wavefunction **R[n, l, r]** for n = 2, 3 for all allowed **l** values and verify that it has **n-l-1** nodes for **r > 0** and vanishes at the origin as well unless **l = 0**.

Exercise 20.2-2

Allowing the angular momentum quantum number **l** to assume noninteger values is useful in some applications. Re-solve the Coulomb radial equation in this case and determine the energy spectrum and eigenfunctions.

Problem 20.2-1

Define the *3D* Fourier transform according to

```
psiFT[kvec] ==
    (2Pi)^(-3/2) Integrate[E^(-I kvec.rvec) psi[rvec],{rvec}]
```

where **{rvec}** symbolizes an integration over all space. Refer to Section 11.1.

(a) Calculate the Fourier transform of the Coulomb potential **-Z e^2/r** by transforming first the *screened Coulomb* or *Yukawa* potential **-Z e^2 E^(-a r)/r**, where **a** is a screening parameter which determines the range of the potential. Show in the limit **a -> 0** that **VFT[kvec] =**

$$- \left(\frac{e^2 \; \mathrm{Sqrt}[\frac{2}{\mathrm{Pi}}] \; Z}{k^2} \right)$$

where **k = Sqrt[kvec.kvec]**. The singularity at **k = 0** is characteristic of the infinite range of the Coulomb potential. Hint: Introduce spherical polar coordinates **t** and **p** with z axis along the wavevector **kvec**. Perform the integral over **s = Cos[t]** with the built-in function **Integrate**, and the integral over **r** with **integExp**.

(b) In a similar fashion, calculate the Fourier transform of the **n = 1** and **n = 2** Coulomb bound-state wavefunctions **psi[{n, l, mz}, r, t, p]**. Check your results against the general closed-form expression from Bethe and Salpeter [6], p. 39,

```
f[n_,l_,k_] :=
    Sqrt[2/Pi (n-l-1)!/(n+1)!] n^2 2^(2(l+1)) l! *
        (n k)^l/(n^2 k^2 + 1)^(l+2) *
        GegenbauerC[n-l-1, l+1, (n^2 k^2 - 1)/(n^2 k^2 + 1)]
```

for the "radial" momentum wavefunctions given in terms of the built-in Gegenbauer polynomials. Note that this formula is in atomic units with **Z = 1** and normalized such that

```
Integrate[k^2 f[n,l,k]^2, {k,0,Infinity}] == 1
```

Problem 20.2-2

Introduce a model central-force potential to study the ground state of the deuteron. This is one of those problems that is straightforward and certainly interesting enough but usually requires too much analysis to be much fun. The computer greatly improves the situation.

The deuteron results from the binding together of a proton and a neutron due to their mutual nuclear "strong" interaction. The simplest empirical potential energy for this system is an exponential one of the form **-A E^(-r/a)**, where **r** is the proton-neutron separation, **A** the "strength" parameter of the order of **32.6 MeV (1 MeV =**
10^6 eV), and **a** the range parameter of the order of **2.16 fm (1 fm = 10^-15 meter)**.

We first set up the radial Schrödinger equation and calculate the ground-state energy variationally. Then we improve our variational estimate by numerically integrating the Schrödinger equation directly. Finally, we compare these approximate solutions with an analytical one in terms of Bessel functions.

(a) Introduce a trial variational function of the form **E^(-b r/(2a))**, where **b** is a variational parameter, and show that the *derivative* of the trial energy **eTrial** with respect to **b** can be expressed as **D[eTrial,b]==**

$$
\frac{-3\ A\ b^2}{(1 + b)^4} + \frac{b\ h^2}{4\ a^2\ m}
$$

Use **integExp** from the package **Quantum`integExp`** to perform the necessary integrals. This result shows we have a fifth-order polynomial, which in general can't be factored analytically but in principle has five roots. Although **Solve** works in this case and calculates the roots analytically (which you might check), it's sufficient to compute the roots numerically.

Given the strength **A** of the interaction, it's convenient to work in **MeV**. Hence, substitute **h -> 6.582 10^-22 MeV sec** for Planck's constant divided by **2Pi** and **m -> mp/2** for the *p-n* reduced mass with **mp -> 938.3 MeV/c^2** the proton mass. (Assume that the neutron and the proton have approximately the same mass.) Here **c** is the speed of light, **c -> 2.998 10^8 meter/sec**.

Show with these values (and **NSolve**) that **D[eTrial,b]== 0** has three real roots and a pair of complex-conjugate roots. The complex roots, however, give rise to complex energies which are not permitted here since the hamiltonian is hermitian. Also, the trivial root at **b = 0** gives a constant wavefunction and can therefore be rejected. Show that the root near **b = 1.3** determines the *minimum* or optimum variational energy to be -2.130 MeV, which is close to but above the *observed* ground-state energy -2.226 MeV, consistent with the variational theorem.   Plot the best variational ground-state wavefunction as a function of r in fm.

Finally, verify that the other root near b = 0.1 gives rise to a positive energy corresponding to an unbound state. Thus, show that the model is consistent with the observation that the deuteron has only one bound state, although we really need to extend our analysis to states with nonvanishing angular momentum L > 0.

(b) Use the shooting method (Section 2.2) to estimate the ground-state eigenfunction and eigenenergy with the built-in function **NDSolve** to integrate the Schrödinger equation numerically.   For convenience, transform to a *reduced radial wavefunction* u[r] = r R[r] and set up a reduced radial equation with energies measured in MeV and distances in fm. Thus, show that

```
-32.6 u[r]
───────────  - 41.499 u''[r] == eps u[r]
 0.462963 r
E
```

where **eps = Energy/MeV** is the scaled eigenenergy. We require R[r] to be everywhere finite (refer to Exercise 20.2-1) and therefore that u[0] = 0.

Now in order to determine eigenfunctions and eigenvalues, we need a boundary condition as r -> Infinity. Numerically, of course, we work with a maximum value **rmax** of r. Since the model exponential potential has a relatively short range, the Schrödinger equation simplifies *asymptotically*, i.e. for large r near **rmax**, where the potential can be ignored. It is straightforward to show that the acceptable asymptotic solution is a decaying exponential (see part **(c)**), and we can simply take u[rmax] = 0, although the method can be refined by building in the asymptotic solution explicitly. Try **rmax = 30.0** but also check the accuracy and stability of your results by varying this value.

Thus, guess a value **eps** for the ground-state energy, one near our variational result for example, and integrate the radial equation from r = rmax towards the origin r = 0 and observe how close the solution u[r] comes to hitting u[eps,0] = 0. Make a table of (unnormalized wavefunction) values u[eps,0] as a function of **eps** and plot your results as in Figure 20.2-1. The point where this line of values crosses the **eps** axis determines the ground-state energy. Interpolate the points (cf. Exercise 3.5, Appendix III) and search for the root of the resulting spline with the built-in function **FindRoot**. The root should be within one percent of the observed ground-state energy and very close to the model's exact value, which we obtain in Part **(c)**.

Normalize the resulting ground-state wavefunction with the built-in function **NIntegrate** and plot it together with the best variational wavefunction from Part **(a)**.

(c) Show that the L = 0 radial equation for the model exponential potential can be transformed to *Bessel's equation*

```
    2    2                      2
(-n   + z ) J[z] + z J'[z] + z   J''[z] == 0
```

by introducing a radial wavefunction of the form R[r] = J[z]/r /. z -> d E^(-r/ (2a)), where J[z] is a *Bessel* function, if we define the (unknown) eigenenergy and the dimensionless parameter d as {e[n],d} ==

```
    2   2      3/2
 -(h   n )    2      a Sqrt[A m]
{───────────, ─────────────────}
      2              h
   8 a   m
```

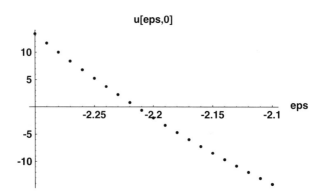

Figure 20.2-1. Wavefunction values (unnormalized) at the origin $r = 0$ as a function of the scaled eigenenergy **eps** for the ground state of the deuteron.

Although **n** is another dimensionless parameter to be determined, show that **d = 3.8289** in the present example.

In order to determine **n**, we now need to impose boundary conditions in the usual way and require that **R[r]** be finite for all values **{r,0,Infinity}**. As with the numerical solution, only the limits **r -> 0** and **r -> Infinity** are of consequence.

As **r -> Infinity**, we have that **z -> 0**, and we require therefore that **J[n,0]** be finite. Thus, our solutions are the built-in Bessel functions of order n, **J[n,z] = BesselJ[n,z]**, which are bounded for all **z** finite if **n** is a *positive* number or a negative integer. In particular, the solution cannot involve the linearly independent built-in function **BesselY[n,z]**, which you should verify by Taylor expansion is unbounded (*irregular*) for **z** small and any **n**. Investigate **BesselJ[n,z]** for n < 0, and with a few examples show that **BesselJ[-n,z] == (-1)^n BesselJ[n,z]** for **n** an integer. Show also by Taylor expansion that **BesselJ[n < 0,z]**, n not an integer, is irregular at **z = 0**. Finally, show that **R[r] = J[n,z]/r ~ E^(-Sqrt[-2m e[n]/h^2]r)** as **r -> Infinity**, i.e. a solution of the asymptotic radial equation.

As **r -> 0**, the variable **z -> d**, and we require **J[n,z -> d] -> 0** in such a way that **R[r] = J[n,z]/r** remains finite. Thus, we require **d** to be a *zero of the Bessel function* and choose the order **n** of the Bessel function accordingly. That is, we have to adjust the order **n** since we already have the numerical value **d = 3.8289**. In this way, we determine the spectrum of eigenenergies **e[n]**.

This is a somewhat nonstandard and interesting state of affairs compared with the usual boundary value problem involving Bessel functions. Life would be especially difficult at this point if we couldn't evaluate **J[n,z]** easily for arbitrary n. Familiar handbooks tabulate **J[n,z]** and its zeros for integer n only. Thus, plot **BesselJ[n,d]** *as a function of* **n** and show graphically that there is only one positive n, around **n = 1**, for which **d** is a root of **BesselJ[n,d]**. We thus again conclude that our model potential supports only one **L = 0** bound state, given the empirical values of **A** and **a**.

Now determine **n** accurately using the secant-method variation of **FindRoot** (refer to Wolfram [62]), since *Mathematica* doesn't give an analytic expression for the derivative of **BesselJ[n,d]** with respect to **n**. (Expressions for this derivative are relatively complicated. Refer to Abramowitz and Stegun [1], eqs. 9.1.64 - 68.) The secant method requires two starting values, which you can take arbitrarily to be on either side of **n = 1**. The results for the ground-state **n** and energy **e[n]** should be close to

```
{n -> 0.997918, e[n] -> -2.21442 MeV}
```

Normalize the resulting ground-state wavefunction (with **NIntegrate**) and plot its difference with the wavefunction you obtained by the shooting method in Part **(b)**.

Parity 20.3

It is useful to know how these wavefunctions transform under an *inversion* of the coordinate axes, **{x,y,z} -> {-x,-y,-z}** or **rvecC -> -rvecC**, i.e. the *parity* of the wavefunctions. In spherical-polar coordinates this transformation consists of the replacements **{r,t,p} -> {r,Pi-t,Pi+p}** (see Figure 4.0-1, Appendix V), and we therefore define a *parity operator*

```
Parity @ psi_ :=
    psi /. {r :> r, t :> Pi-t, p :> Pi+p} //Expand
```

in analogy with the operator we defined in Cartesian coordinates in Chapter 5. We readily check that this function generates the desired inversion in spherical coordinates by noting

```
rvecC = r er /.
    er -> {Cos[p] Sin[t], Sin[p] Sin[t], Cos[t]};

Parity @ rvecC == -rvecC
```

```
    True
```

Recall that an eigenfunction of the parity operator with eigenvalue **+1** is said to have *even parity*, and one with eigenvalue **-1** *odd parity*.

Of course, if an observable commutes with the parity operator, eigenfunctions of the observable can always be found which are simultaneous eigenfunctions of parity (cf. Section 15.1). For example, it's easy to show that the spherical harmonics **Y[l,mz]**, which are eigenfunctions of **lsq** and **lz**, have the definite parity **(-1)^l**, independent of **mz**. Thus, for **l = 0** and **1**,

```
Y[l_,m_,t_,p_] := SphericalHarmonicY[l,m,t,p]

Table[
    Parity @ Y[l,m,t,p] == (-1)^l Y[l,m,t,p],
    {l,0,1},{m,-1,1}
]
```

```
    {{True}, {True, True, True}}
```

Hence, the parity operator must commute with **lsq** and **lz**. This is evident from the observation that under parity **lvec -> -I h (-rvec) ×** **(-grad) == lvec**, and thus formally **Parity @ lvec @ psi == lvec @**

Parity @ psi. For this reason, by the way, **lvec** is also called an *axial vector*. (Note that our above definition of **Parity** would have to be extended if we actually wanted to apply it to derivative operators such as **lvec**. For example, we cannot directly substitute an algebraic expression for the variable **x** in Dt[**f,x**].)

Exercise 20.3-1

(a) Deduce the parity of the spherical harmonics by first determining the parity of the Legendre polynomials. Refer to Problem 19.1-2. Introduce the parity operator in Cartesian coordinates we defined in Chapter 5 and check the parity of the *solid harmonics* defined in Exercise 19.1-2.

(b) Matrix elements of a product of three spherical harmonics **Y[11,m1] Y[12,m2] Y[13,m3]** arise in many applications. Derive the *selection rule* which states that these matrix elements vanish identically unless **11+12+13 ==** *even*.

Clearly, coordinate inversion commutes with any function of **r = Sqrt[x^2 + y^2 + z^2]** and in particular with the Coulomb potential. That is, after re-entering our commutator rule from Section 18.2,

```
Commutator[A_, B_] @ psi_ :=
    A @ B @ psi - B @ A @ psi //Expand

Commutator[Parity, V[r]] @ psi == 0
```

```
    True
```

Hydrogen-atom wavefunctions can therefore always be constructed which have definite parity. Such eigenfunctions are of course the **psi[{n,l,mz}]** with parity **(-1)^l** since they are proportional to spherical harmonics and the radial wavefunctions clearly have even parity. For example, for all **n = 1** and **n = 2** states

```
Table[
    Parity @ psi[{n,l,m},r,t,p] ==
        (-1)^l psi[{n,l,m},r,t,p],
    {n,1,2},{l,0,n-1},{m,-1,1}
] //ExpandAll //MapAll[Factor,#]&
```

```
    {{{True}}, {{True}, {True, True, True}}}
```

One thus loosely refers to parity as being a good quantum number for these states.

Continuum Wavefunctions   20.4

Finally, we point out that if the principal quantum number **n** is replaced by **-I/(ao/Z k)**, where **k** is a continuous parameter, then the discrete bound-state energy spectrum **e[n]** transforms to that corresponding to unbounded motion with *continuum* **Energy > 0**:

```
e[k] = -Z^2 Ry/n^2 /.n -> -I/(ao/Z k) //.aoRy

    2  2
   h  k
   ─────
    2 m
```

Hence, **k > 0** is the electron's wavenumber with units of inverse length and **2Pi/k** is the electron's deBroglie wavelength. Following our derivation of the wavefunction through again, we find that the bound-state radial wavefunctions similarly transform to the (unnormalized) radial wavefunctions describing unbounded motion. Thus, replacing **n** by **-I/(ao/Z k)**, we obtain

```
RC[k_,l_,r_] =
    r^l E^(-Z r/(ao n)) w1F1[n,l,2 Z r/(ao n)] /.
        n -> -I/(ao/Z k)
  -I k r  l                          I Z
 E        r  Hypergeometric1F1[1 + l + ───, 2 + 2 l, 2 I k r]
                                      ao k
```

Because these are functions of the continuous parameter **{k,0,Infinity}**, the usual normalization integral is proportional to a Dirac delta function, which at the same time expresses the orthogonality of eigenfunctions belonging to different **k** values (cf. Merzbacher [45], Section 6.3). In practice, other normalizations are employed, as for example a scattered beam of particles with unit incident flux (see Section 14.2 and e.g. Schiff [59], Section 21). It should be clear physically that the bound and continuum states together are required for a complete, orthogonal set. We note in this connection that a wide class of matrix elements involving bound-state and continuum Coulomb functions can be evaluated generally with the formulas and method introduced in Section 13.5 to normalize the Morse-oscillator wavefunctions.

The continuum radial functions **RC[k,l,r]** are referred to as *regular Coulomb functions*, since they are finite at the origin, and are the only functions which can arise in problems involving pure Coulomb potentials. The so-called *irregular Coulomb functions* are required, however, in constructing general solutions in applications involving central potentials which deviate from a pure Coulomb potential near the origin. These linearly independent solutions are of course proportional to the **HypergeometricU** we discarded for the bound states and become important for instance in describing charged particles near nuclei.

We conveniently construct *3D* plots of the probability distribution by introducing the Cartesian coordinates **x = r Sin[t]** and **z = r Cos[t]** used in

Figure 20.1-1, since the square of the wavefunction is cylindrically symmetric about the z axis of quantization. The wavefunctions are easily transformed (cf. Exercise 19.1-2). (Although we could also plot directly as a function of **r** and **t** using **SphericalPlot3D** as in Figure 19.1-1, the coordinates **x** and **z** have the advantage that the z axis is explicitly represented.) Two excited-state probability distributions of the hydrogen atom are shown in Figure 20.4-1, the *2p*(**mz=0**) bound state on the left and a *p*(**mz=0**) continuum wave on the right with **k = Sqrt[2]** and hence **e[k] = 1** a.u. The definite parity of the wavefunctions is evident (i.e. the even parity of the probability). Note that the central maximum in the continuum wave has been arbitrarily chopped to make the secondary maxima more visible.

For comparision, the *radial* components of these wavefunctions are isolated in Exercise 20.4-1 and plotted in Figure 20.4-2 along with **Veff**.

Exercise 20.4-1

(a) Make plots of the wavefunctions squared like the ones shown in Figure 20.4-1. (The bound state plots promptly, but the continuum wave takes about 50 minutes.) If you have the time or computer resources, do this for **n = 1, 2, 3** and for **k = 0, 1, 10**. For a given **n** or **k**, make an animation to display different **1** and **mz** values.

(b) Make plots, as in Figure 20.4-2, of the *radial* probability distributions for **n = 1, 2, 3** and of the **Re** and **Im** parts of the continuum radial wavefunction for **k = 0, 1, 10** and **1 = 0, 1, 2**. For convenience use atomic units. Note how the centrifugal barrier pushes the wavefunction away from the origin and how the wavelength of the continuum levels varies with energy (cf. Exercise 19.1-3). Figure 20.4-2 is schematic since the probabilities have been arbitrarily scaled and shifted by their eigenenergy and plotted along with the effective potential **Veff**.

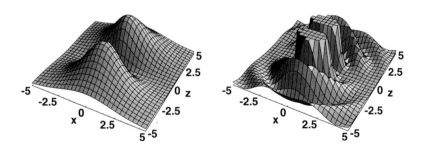

Figure 20.4-1. Hydrogenic (**z = 1**) probability distributions (wavefunction squared) as functions (in a.u.) of the Cartesian coordinates **x = r Sin[t]** and **z = r Cos[t]** introduced in Figure 20.1-1. Shown on the left is a *2p*(**mz=0**) bound state and on the right a *p*(**mz=0**) continuum state with **k = Sqrt[2]** and hence **e[k] = 1** a.u.

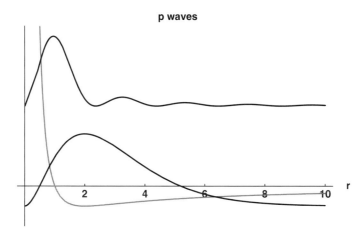

Figure 20.4-2. The *radial p*-state probability distributions from Figure 20.4-1 shifted by their energies and scaled for comparison with the effective potential **Veff** from Figure 20.1-1 shown in gray.

Exercise 20.4-2

(a) Evaluate the hydrogenic expectation values **<r>** and **<r^2>** for the five states of the **n = 1** and **2** levels. Also, compute **<1/r>** and use it and the virial theorem to compute the energy of these states (cf. Problem 18.2-1).

(b) The *positron* is a particle with the same mass as an electron but with charge +1 (relative to the electron charge). *Positronium* is a short-lived system observed when an electron and a positron bind via the Coulomb interaction about their common center of mass. It decays when the positron and electron mutually annihilate. Compute its energy and **<r>** in the ground state. What is the wavelength of light emitted when the system decays from an excited **n = 2** level to its **n = 1** ground state?

(c) Muons are particles with mass 206.8 times greater than that of electrons and with charge ±1 (particle, antiparticle). *Muonic atoms* are observed when one of the electrons is replaced by a negative muon. Estimate the energy of the x-ray emitted when a muon in lead (**Z = 82**) decays from an excited **n = 2** level to its **n = 1** ground-state. What is the wavelength of light emitted when the system decays from an excited **n = 2** level to its **n = 1** ground state? The observed value is 10.48 MeV. Compute **<r>** for these two levels. Why is it reasonable to assume that the muonic atom consists of only a bare lead nucleus and a muon?

Problem 20.4-1

Show that for **Energy = 0** the hydrogenic-atom radial wavefunctions are proportional to **BesselJ[2l+1, Sqrt[c r]]/Sqrt[c r]** and determine the constant c. Plot this function and compare it with **RC[k,l,r]**. Make an animation to check the limit **k -> 0**. Compare with results from Exercise 19.1-3.

20.5   Separation in Parabolic Coordinates

Before turning to the construction of the wavefunctions from the Runge-Lenz algebra, we should examine the solution of the Coulomb problem in the parabolic coordinates $u = r + z$, $v = r - z$ and p defined in Problem 8, Appendix V. This approach also permits a separation of variables in the Schrödinger equation and is analogous to the solution in spherical coordinates. We relegate the investigation, however, to Problem 20.5-1 and for the moment remark that the resulting wavefunctions are not eigenfunctions of lsq and lz. Rather, as we shall see later on, they are simultaneous eigenfunctions of the components lz and Az of the angular-momentum and Runge-Lenz vectors.

Recently, the parabolic quantum numbers have also been found to provide useful clues in the hunt for a complete set of commuting observables in the two-electron atom (three-body) Coulomb problem. When two electrons are excited but nevertheless slightly bound one electron tends to be drawn into a kind of Stark effect caused by the long-range dipole field of the other electron and the nucleus. One finds that such atoms are partially characterized by parabolic quantum numbers $n1$ and $n2$, in particular, the so-called *electric quantum number* $K = n2 - n1$, which determines the shift in an energy level caused by an external electric field (see Section 21.5 below). This is quite interesting when one considers that a separation of variables in a system as fundamental as helium does not appear possible. In fact, it's not even clear what the proper choice of coordinates is, for presumably if we knew, we could fully separate the two-electron Schrödinger equation. As a result, the full spectroscopy of helium has never been systematically explained. The occurrence of approximate quantum numbers encourages one to seek at least a partial separation of the problem (refer to Rau [55] and also to Fano and Rau[19] and references therein).

The parabolic or one-center symmetries are actually just one aspect of a more general description of two-electron atoms based on two-center *elliptical* symmetries, which recognizes the basic likeness of two electrons. We see from Problem 9, Appendix V, that parabolic coordinates are just the large-R limit of elliptical coordinates for fixed nuclei. An analogy with molecules thus arises because the Schrödinger equation for the hydrogen molecular ion (two protons and one electron) separates in elliptical coordinates if the nuclei are assumed fixed. The two electrons in say helium are given the role of the two nuclei in the molecule and are thus placed at the two centers with a separation R. We employ this point of view in Problem 20.5-5 to calculate the ground states of helium and H^- variationally. The justification of this unconventional approach is its ability to account for observed similarities in the energy spectrums of systems as seemingly different as the hydrogen molecular ion and the helium atom (see Feagin and Briggs [20] and the review by Rost and Briggs [57]).

It's interesting in this connection to note that the hydrogen-atom Schrö-dinger equation also separates in elliptical coordinates, a fact rarely referred to. One thus imagines a molecule with the hydrogen nucleus (the proton) at one center and a phantom nucleus with zero charge at the other. We examine this possibility in Problem 20.5-2.

Problem 20.5-1

Show that the Schrödinger equation for a hydrogenic atom also separates and can be solved in *parabolic* coordinates u = r + z, v = r - z and p, where both u and v are defined for {0,Infinity} and p for {0,2Pi}. Refer to Problem 8, Appendix V. Derive the Balmer-Bohr energy spectrum and the normalized wavefunctions with principal quantum number n ≤ 3. Verify that the degeneracy of the *n*th level is n^2.

(a) Assume that the system wavefunction is of the form chi = f[u]g[v]E^(I mz p) and separate variables. Arrange terms into *two* expressions uPart and vPart *identical* in form. In order to facilitate comparison with solutions in the text, write the energy as -Z^2 Ry/n^2 and introduce K Z/(2n ao) == uPart == -vPart as the separation constant, where n and K are numbers to be determined. (You might find **Select** convenient for separating variables. See Exercise 1.9, Appendix III.)

(b) Solve the f and g equations near the origin and asymptotically and show that

```
f[u]  == u^(Abs[mz]/2) E^(-Z u/(2ao n)) *
              LaguerreL[n1,Abs[mz],Z u/(ao n)]
g[v]  == v^(Abs[mz]/2) E^(-Z v/(2ao n)) *
              LaguerreL[n2,Abs[mz],Z v/(ao n)]
```

where n1 and n2 are positive integers or zero and n = n1 + n2 + Abs[mz] + 1. This identifies n as the principal quantum number and recovers the Balmer-Bohr energy spectrum. In addition, show that K = n2 - n1. This relates the separation constant to the so-called *electric quantum number* (see Bethe and Salpeter [6], p. 230, and Section 21.2 below).

Make a series of plots to show that n1 and n2 specify the number of nodes in f[u] and g[v], respectively.

(c) Collect your results in a definition of the function chi[{n1,n2,mz},u,v,p]. In order to obtain phases in agreement with the Runge-Lenz solutions in the next section, include a factor (-1)^(n1+mz) and define chi for negative mz by the rule chi[-mz]== (-1)^mz chi[mz] in analogy with the spherical harmonics. Normalize chi explicitly for n1, n2, mz ≤ 2 (using integExp). (We calculate the normalization generally in Problem 21.5-2 below.)

(d) As we shall see in Section 21.6, the parabolic states chi[{n1,n2,mz}] represent in general linear combinations of the spherical states psi[{n,1,mz}], and vice versa. However, the states chi and psi at the top and bottom of the angular momentum ladder with mz = ±1 = ±(n-1) have a one-to-one correspondence with each other. Show that the top- and bottom-rung parabolic states are given by n1 = n2 = 0, which includes the ground state. Transform to spherical coordinates and verify that chi[{0,0,(n-1)}] == psi[{n,n-1,(n-1)}] for n ≤ 3. Transform the two middle-rung states to spherical coordinates and show that {chi[{1,0,0}], chi[{0,1,0}]} ==

$$\left\{ \frac{Z^{3/2} (-2\ ao + r\ Z + r\ Z\ Cos[t])}{8\ ao^{5/2}\ E^{(r\ Z)/(2\ ao)}\ Sqrt[Pi]} \right., $$

$$\left. \frac{Z^{3/2} (2\ ao - r\ Z + r\ Z\ Cos[t])}{8\ ao^{5/2}\ E^{(r\ Z)/(2\ ao)}\ Sqrt[Pi]} \right\}$$

What linear combinations of these functions give `psi[2s]` and `psi[2p(mz=0)]`? Finally, calculate all *nine* n = 3 normalized states `chi`.

When comparing with other references, note that the definitions of u and v and of n1 and n2 as well are sometimes the reverse of those used here, which follow Bethe and Salpeter [6].

(e) From n1 + n2 = n - Abs[mz] - 1, we have that `Max[n1] = n - Abs[mz] - 1 = Max[n2]` for given values of n and mz. This thus corresponds to n - Abs[mz] combinations of n1 and n2. The degeneracy is obtained by summing this quantity over mz from 0 to n - 1:

```
n + 2(n-1) + 2(n-2) + ... + 2(n-(n-1))
```

where the factors of 2 arise from the two possibilities ±mz for mz > 0. Load the package `Algebra`SymbolicSum` to show that the degeneracy equals n^2 in agreement with the result from Problem 18.7-1.

Problem 20.5-2

Show that the Schrödinger equation for a one-electron atom also separates in two-center molecular or *elliptical* coordinates, as defined in Problem 9, Appendix V. Imagine the atomic nucleus charge Z e at one center and another *phantom* nucleus with zero charge at the other.

The possibility of this separation is also clear from the general description of the molecular two-center Schrödinger equation. For example, G. Hunter, B. .F. Gray, and H. O. Pritchard, J. Chem. Phys. **45**, 3806 (1966), separated variables for two arbitrary nuclear charges Z1 e and Z2 e in terms of a charge ratio q = Z2/Z1. The hydrogen-atom separation is thus obtained from their results simply by setting q equal to zero.

Problem 20.5-3

Determine the eigenfunctions and eigenenergies of the *2D* isotropic harmonic oscillator by solving the Schrödinger equation in plane polar coordinates r and p. Assume simultaneous eigenfunctions of the angular momentum lz (perpendicular to the orbital plane) of the form `psi == f[r]E^(I mz p)` and show that `f[r]` is proportional to an associated Laguerre polynomial in the variable m w r^2/h.

Writing the energy as h w n2D and requiring that the wavefunction be everywhere finite, show that the quantum number n2D must be a *nonzero* positive integer of the form n2D = 2q + Abs[mz] + 1 with q = 0, 1, 2, ... Hence, {mz,-n2D + 1,n2D - 1,2} *in steps of 2!* Compare with the solution from Problem 18.7-2.

Problem 20.5-4

Transform the hydrogenic-atom radial equation to that of a *2D* isotropic harmonic oscillator, and use the energy spectrum of the oscillator to obtain the Balmer-Bohr spectrum. Thus, change the independent and dependent variables according to

```
r -> r2D^2/(Sqrt[-e[n]/(Z^2 Ry)]) Z/ao
R -> Sqrt[Sqrt[-e[n]/(Z^2 Ry)]]/r2D f[r2D]
```

where `r2D` is the polar coordinate of the oscillator and `f[r2D]` its radial wavefunction, and show that `w = 2 h/m (Z/ao)^2` is the frequency of the oscillator. (This transformation is already evident from a comparision of the H-atom ground-state wavefunction with that of a *1D* oscillator.) Refer to the previous problem and to Problem 18.7-2.

Problem 20.5-5

Let's estimate variationally the ground-state energies of two fundamental atomic systems, helium and the hydrogen negative ion H^-. We shall do this in two ways. The first approach can be found in most textbooks and is based on a trial wavefunction of the form `E^(-a(r1 + r2))`, where `r1` and `r2` are the electron-ion separations and `a` the variational parameter. This approach is moderately accurate for helium but inadequate for H^-. The second approach starts with the same trial function and introduces elliptical or molecular coordinates `R, l, m` (refer to Problem 9, Appendix V) in order to define an effective *1D* potential for the electron pair as a function of the interelectronic separation `R`. This *molecular model* is remarkably accurate for both systems and easily extended to excited states (refer to J.M. Rost and J.S. Briggs, J. Phys. B**21**, L233 (1988)).

(a) Thus, introducing the trial wavefunction `E^(-a(r1 + r2))` and integrating with respect to the spherical coordinates `r1` and `r2`, evaluate the variational energy `eTrial` of the two-electron hamiltonian (in atomic units) `H ==`

$$\frac{1}{R} - \frac{Z}{r1} - \frac{Z}{r2} - \frac{laplacian[r1]}{2} - \frac{laplacian[r2]}{2}$$

Here `z` is the atomic number, viz. `Z = 2` for helium and `Z = 1` for H^-, and `1/R` the electron-pair interaction with `R = Abs[r1vec - r2vec]`. The spherical-coordinate laplacian is defined in Section 4.6, Appendix V.

The required integrations are straightforward except for the expectation value of `1/R`. This electron-pair *correlation interaction* is nevertheless elementary to evaluate if spherical symmetry and Gauss's law are invoked. Thus, first calculate the effective potential experienced by electron *1* due to electron *2* with charge density proportional to `E^(-2a r2)` and show that

$$\frac{1 - \dfrac{1 + a\ r1}{E^{2\ a\ r1}}}{r1}$$

(Be sure to normalize the trial function first.) Use this result to calculate the electron-pair interaction energy. Collecting results, show that the *minimum* total energy `eTrial =`

$$- (- (\frac{5}{16}) + Z)^2$$

corresponding to an optimum variational parameter $a \to Z - 5/16$. Physically, this means that each electron is partially screened from the full charge of the nucleus by the other electron.

Compare your predicted ground-state energies in helium and H^- with the generally accepted best variational estimates e[0][Z = 2] = -2.90372 and e[0][Z = 1] = -0.52775 (from C.L. Pekeris, Phys. Rev. **126**, 1470 (1962)). Explain that while the value eTrial for helium is acceptable, it predicts for H^- the nonexistence of a bound state. (Hint: Compare with the ground-state energy of one-electron hydrogen.) It's interesting to note that the Sun's opacity results from the absorption of light by the ground state of H^- near the surface of the Sun.

(b) Now introduce elliptical coordinates R, 1 = (r1+r2)/R, and m = (r1-r2)/R such that E^(-a(r1 + r2)) == E^(-a R 1). Restricting our considerations to states with total angular momentum L = 0, we can confine our evaluations to the plane of three classical particles and the coordinates 1, m and R.

One can think of the function phi = E^(-a R 1)/Sqrt[n[R]], with n[R] a normalization, as describing the motion of the electron-pair center of mass for fixed R (refer to Problem 15.2-1). We therefore greatly improve the accuracy of the trial wavefunction by introducing a factor f[R] to describe the relative (vibrational) motion of the two electrons. Integrating over 1 and m and requiring for fixed R that phi be normalized to unity, show that n[R] =

$$\frac{3 + 3 w + w^2}{48 \, a^3 \, E^w}$$

where w = 2 a R. (Be sure to include the volume element R^3/8 (1^2 - m^2) from Problem 9, Appendix V.) Show then that the system wavefunction f[R]/R phi transforms the two-electron Schrödinger equation to the *1D* radial equation

$$f[R] \; (Energy - \frac{1}{R} - u[R]) + f''[R] == 0$$

where u[R] is an effective electron-pair potential energy defined by the expectation value of H - 1/R with respect to phi integrated over 1 and m for fixed R. Show that u[R] =

$$a^2 \; (1 - \frac{w^2 + 2 w^3 + w^4}{(3 + 3 w + w^2)^2}) - \frac{12 \, a \, (1 + w) \, Z}{3 + 3 w + w^2}$$

This potential is plotted in Figure 20.5-1 including the 1/R electron-pair correlation. Note that u[R] + 1/R -> 0 as R -> Infinity corresponding to both electrons at infinity, i.e. the *double-ionization threshold*.

Hint: You can calculate u[R] in one of two ways. The most straightforward is to transform the hamiltonian H to elliptical coordinates by combining the results of Problems 15.2-1, 19.0-2, and Problem 10, Appendix V. Operating on just the trial function f[R]E^(-a R 1)/(R Sqrt[n[R]]) greatly reduces the algebra. Alternatively, you

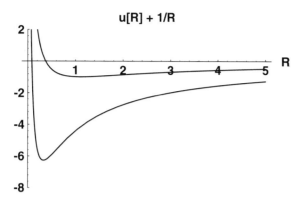

Figure 20.5-1. Effective electron-pair variational potentials for helium (lower curve with **a** = **Z** = 2)and H^- (upper curve with **a** = **Z** = 1) as a function of the interelectronic separation **R**. All values are in atomic units.

can calculate the local energy with the original hamiltonian **H** from part **(a)** as a function of **r1** and **r2** using **phi == E^(-a (r1 + r2))/Sqrt[n[R]]** and then transform to elliptical coordinates to evaluate the expectation value with respect to **l** and **m** for fixed **R**. However, you will have to add in derivatives of **1/Sqrt[n[R]]** from the radial component **-D[R^2 D[psi,R],R]/R^2** of the transformed hamiltonian.

Finally, use the shooting method subject to the boundary condition that **f[0] = 0** to obtain the ground-state wavefunction **f[R]** for several values of the variational parameter **a**. Minimize the resulting eigenenergies as a function of **a** to determine the optimum value of **a** and hence the best ground-state energy of both systems (cf. Problem 20.2-2). Your value for helium should be within one percent of the best variational estimate given in part **(a)**, and for H^- within a few percent.

An advantage of this approach is that once **a** has been optimized the same potential **u[R]** gives good estimates of the entire series of **L** = 0 doubly-excited states up to the double-ionization threshold in both systems.

21
Wavefunctions from the Runge-Lenz Algebra

We now construct the Coulomb bound-state wavefunctions from the raising and lowering operators of the *pseudo* angular momenta we defined in Problem 18.7-1. We proceed in close analogy with the derivation of the spherical harmonics in Problem 19.1-3 from the raising and lowering operators of the orbital angular momentum. As we shall see, the approach is powerful and provides insight into the properties and various forms of the Coulomb wavefunctions.

We shall thus work in the subspace of energy eigenfunctions and replace the hamiltonian `H` by its eigenvalue `e[n]`. We assume that the energy spectrum is known and for bound states given by `e[n] = -Z^2 Ry/n^2`, where the principal quantum number `n` is a nonzero positive integer. Hence, we often refer to the `n` subspace or `n` *manifold* for short.

Raising and Lowering Operators 21.1

We begin by transforming the Runge-Lenz vector **Avec** to spherical coordinates. The simplest way to do this is to recompute its Cartesian components using those of **pvec** and **lvec** expressed in spherical coordinates. We calculated **lvecC** in Section 19.0, and we can calculate **pvecC** here by analogy. (We also need to declare the spherical coordinates to be independent with our function **IndependentVariables** from Section 18.1.) Hence, we re-compute **AvecC** from Section 18.6 in spherical coordinates and simplify (this takes about 5 minutes):

```
IndependentVariables[x_,y_,z_] :=
    Module[{},
        x/: Dt[x,y] := 0;    x/: Dt[x,z] := 0;
        y/: Dt[y,x] := 0;    y/: Dt[y,z] := 0;
        z/: Dt[z,x] := 0;    z/: Dt[z,y] := 0
    ]


Needs["Quantum`Trigonometry`"];

pvecC @ psi_ := -I h grad[psi] /. eCrep

IndependentVariables[r,t,p];
SetAttributes[{Z,e,m,h},Constant]

AvecC @ psi_ =
    Expand[
        Table[
            Sum[Signature[{i,j,k}]
                (pvecC[#][[i]]& @ lvecC[#][[j]]& @ psi -
                lvecC[#][[i]]& @ pvecC[#][[j]]& @ psi),
                {i,1,3},{j,1,3}
            ],
            {k,1,3}
        ]/(2m) - Z e^2 (er /.eCrep) psi
    ] /.
        {m->1, h->1, e->1} //Expand //TrigReduce;

{Ax @ psi_ = AvecC[psi][[1]],
 Ay @ psi_ = AvecC[psi][[2]],
 Az @ psi_ = AvecC[psi][[3]]}
```

For convenience in the following, we've transformed to atomic units and set $h = m = e = 1$. We leave the atomic number Z, however, unspecified. We have, for example, that

```
Az @ psi
```

$$-(\text{psi Z Cos[t]}) - \text{Cos[t] Dt[psi, r]} - \frac{\text{Cos[t] Cot[t] Dt[psi, t]}}{r} -$$

$$\frac{\text{Cot[t] Csc[t] Dt[psi, \{p, 2\}]}}{r} - \frac{\text{Cos[t] Dt[psi, \{t, 2\}]}}{r} -$$

$$\text{Dt[psi, r, t] Sin[t]}$$

The wavefunctions we seek are eigenfunctions of the pseudo angular momentum operators **Jisq** and **Jiz** with $i = 1, 2$ we defined in Section 18.7, viz. relations (18.7-r2). We thus construct raising and lowering operators **JiR** and **JiL** in analogy with ordinary angular momentum.

Restricting ourselves to bound-states, we first note that the factor from relation (18.7-r1) which defines the scaled Runge-Lenz vector **Mvec** in the **n** subspace is simply **n/Z** in atomic units:

```
Sqrt[m/(-2 e[n])] == n/Z /. e[n] -> -Z^2 Ry/n^2 /.
    {m->1, h->1, e->1, Ry->1/2} //PowerExpand
```

```
    True
```

We thus define the z component of this vector and its raising and lowering operators and label them with the principal quantum number **n**. The trig rules introduced here simplify the results considerably.

```
Mz[n_] @ psi_ = n/Z Az[psi];

MR[n_] @ psi_ =
    n/Z (Ax[psi] + I Ay[psi]) //.
        {Cos[p] -> TrigToComplex[Cos[p]],
         Sin[p] -> TrigToComplex[Sin[p]]} //
            Expand;

ML[n_] @ psi_ =
    n/Z (Ax[psi] - I Ay[psi]) //.
        {Cos[p] -> TrigToComplex[Cos[p]],
         Sin[p] -> TrigToComplex[Sin[p]]} //
            Expand;
```

Thus, for example

```
MR[1] @ psi //Collect[#,E^(I p)]&
```

$$
E^{I\,p}\,\left(-\left(\frac{\text{Cos}[t]\ \text{Dt}[psi,\,t]}{r\ Z}\right) - \frac{\text{Csc}[t]\ \text{Dt}[psi,\,\{p,\,2\}]}{r\ Z} + \right.
$$

$$
\frac{I\ \text{Csc}[t]\ \text{Dt}[psi,\,p,\,r]}{Z} + \frac{\text{Cos}[t]\ \text{Dt}[psi,\,r,\,t]}{Z} - psi\ \text{Sin}[t] -
$$

$$
\left. \frac{\text{Dt}[psi,\,r]\ \text{Sin}[t]}{Z} - \frac{\text{Dt}[psi,\,\{t,\,2\}]\ \text{Sin}[t]}{r\ Z}\right)
$$

We can now define the raising and lowering operators for the pseudo angular momenta. It's appropriate in addition to define the orbital angular momentum **LvecC** in atomic units from **lvecC**. We thus have

```
Lz @ psi_ = lz @ psi /. h->1;
LR @ psi_ = lR @ psi /. h->1;
LL @ psi_ = lL @ psi /. h->1;
```

```
J1R[n_] @ psi_ = (LR[psi] + MR[n][psi])/2 //Expand;
J1L[n_] @ psi_ = (LL[psi] + ML[n][psi])/2 //Expand;
J2R[n_] @ psi_ = (LR[psi] - MR[n][psi])/2 //Expand;
J2L[n_] @ psi_ = (LL[psi] - ML[n][psi])/2 //Expand;
```

Exercise 21.1-1

Verify that the operators **Lvec**, **Mvec**, **J1vec** and **J2vec** have the right properties from Section 18.6 and satisfy the appropriate commutation relations from Problem 18.7-1.

21.2 Top-Rung States

We turn to the construction of the bound-state eigenfunctions **phi[{j,m1, m2},r,t,p]** of **J1sq[n]** and **J2sq[n]** and of **J1z[n]** and **J2z[n]** with quantum numbers **j1 = j2 = j** and **m1** and **m2**, respectively. From Problem 18.7-1, we have that the principal quantum number **n = 2j+1** or **j = (n-1)/2**. Hence, **j = 0, 1/2, 1, 3/2, ...**

It's evident that the functions **phi[{j,m1,m2}]** are also eigenfunctions of **Lz** and **Mz[n]** (but not in general of **Lsq** and **Msq[n]**) with *integer* eigenvalues **mz = m1+m2** and **K = m1-m2**, respectively. For a given **n**, it follows that **Max[mz]=Max[K]=2j=n-1**, since **Abs[m1]**, **Abs[m2]** $\leq$ **j**. Hence, **mz** and **K** each take on the **2n-1** values **-n+1, -n+2, ..., n-2, n-1**. We use the label **K** for the eigenvalue of **Mz[n]** since we shall identify it shortly with the *electric quantum number*, i.e. the parabolic separation constant from Problem 20.5-1.

Working in analogy with the orbital angular momentum and Problem 19.1-3, we first calculate for an arbitrary level **n** the "top rung" **phi[{j,j,j}]** for **m1 = m2 = j**. This function is defined as the normalized solution of

$$J1R[n] \ @ \ phi[\{j,j,j\}] \ == \ 0 \tag{21.2-r1}$$

(or equivalently of the same equation with **J2R**), which requires that

$$\begin{aligned} LR \quad & @ \ phi[\{j,j,j\}] \ == \ 0 \\ MR[n] \quad & @ \ phi[\{j,j,j\}] \ == \ 0 \end{aligned} \tag{21.2-r2}$$

From Problem 19.1-3, **phi[{j,j,j},r,t,p]** is thus proportional to the top-rung spherical harmonic **Y[2j,2j,t,p]** with **mz = Max[mz] = 2j**. The proportionality factor is a function of **r** to be determined. Hence, **phi[{j,j,j}]** is also an eigenfunction of **Lsq** with eigenvalue **l = Max[l] = 2j = n-1**. We thus set

```
phi[{j_,j_,j_},r_,t_,p_] = f[r] Y[2j,2j,t,p];

    SetAttributes[{n,l,j},Constant];
```

As a simultaneous eigenfunction of **Lsq** and **Lz**, this function when normalized must be equivalent to our spherical solution of the Schrödinger equation **psi[{n,n-1,n-1},r,t,p]** for **l = mz = n-1** from Section 20.2. To verify the equivalence, we calculate **phi[{j,j,j}]** for arbitrary **j** starting with the general expression for the top-rung spherical harmonic from Problem 19.1-3:

```
Y[l_,l_,t_,p_] = (-1)^l 2^(-1 - l)/l! *
                 E^(I l p) Sqrt[(1 + 2 l)!/Pi] Sin[t]^l
```

$$\frac{(-1)^l \; 2^{-1-l} \; E^{I l p} \; Sqrt[\dfrac{(1 + 2 l)!}{Pi}] \; Sin[t]^l}{l!}$$

We easily check that

```
LR @ phi[{j,j,j},r,t,p] == 0
```

```
    True
```

as required. We determine the radial function **f[r]** by applying the operator **MR[n]** and simplifying:

```
Expand[
    MR[n] @ phi[{(n-1)/2,(n-1)/2,(n-1)/2},r,t,p]/
        Y[n-1,n-1,t,p]
] //.
    Cos[x_] Cot[x_] :> Csc[x] - Sin[x] //
    Expand //Factor
```

$$\frac{E^{I p} \; n \; Sin[t] \; (-(n \; f[r]) + n^2 \; f[r] - r \; Z \; f[r] - n \; r \; f'[r])}{r \; Z}$$

Equating this expression to zero, we obtain a *first-order* differential equation in **f[r]**. In effect, we've "factored" the second-order Schrödinger equation with the Runge-Lenz algebra and thereby considerably reduced the effort required to extract a solution. We obtain the *unnormalized* top-rung radial wavefunction with **DSolve**:

```
f[{j_,j_,j_},r_] = f[r] /.
    DSolve[ % == 0, f[r], r ][[1]] /. n -> 2j+1
```

$$\frac{r^{2j} \; C[1]}{E^{(r \; Z)/(1 + 2 \; j)}}$$

We conveniently normalize these functions and evaluate **C[1]** for *arbitrary* **j** using **integExp**, thus defining normalized functions **F[{j,j,j},r]**. The function **PowerContract** from the package **Quantum`PowerTools`** compacts the result.

```
Needs["Quantum`integExp`"]
Needs["Quantum`PowerTools`"]

Crep =
    Solve[
        integExp[r^2 f[{j,j,j},r]^2,{r,0,Infinity}] == 1,
        C[1]
    ][[1]];

F[{j_,j_,j_},r_] = f[{j,j,j},r] /.Crep //
                PowerContract //MapAll[Factor,#]&
```

$$\frac{2^{3/2 + 2j}\, Z^{3/2} \left(\dfrac{r\, Z}{1 + 2\, j}\right)^{2\, j}}{E^{(r\, Z)/(1 + 2\, j)}\, \mathrm{Sqrt}[(1 + 2\, j)^3\, (2\, (1 + 2\, j))!]}$$

Specializing to **j = 0** and **j = 1/2**, we obtain for the *1s* ground state and the *2p* **(mz=1)** excited state, respectively,

```
{F[{0,0,0},r], F[{1/2,1/2,1/2},r]}
```

$$\left\{\frac{2\, Z^{3/2}}{E^{r\, Z}\, r\, Z}\,,\ \frac{r\, Z^{5/2}}{\mathrm{Sqrt}[24]\, E^{(r\, Z)/2}}\right\}$$

which we can compare with our spherical solutions from Section 20.2, *converted to atomic units*:

```
% == {R[1,0,r], R[2,1,r]} /. ao -> 1

    True
```

Exercise 21.2-1

Verify **f[{j,j,j}]** by requiring **f[r]Y[2j,2j]** to be an eigenfunction of **Mz[n]**.

Putting back in the top-rung spherical harmonic **Y[2j,2j]**, we obtain the full *normalized* top-rung wavefunction

```
phi[{j_,j_,j_},r_,t_,p_] = F[{j,j,j},r] Y[2j,2j,t,p];
```

which is identical with our spherical solution **psi[{n,n-1,n-1}]** from Section 20.2. For example,

```
phi[{1/2,1/2,1/2},r,t,p] == psi[{2,1,1},r,t,p] /. ao -> 1

    True
```

Clearly, the radial function `F[{j,j,j},r]` is independent of the orientation of the z axis. We would achieve the same solution starting with the bottom rung for `m1 = m2 = -j` and applying `ML[n]` instead. We thus obtain the full bottom-rung wavefunction `phi[{j,-j,-j}]` simply by putting in the bottom-rung spherical harmonic `Y[2j,-2j]` for `mz = Min[mz] = -2j`:

```
Needs["Quantum`QuickReIm`"]


phi[{j_,m1_,m2_},r_,t_,p_] :=
    F[{j,j,j},r] (-1)^(2j) Conjugate[Y[2j,2j,t,p]] /;
        m1 == m2 == -j
```

using the relation `Y[l,-mz] == (-1)^mz Conjugate[Y[l,mz]]` from Section 19.1. Again comparing with our spherical solution, we have for example that

```
phi[{1/2,-1/2,-1/2},r,t,p] == psi[{2,1,-1},r,t,p] /.
    ao -> 1

    True
```

For the ground state with `j = 0` there is only one rung, and the ground state is therefore nondegenerate.

Exercise 21.2-2

Verify that `phi[{j,j,j}]` and `phi[{j,-j,-j}]` are eigenstates of `Jisq[n]` for arbitrary `j = (n-1)/2`.

Down the Ladder 21.3

Let us now generate the remaining states in the **n = 2** manifold starting with the top rung and applying `J1L[n]` and `J2L[n]` successively until the bottom rung is reached. Since there are only `n^2 = 4` states altogether and we've already computed the bottom rung, we have just to calculate the two middle rungs and simplify:

```
{phi[{1/2,-1/2, 1/2},r_,t_,p_] =
    J1L[2] @ phi[{1/2,1/2,1/2},r,t,p] //
        Expand //TrigReduce //Factor,

 phi[{1/2, 1/2,-1/2},r_,t_,p_] =
    J2L[2] @ phi[{1/2,1/2,1/2},r,t,p] //
        Expand //TrigReduce //Factor}
```

$$\left\{ \frac{Z^{3/2}\ (-2 + r\ Z + r\ Z\ \mathrm{Cos}[t])}{8\ E^{(r\ Z)/2}\ \mathrm{Sqrt}[\mathrm{Pi}]},\ \frac{Z^{3/2}\ (2 - r\ Z + r\ Z\ \mathrm{Cos}[t])}{8\ E^{(r\ Z)/2}\ \mathrm{Sqrt}[\mathrm{Pi}]} \right\}$$

Note that the factor **Sqrt[(j+mi)(j-mi+1)]** (from Problem 16.1-1, Part (c)) generated by application of **JiL** equals unity when **j = mi = 1/2**. Hence, the above wavefunctions are normalized, since the top rung **phi[{j,j,j}]** is. Also, we can check the bottom rung by one more application of **J1L[2]** or **J2L[2]** to the appropriate middle rung (this takes about 2 minutes):

```
J1L[2] @ phi[{1/2, 1/2,-1/2},r,t,p] ==
J2L[2] @ phi[{1/2,-1/2, 1/2},r,t,p] ==
      phi[{1/2,-1/2,-1/2},r,t,p] //ExpandAll
```

    True

We easily verify that these solutions are simultaneous eigenfunctions of **Lz** and **Mz[2]** with *integer* eigenvalues **mz = m1+m2** and **K = m1-m2**, respectively:

```
{Lz @ phi[{1/2, 1/2, 1/2},r,t,p]/
     phi[{1/2, 1/2, 1/2},r,t,p],
 Lz @ phi[{1/2, 1/2,-1/2},r,t,p]/
     phi[{1/2, 1/2,-1/2},r,t,p],
 Lz @ phi[{1/2,-1/2, 1/2},r,t,p]/
     phi[{1/2,-1/2, 1/2},r,t,p],
 Lz @ phi[{1/2,-1/2,-1/2},r,t,p]/
     phi[{1/2,-1/2,-1/2},r,t,p]}//
       Expand //TrigReduce
```

   {1, 0, 0, -1}

```
{Mz[2] @ phi[{1/2, 1/2, 1/2},r,t,p]/
        phi[{1/2, 1/2, 1/2},r,t,p],
 Mz[2] @ phi[{1/2, 1/2,-1/2},r,t,p]/
         phi[{1/2, 1/2,-1/2},r,t,p],
 Mz[2] @ phi[{1/2,-1/2, 1/2},r,t,p]/
         phi[{1/2,-1/2, 1/2},r,t,p],
 Mz[2] @ phi[{1/2,-1/2,-1/2},r,t,p]/
          phi[{1/2,-1/2,-1/2},r,t,p]}//
            Expand //TrigReduce //Together
```

   {0, 1, -1, 0}

Note that the top- and bottom-rung states have **K = j-j = 0**.

We readily find that the Runge-Lenz vector **Avec** and therefore **Mz** change sign under a coordinate inversion and hence don't commute with **Parity**. (The Runge-Lenz vector, like the position and momentum vectors, is thus a *polar vector*, unlike the angular momentum which is an *axial vector*. See Section 20.3) Therefore, the eigenfunctions **phi[{j,m1,m2}]** of **Mz** do not in general have definite parity. However, linear combinations of the **phi** can be arranged which do (see Problem 20.5-1, Part (d)). As we shall see shortly, such functions are just the spherical solutions **psi[{n,l,mz}]**.

Of course, the top and bottom rungs are exceptions and do have a definite parity. This in fact explains why the eigenvalue **K** vanishes for these states, which you can prove in the following exercise.

Exercise 21.3-1

(a) Show that `psi[{1/2,1/2,-1/2}]` and `psi[{1/2,-1/2,1/2}]` are eigenstates of `J1sq` and `J2sq` but not of `Lsq`. Verify by direct calculation that these wavefunctions are normalized and that they are degenerate eigenstates of the hamiltonian in spherical coordinates. Show that they are not eigenstates of `Parity`, but rather transform into one another under parity.

(b) What is the parity of the top and bottom rungs? Prove that `K` has to vanish for these states.

Connection with the   21.4
Parabolic Separation

Collecting our results for the **n = 2** states, we see that the Runge-Lenz eigen-functions `phi[{m1,m2,mz}]` bear a one-to-one correspondence with the parabolic solutions `chi[{n1,n2,mz}]` of the Schrödinger equation from Problem 20.5-1. For the top- and bottom-rungs, this follows from the equivalence of the parabolic and spherical solutions, which we demonstrated in Problem 20.5-1, and then the equivalence of the spherical and Runge-Lenz solutions, which we demonstrated in Section 21.2. For the **n = 2** middle rungs, we have in addition from Problem 20.5-1, part **(d)**, that

```
{phi[{1/2,  1/2,-1/2},r,t,p] == chi[{0,1,0},r,t,p],
 phi[{1/2,-1/2, 1/2},r,t,p] == chi[{1,0,0},r,t,p]} /.
    ao->1
```

```
{True, True}
```

These results also indicate a direct relation between the eigenvalue `K =` `m1-m2` of `Mz[n]` and the parabolic quantum numbers `n1` and `n2`.

We in fact show in Problem 21.4-1 that the parabolic solutions `chi[{n1,` `n2,mz}]` provide simultaneous eigenfunctions of the Runge-Lenz operator `Az` with eigenvalue `Z/n (n2-n1)` (in atomic units) and thus of the scaled Runge-Lenz operator `Mz[n]` with eigenvalue `n2-n1`. Therefore, `K = m1-m2` `= n2-n1`. Hence, the parabolic solutions also provide eigenfunctions of the `Jiz == (Lz ± n/Z Az)/2` with eigenvalues `mi = (mz ±(n2-n1))/2`.

We link in this way the separation constant in parabolic coordinates from Problem 20.5-1 to the eigenvalue of `Az`. This reminds us of the separation in spherical coordinates in Section 20.1 in which the eigenvalue `l(l+1)` of `Lsq` is the separation constant.

The proof in Problem 21.4-1 relies on the transformation of `Az` to parabolic coordinates **u**, **v**, and **p**. The result in Part **(a)** of the problem already suggests that the parabolic coordinates are superior for expressing the symmetries of the hydrogen atom. In fact, we show in Problem 21.4-2 that `Az[u,v]` can be further simplified as `Az[u,v] == S[u] - S[v]`, i.e. the difference of two operators identical in form but one a function of **u** and the other a function of **v**. Furthermore, **S[u]** and **S[v]** can be directly related to the separation of variables in parabolic coordinates with the solutions **chi[{n1,n2,mz}]** as their eigenfunctions and with eigenvalues determined by **n1** and **n2**, respectively.

In the next section, we interpret the eigenvalue **K** physically in terms of the *linear Stark effect* and thus relate it to the energy splittings of the *nth* level when the atom is placed in a uniform electric field. For this reason, **K** is known as the *electric quantum number*. (The definition **K = n2 - n1** here is consistent with that used in the recent literature on approximate symmetries of two-electron atoms (see Feagin and Briggs [20] and also Rost and Briggs [57]), although Bethe and Salpeter [6], p. 230, call **n1 - n2** the electric quantum number. See also the remark in Problem 20.5-1, Part **(d)**.)

Exercise 21.4-1

Show that if the fundamental constants are restored the eigenvalue of **Az** is **Z e^2 K/n** while that of **Mz[n]** is **ℏ K**.

We have now several quantum numbers floating around to describe the same states, so before we go on it might be helpful if we summarize and give some perspective. The situation here is a good example of choosing alternative complete sets of commuting observables among several possible constants of the motion.

A general two-body problem has *six* coordinate degrees of freedom, three of which are accounted for by the center-of-mass motion. The reduced-mass motion provides the remaining three, which we describe in the hydrogen atom by the spherical coordinates **r**, **t** and **p** or by the parabolic coordinates **u**, **v** and **p**. Thus, there exist at most *six* independent commuting observables whose eigenvalue spectrums fully and uniquely specify linearly independent solutions of the Schrödinger equation. If this set includes the (time-independent) hamiltonian, then all of the observables must be *constants of the motion*. The components of the center-of-mass momentum can provide of course three constants of the motion and the hamiltonian a fourth. Whereas in the general problem two additional constants of the motion cannot be found (cf. the discussion on dynamical symmetry in Section 18.5), we now have several possibilities for the Coulomb problem. Of course, they are not all independent.

For example, we might use the pseudo angular momentum operators **J1sq == J2sq**, **J1z** and **J2z** as three commuting observables and constants of the motion. They provide the complete set of quantum numbers {**j,m1,m2**} with which to label solutions describing the reduced-mass motion. Alternatively, we might take the reduced-mass hamiltonian **H**, the Runge-Lenz component **Mz[n]** and the angular-momentum component **Lz** and label solutions with the three quantum numbers {**n,K,mz**}. Also, in parabolic coordinates (see Problem 21.4-2) one can replace **H** and **Mz[n]** by two new commuting observables and constants of the motion **S[u]** and **S[v]** whose eigenvalues are directly related to the quantum numbers **n1** and **n2**, respectively, and label solutions with the set **{n1,n2,mz}**.

Finally, recall that the spherical-coordinate labels **{n,l,mz}** correspond to the three operators **H**, **Lsq** and **Lz**. These labels, however, do not have a one-to-one correspondence with any of the above sets, except for the top- and bottom-rung states. As we shall demonstrate below in Section 21.6, this is because the spherical wavefunctions **psi[{n,l,mz}]** represent linear combinations of the parabolic ones **chi[{n1,n2,mz}]** (see also Problem 20.5-1, part **(d)**).

We can summarize and relate these different sets of labels by defining rules for converting from one set to another. Consider for example the lists **{j,m1,m2}** and **{n,K,mz}**, which we tag **jm1m2** and **nKmz** to distinguish explicit cases as, for example, **jm1m2[{0,0,0}]** and **nKmz[{1,0,0}]**. The following two rules convert back and forth between the lists using the connections we deduced at the beginning of this section:

```
nKmz   @ jm1m2[{j_,m1_,m2_}] :=
             nKmz[{2j+1,m1-m2,m1+m2}]
jm1m2 @   nKmz[{n_, K_,mz_}] :=
             jm1m2[{(n-1)/2,(mz+K)/2,(mz-K)/2}]
```

Thus, for example,

```
nKmz @ jm1m2[{1/2,1/2,1/2}]

   nKmz[{2, 0, 1}]
```

and back again

```
jm1m2 @ %

            1   1   1
   jm1m2[{-, -, -}]
            2   2   2
```

We can also apply **nKmz** to the entire **n = 2** manifold at once using **/@** (**Map**):

```
nKmz /@ {jm1m2[{1/2, 1/2,1/2}], jm1m2[{1/2, 1/2,-1/2}],
          jm1m2[{1/2,-1/2,1/2}], jm1m2[{1/2,-1/2,-1/2}]}

   {nKmz[{2, 0, 1}], nKmz[{2, 1, 0}],

      nKmz[{2, -1, 0}], nKmz[{2, 0, -1}]}
```

Note there are **2n-1** separate values of **K** and of **mz**.

In all, we can define six such rules connecting the three sets of labels **{j,m1,m2}**, **{n,K,mz}** and **{n1,n2,mz}**. Their construction is left as an exercise.

Exercise 21.4-2

Add the rules **nKmz @ n1n2mz**, **jm1m2 @ n1n2mz**, **n1n2mz @ nKmz** and **n1n2mz @ jm1m2** for converting between all three equivalent sets of quantum numbers **{j,m1,m2}**, **{n,K,mz}** and **{n1,n2,mz}kern -2pt**. Refer to Problem 20.5-1. Check your rules by showing for the **n = 2** manifold that **jm1m2 -> nKmz -> n1n2mz** and back to **jm1m2**.

Problem 21.4-1

Transform the Runge-Lenz vector **Avec** to *parabolic coordinates* and show that the parabolic wavefunctions **chi[n1,n2,m]** from Problem 20.5-1 are eigenfunctions of **Az** with eigenvalues **Z K/n** in atomic units, where **K = n2 - n1**. Refer to Problem 19.0-1 and to Problem 8, Appendix V.

(a) Thus, derive the *Cartesian* components of the Runge-Lenz vector **Avec** in *parabolic coordinates* **u**, **v**, and **p** and show that **Az @ psi ==**

$$-\left(\frac{psi\ u\ Z}{u\ +\ v}\right) + \frac{psi\ v\ Z}{u\ +\ v} + \frac{2\ v\ Dt[psi,\ u]}{u\ +\ v} - \frac{2\ u\ Dt[psi,\ v]}{u\ +\ v} +$$

$$\frac{Dt[psi,\ \{p,\ 2\}]}{2\ u} - \frac{Dt[psi,\ \{p,\ 2\}]}{2\ v} +$$

$$\frac{2\ u\ v\ Dt[psi,\ \{u,\ 2\}]}{u\ +\ v} - \frac{2\ u\ v\ Dt[psi,\ \{v,\ 2\}]}{u\ +\ v}$$

in atomic units. Verify that all the components have the right properties and satisfy the appropriate commutation relations. Refer to Exercise 21.1-1.

(b) Introduce the separation *ansatz* **chi[u,v,p] = f[u] g[v] E^ (I m p)** and relate **(Az @ chi)/chi** to the expressions **uPart** and **vPart** generated by the separation of variables in Problem 20.5-1, part **(a)**. Then introduce the separation constant **K Z/ (2n)** (in a.u.) and the fact that **uPart == -vPart == K Z/ (2n)** to show generally that **Az** has the eigenvalue **Z K/n**.

(c) Verify by direct computation for **n ≤ 3** that the parabolic states **chi[n1,n2,m]** are eigenfunctions of **Az**.

Problem 21.4-2

(a) Show that the hamiltonian **H** and the Runge-Lenz operator **Az** expressed in parabolic coordinates can be replaced by the pair of operators **S[u]** and **S[v]** where (in atomic units)

```
S[u_] @ psi =
      Energy/2 u psi + 1/(4u) Dt[psi,{p,2}] + Dt[u Dt[psi,u],u]
```

Refer to the previous problem. Hence, show that **Az[u,v] == S[u] - S[v]** and that the Schrödinger equation can be written as **S[u] @ chi + S[v] @ chi + Z chi == 0**. (Hint: Replace the eigenvalue **Energy** in **S[u]** and **S[v]** by the hamiltonian **H**.) Clearly, **S[u]** and **S[v]** commute. Show that they separately commute with **H**.

(b) The form of these operators is suggested by the expressions **uPart** and **vPart** generated in the parabolic separation of variables from Problem 20.5-1. Relate **S[u]** to

uPart and **S[v]** to **vPart** and show that **chi[{n1,n2,mz}]** is also an eigenfunction of **S[u]** and **S[v]** separately. Show that their eigenvalues **su** and **sv** scaled by **n/Z** are, respectively,

```
     -(1 + 2 n1 + Abs[mz])   -(1 + 2 n2 + Abs[mz])
   {---------------------- , ----------------------}
            2                         2
```

Thus, recover the eigenvalue of **Az** derived in the previous problem.

Problem 21.4-3

Starting with the top rung and applying the lowering operators **JiL[n]**, work out all nine *normalized* states **phi[{j,m1,m2}]** of the **n = 3** manifold and compare them with the solutions **chi[{n1,n2,mz}]** from Problem 20.5-1.

Linear Stark Effect   21.5

Let's calculate now matrix elements of an electric dipole potential between Runge-Lenz eigenstates **phi[{j,m1,m2}]** and thus evaluate the *linear* response of the atom to a uniform external electric field **Evec**. In the language of perturbation theory, we determine the first-order shifts in the energy of the *nth* level and the corresponding zeroth-order wavefunctions. The resulting variation in the spectroscopy of the atom is called the linear Stark effect.

We greatly reduce our efforts by taking the positive z axis in the direction of the electric field (cf. Problems 18.4-1 and 20.2-1). Then the interaction of the atom with the field is given by the perturbation potential

```
vD = e Eo ao z /. z -> r Cos[t]

   ao e Eo r Cos[t]
```

Here **e** is the (positive) magnitude of the electron charge and **Eo** the field strength. In order to obtain results in terms of the basic constants **e** and **ao**, we've expressed **r** in atomic units, since the wavefunctions are. (In effect, we set **z = ao(z/ao) -> ao z**.) We see that this potential has the proper sign by noting that it gives the correct direction of the classical force for a negatively charged electron, viz. **-D[vD,z] == -e Eo ao** is opposite the applied field.

The dipole interaction clearly has odd parity, since under coordinate inversion **z -> -z**. It follows (from our arguments at the end of Chapter 5) that matrix elements **vDme** of the dipole interaction between wavefunctions of the *same* parity vanish identically. We thus conclude on the basis of parity alone that all top- and bottom-rung states, which include the ground state, show no linear Stark effect, since these states have definite parity. Specifically,

```
vDme[{j_,m1_,m2_},{j_,m1_,m2_}] := 0 /;
     m1 == m2 == j || m1 == m2 == -j
```

This means that the hydrogen atom in its ground state doesn't have a *permanent electric dipole moment*, i.e. it doesn't exhibit any energy change linear in the electric field strength **Eo**. It is possible, however, to *induce* an electric dipole moment proportional to **Evec** and therefore an energy change proportional to **Eo^2**. This is referred to as the *quadratic* or, from the point of view of perturbation theory, *2nd-order Stark effect*. One in fact observes in Nature quite generally that atoms and nuclei in their ground states exhibit at best extremely small permanent electric dipole moments. (Nonvanishing moments could in principle arise from the *weak interaction*, which does not have a definite parity.)

As we have seen, the eigenstates **phi[{j,m1,m2}]** which constitute the middle rungs do not in general have a definite parity because the **Parity** operator doesn't commute with the Runge-Lenz vector. As we shall show in a moment, these states of the hydrogen atom exhibit a linear Stark effect. For excited states of atoms in general, however, this isn't the case since for more than two charges the special symmetry of a *point* Coulomb interaction is lost and parity is a good quantum number.

We can further simplify setting up the interaction matrix by noting that the dipole potential preserves rotational symmetry about the z axis and therefore commutes with the z component of the angular momentum (see Section 18.4 and Problem 18.4-1). Since the Runge-Lenz eigenstates **phi[{j,m1,m2}]** are also eigenstates of **Lz** with eigenvalue **m1 + m2**, we then obtain the *selection rule* that the matrix elements **vDme** vanish unless the **Lz** eigenvalues of the two states are identical, i.e.

```
vDme[{jp_,m1p_,m2p_},{j_,m1_,m2_}] := 0 /; m1p+m2p != m1+m2
```

Exercise 21.5-1

Show with paper and pencil that a similar selection rule holds for any interaction **V** if the interaction matrix elements are defined by eigenstates of a hermitian operator **Q** that commutes with **V**.

We now add a rule for calculating the remaining matrix elements **vDme** in the **jm1m2** basis. Because of the azimuthal symmetry in the remaining functions, the integral over **p** simply contributes a factor **2Pi**. We can integrate over **t** using the built-in function **Integrate**, but we shall integrate over **r** with **integExp** for convenience. In any case, it turns out to be faster to perform the **r** integration first. The following procedure should make these steps clear:

```
vDme[{jp_,m1p_,m2p_},{j_,m1_,m2_}] :=
    vDme[{jp,m1p,m2p},{j,m1,m2}] =
    Integrate[ 2Pi Sin[t] *
            integExp[ r^2 *
                Conjugate[phi[{jp,m1p,m2p},r,t,p]] *
                vD phi[{ j, m1, m2},r,t,p],
                {r,0,Infinity}
           ],
```

```
        {t,0,Pi}
   ]
```

The dynamic programming construct in the first two lines ensures that matrix elements are automatically stored if they're computed (see Exercise 3.1, Appendix III). As a simple check, we can verify that the diagonal matrix element for the ground state $j = 0$ vanishes:

```
vDme[{0,0,0},{0,0,0}] == 0
```

```
   True
```

For $n = 2$ we can conveniently calculate the matrix of **vD** as a **Table** if we first form a list of all possible pairs of **m1** and **m2**, as we did before constructing the angular momentum matrices in Section 17.1. Here we include the label j to correspond to the wavefunctions **phi[{j,m1,m2}]**:

```
m1m2 =
    Table[{1/2,-m1,-m2},{m1,-1/2,1/2},{m2,-1/2,1/2}] //
        Flatten[#,1]&
```

$$\left\{\left\{-\frac{1}{2},\ -\frac{1}{2},\ -\frac{1}{2}\right\},\ \left\{-\frac{1}{2},\ -\frac{1}{2},\ -\left(-\frac{1}{2}\right)\right\},\ \left\{-\frac{1}{2},\ -\left(-\frac{1}{2}\right),\ -\frac{1}{2}\right\},\ \left\{-\frac{1}{2},\ -\left(-\frac{1}{2}\right),\ -\left(-\frac{1}{2}\right)\right\}\right\}$$

We thus obtain the full dipole-interaction matrix in the $n = 2$ manifold according to (this takes about 4 minutes)

```
Table[ vDme[m1m2[[i]],m1m2[[j]]], {i,1,4},{j,1,4} ] //
        TableForm[#,TableAlignments -> Center]&
```

0	0	0	0
0	$\dfrac{-3\ \text{ao e Eo}}{Z}$	0	0
0	0	$\dfrac{3\ \text{ao e Eo}}{Z}$	0
0	0	0	0

A diagonal matrix! That is, the eigenfunctions **phi[{j,m1,m2}]** of a one-electron atom diagonalize not only the Coulomb interaction (the unperturbed hamiltonian) but also the electric dipole interaction, at least for a given energy or n value. The Runge-Lenz functions **phi[{j,m1,m2}]** are the *zeroth-order* wavefunctions of the perturbed subspace. The corresponding *first-order* perturbed energies are given (from Section 7.4) by the diagonal matrix elements according to

$$-Z\verb|^|2\ Ry/n\verb|^|2\ +\ vDme[\{j,m1,m2\},\{j,m1,m2\}] \qquad (21.5\text{-r1})$$

The dipole interaction thus partially lifts the degeneracy of the $n = 2$ level.

Since the **phi[{j,m1,m2}]** are equivalent to the parabolic wavefunctions **chi[{n1,n2,mz}]**, a diagonal matrix suggests that the hydrogen-atom Schrödinger equation in parabolic coordinates also separates in the presence of a uniform electric field. We in fact show this in Problem 21.5-1. Moreover, the separation suggests that the Runge-Lenz vector commutes with the dipole interaction **vD**, a result which turns out to be true *in a given energy subspace* (see Biedenharn and Louck [7], p. 360). The **phi[{j,m1,m2}]** or **chi[{n1,n2,mz}]** are thus also referred to as *Stark states*.

That these properties are confined to a particular subspace becomes evident when we note how matrix elements connecting different subspaces are in general *nonvanishing*. For example, considering the **n = 1** ground state and the two **n = 2, mz = 0** excited states, we obtain the matrix element values

```
{vDme[{0,0,0},{1/2,1/2,-1/2}],
 vDme[{0,0,0},{1/2,-1/2,1/2}]}
    128 ao e Eo   128 ao e Eo
  {------------, ------------}
      243 Z         243 Z
```

In fact, these matrix elements contribute in perturbation theory to the *2nd-order* Stark effect of the ground state. Nevertheless, we also show in Problem 21.5-1 that a generalized Runge-Lenz vector can be constructed which includes the electric field and commutes with **vD**. We also note in this regard that the *1st- and 2nd-order* Stark effects in hydrogen can be calculated in closed form in a purely algebraic way (see H.G. Becker and K. Bleuler, Z. Naturforsch. **31a**, 517 (1976)).

Exercise 21.5-2

It's much simpler to calculate the dipole matrix elements in parabolic coordinates **u**, **v** and **p**. Transform the wavefunctions **phi[j,m1,m2}]** and redefine **vDme** in parabolic coordinates. Refer to Problem 20.5-1. Verify our above results, then see if you can evaluate the top- and bottom-rung diagonal matrix elements generally using **integExp**.

It's evident that the diagonal matrix elements are proportional to the electric quantum number **K = m1−m2**. To make this clear, we need to match up the diagonal elements with the labels **{n,K,mz}**, which we conveniently do using our conversion rule **nKmz** from the previous section. Thus, mapping first the list name **jm1m2** onto the label list **m1m2** and then the rule **nKmz**, we see that the diagonal elements (from left to right)

```
jm1m2 /@ m1m2
            1   1   1              1   1      1                  1      1   1
 {jm1m2[{-, -, -}], jm1m2[{-, -, -(-)}], jm1m2[{-, -(-), -}],
            2   2   2              2   2      2                  2      2   2
            1      1      1
   jm1m2[{-, -(-), -(-)}]}
            2      2      2
```

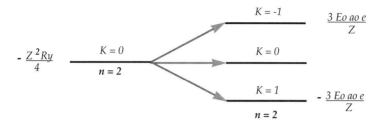

Figure 21.5-1. Splitting of the **n** = 2 hydrogenic level by an electric dipole interaction. Here **Eo** is the electric field strength and **K** the electric quantum number, proportional to the eigenvalue of the z component of the Runge-Lenz vector.

correspond to

```
nKmz /@ %
   {nKmz[{2, 0, 1}], nKmz[{2, 1, 0}],
      nKmz[{2, -1, 0}], nKmz[{2, 0, -1}]}
```

Thus, the first and the last diagonal elements have **K = 0** and therefore correspond to the top and bottom rungs states. Hence, these two sublevels remain degenerate, as we predicted on the basis of parity. The second matrix element has **K = +1** and lowers the energy by **3Eo ao e/Z**, whereas the third element has **K = −1** and raises the energy by **3Eo ao e/Z**. Note that both **K** $\neq$ **0** states have the same value **mz = 0**. For these two levels, the perturbation has lifted the degeneracy. This splitting of the **n = 2** level is depicted in Figure 21.5-1.

We can in fact show generally that the diagonal matrix elements are given by **−3e Eo ao n K/(2Z)** such that the *nth* energy level splits independently of **mz** into **2n−1** sublevels corresponding to the **2n−1** possible values of the electric quantum number **K**. The derivation of this result is the subject of Problem 21.5-2.

The **n = 2** middle-rung (**Z = 1**) probability distributions are plotted in Figure 21.5-2 as functions of **x** and **z**, as in Figure 20.4-1. The figures make clear that these two states transform into one another under coordinate inversion, viz. **x -> −x, z -> −z** (see Exercise 21.3-1). The parabolic shape of the nodal lines is also evident. Moreover, it's apparent that the *2p* (**mz = 0**) distribution in Figure 20.4-1 is a linear combination of these two middle-rung distributions (see Problem 20.5-1 and Section 21.6 below).

For the case **K > 0**, it's evident from the left-hand plot in Figure 21.5-2 that the electron favors the negative z axis (nucleus at the origin) and hence the "downhill" side of the dipole potential. (The broad shoulder on the negative side contains more probability than the relatively sharp peak on the positive side.) Likewise, it's apparent from the right-hand plot in Figure 21.5-2 that for **K < 0** the electron favors the "uphill" side. Thus, the energy of the **K > 0**

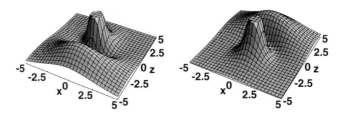

Figure 21.5-2. Hydrogenic ($Z = 1$) $n = 2$ probability distributions corresponding to the $K = 1$ (on the left) and $K = -1$ (on the right) levels depicted in Figure 21.5-1. (The maximum in these figures has been arbitrarily chopped to emphasize the background.)

state is lowered by the dipole interaction, while that of the $K < 0$ state is raised, as depicted in Figure 21.5-1.

We used essentially the same argument in setting up the dipole interaction matrix to show that the top- and bottom-rung $K = 0$ states of definite parity and therefore with symmetric charge distribution about the origin have zero dipole potential energy (diagonal matrix elements). Hence, the energy of these states is unchanged in lowest order by the dipole interaction, as depicted in Figure 21.5-1.

We thus see that the $K \neq 0$ states behave as though the atom has a permanent electric dipole moment of magnitude $3\,e\,a0\,/\,Z$ oriented for $K < 0$ parallel to the external field and for $K > 0$ antiparallel. Because of their symmetry, the $K = 0$ states have no permanent electric dipole moment. (Recall that the dipole moment is defined as the integral of **rvec** over the charge distribution $-$**e Abs[phi]^2**.)

Exercise 21.5-3

Compare the plots in Figure 21.5-2 with similar plots of the distributions corresponding to the solutions **psi[2s]** and **psi[2p(mz=0)]** in spherical coordinates. (See Problem 20.5-1, Part **(d)**.) Also, make contour (**ContourPlot**) and density plots corresponding to Figure 21.5-2 and note the underlying parabolic symmetry.

The Stark effect is usually discussed in terms of the spherical solutions **psi[{n,l,mz}]**. As we have seen, however, these wavefunctions are not eigenfunctions of the Runge-Lenz operator **Az** and therefore do not diagonalize the dipole interaction **vD**. Rather, matrix diagonalization has to be performed in an extra step. Of course, the resulting (matrix) eigenvectors determine as usual the coefficients of linear combinations of the spherical solutions **psi[{n,l,mz}]** which are also eigenfunctions **phi[{j,m1,m2}]** of **Az**. We investigate this approach in Problem 21.5-3. We show in the next section that these expansion coefficients are just the Clebsch-Gordan coefficients which provide the coupling of the pseudo angular momenta `J1vec` and `J2vec` to form the orbital angular momentum `lvec`.

Problem 21.5-1

(a) Show that the hydrogen-atom Schrödinger equation separates in parabolic coordinates even in the presence of a uniform electric field `Eo nvec` (cf. Problem 18.4-1). Assume that the field points along the *z* axis so that the perturbation (dipole) potential is given by `vD`. You don't have to solve the resulting equations in **u** and **v**, which is usually done numerically or by a perturbation procedure.

(b) Verify that the *generalized* Runge-Lenz vector

```
Cvec ==
    (pvec×lvec - lvec×pvec)/(2m) -
        Z e^2 rvec/r + e/2 rvec×Evec×rvec
```

commutes with the Coulomb hamiltonian *including* an electric dipole interaction. Refer to Redmond [56].

Problem 21.5-2

Calculate the diagonal matrix elements of the dipole interaction **vD** using the parabolic solutions `chi[{n1,n2,mz}]` of the Schrödinger equation from Problem 20.5-1. Show for *arbitrary* `n1` and `n2` that a general element is given by `-3e Eo ao n K/ (2Z)`, where `K = n2-n1`.

Hint: You'll first have to calculate the normalization `chiNorm` of the parabolic states for arbitrary `n1`, `n2` and `mz`. Use the following general result (derived in Bethe and Salpeter [6], Section 3) for integrals involving Laguerre polynomials:

```
Integrate[E^-r r^t LaguerreL[n,m,r]^2,{r,0,Infinity}] ==
    j[n+m,m,t-m]
```

where

```
j[l_,m_,s_] :=
    (-1)^s l! s!/(l-m)! *
    Sum[
        (-1)^b Binomial[s,b] Binomial[l+b,s] Binomial[l+b-m,s],
        {b,0,s}
    ] /; s >= 0
```

Check your rule by comparing with the normalization integral given in Exercise 20.2-1. (This expression differs somewhat from that given in Bethe and Salpeter, since here `LaguerreL` is the *Mathematica* function. See Exercise 20.2-1). Show that in atomic units `chiNorm[n1,n2,mz] ==`

$$
n^{-2 - \text{Abs}[mz]} \; Z^{3/2 + \text{Abs}[mz]}
$$

$$
\text{Sqrt}\left[\frac{n1! \; n2!}{\text{Pi} \; (n1 + \text{Abs}[mz])! \; (n2 + \text{Abs}[mz])!}\right]
$$

Problem 21.5-3

Calculate the **n = 2** matrix **vDme** of the dipole interaction **vD** using the *spherical* eigenfunctions **psi[{n,1,mz}]** from Section 20.2. Diagonalize **vDme** and verify the Stark shifts obtained in the text. Use the matrix of eigenvectors to obtain the eigenfunctions **phi[{j,m1,m2}]** as linear combinations of the **psi[{n,1,mz}]**.

Problem 21.5-4

Extend our Stark-effect calculation in the text to the **n = 3** manifold and calculate the matrix **vDme** of the dipole interaction **vD** in the Runge-Lenz **jm1m2** basis. Verify that the matrix is diagonal with diagonal elements given by **-3e Eo ao n K/(2Z)**, in agreement with the general result from Problem 21.5-2.

21.6   Connection with the Spherical Separation

Finally, we connect the eigenfunctions **phi[{j,m1,m2}]** of the pseudo angular momenta **Jisq** and **Jiz** with the eigenfunctions **psi[{n,1,mz}]** of the orbital angular momentum **1sq** and **1z**. Since **1vec == J1vec + J2vec** from relation (18.7-r2), we require simply an angular momentum coupling, a problem we solved generally with the Clebsch-Gordan coefficients in Section 17.3. Recalling that **j = (n-1)/2**, the connection between the two sets of eigenfunctions is therefore given by

```
psiCG[{n_,1_,mz_},r_,t_,p_] :=
    Sum[
        c[(n-1)/2,(n-1)/2,{m1,m2},{1,mz}] *
            phi[{(n-1)/2,m1,m2},r,t,p],
        {m1,-(n-1)/2,(n-1)/2}, {m2,-(n-1)/2,(n-1)/2}
    ]
```

where the expansion coefficients are just the Clebsch-Gordan coefficients from Section 17.3

```
Needs["Quantum`Clebsch`"]
```

```
c[j1_,j2_,{m1_,m2_},{j_,m_}] :=
    Clebsch[{j1,m1},{j2,m2},{j,m}]
```

Here, we could also use the built-in function **ClebschGordan**. We introduce the function name **psiCG** to distinguish this superposition from our previous spherical solutions **psi[{n,1,mz}]** from Section 20.2. Thus, for the **n = 1** ground state we verify that

```
psiCG[{1,0,0},r,t,p] == psi[{1,0,0},r,t,p] /. ao->1
```

```
        True
```

The connection in the **n = 2** manifold amounts to *spin-$1/2$* coupling (cf. Exercise 17.3-2). Hence, we have the "singlet"

```
psiCG[{2,0,0},r,t,p] == psi[{2,0,0},r,t,p] /. ao->1 //
    ExpandAll
```

    True

and the "triplets"

```
{psiCG[{2,1, 1},r,t,p] == psi[{2,1, 1},r,t,p],
 psiCG[{2,1, 0},r,t,p] == psi[{2,1, 0},r,t,p],
 psiCG[{2,1,-1},r,t,p] == psi[{2,1,-1},r,t,p]} /.
    ao->1 //ExpandAll
```

    {True, True, True}

The **psiCG[{n,l,mz}]** are normalized, as long as the **phi[{j,m1, m2}]** are, since the Clesbsch-Gordan transformation is unitary. Moreover, we easily obtain the inverse transformation and therefore the Runge-Lenz or Stark-state solutions **phi[{j,m1,m2}]** as linear combinations of the spherical solutions **psi[{n,l,mz}]** summed over **l** and **mz**. Formally, we have for example that

```
ketCG[{j_,m1_,m2_}] :=
    Sum[
        c[j,j,{m1,m2},{l,mz}] ket[{2j+1,l,mz}],
        {l,0,2j},{mz,-l,l}
    ]
```

where we have substituted **n = 2j+1**. The formal expansion lets us see which **{n,l,mz}** states contribute to a particular linear combination. For example, the two middle rungs in the **n = 2** manifold are found to be

```
{ketCG[{1/2,-1/2,1/2}], ketCG[{1/2,1/2,-1/2}]}
```

$$\left\{-\left(\frac{ket[\{2, 0, 0\}]}{Sqrt[2]}\right) + \frac{ket[\{2, 1, 0\}]}{Sqrt[2]}\,,\right.$$

$$\left.\frac{ket[\{2, 0, 0\}]}{Sqrt[2]} + \frac{ket[\{2, 1, 0\}]}{Sqrt[2]}\right\}$$

that is, linear combinations of the *2s* and *2p* (**mz = 0**) spherical states (cf. Problem 20.5-1 and Exercise 21.6-1). Of course, we can always substitute in wavefunctions and obtain the coordinate representation of these states. Thus,

```
{phiCG[{1/2,-1/2,1/2}], phiCG[{1/2,1/2,-1/2}]} =
{ketCG[{1/2,-1/2,1/2}], ketCG[{1/2,1/2,-1/2}]} /.
    ket[{n_,l_,mz_}] -> psi[{n,l,mz},r,t,p] /.
        ao -> 1 //ExpandAll //Map[Factor,#]&
```

```
     3/2                                    3/2
    Z     (-2 + r Z + r Z Cos[t])     Z     (2 - r Z + r Z Cos[t])
  { ---------------------------------, --------------------------------- }
              (r Z)/2                           (r Z)/2
          8 E          Sqrt[Pi]             8 E          Sqrt[Pi]
```

which agree with our previous solutions. That is, from Section 21.3

```
% == {phi[{1/2,-1/2,1/2},r,t,p],
      phi[{1/2,1/2,-1/2},r,t,p]}
```

```
True
```

The formal expansion is also handy for dissecting and calculating higher states. The **phiCG[{3/2,-1/2,3/2}]** Stark-state probability distribution has been determined in this way and plotted for **Z = 1** in Figure 21.6-1, as a function of the Cartesian coordinates **x** and **z**, as in Figure 21.5-2, and as a function of the parabolic coordinates **u == r + z, v == r - z**. This state corresponds to {**n,K,mz**} = {**4,-2,1**} or {**n1,n2,mz**} = {**2,0,1**}, which we can check using the additional conversion rules from Exercise 21.4-2:

```
{nKmz @ jm1m2[{3/2,-1/2,3/2}],
 n1n2mz @ jm1m2[{3/2,-1/2,3/2}]}
```

```
{nKmz[{4, -2, 1}], n1n2mz[{2, 0, 1}]}
```

It is evident from the **x-z** plot on the left in Figure 21.6-1 that the electron favors the positive z axis consistent with **K < 0**. The parabolic nodal patterns in this figure are striking, and it's useful to replot the distribution in parabolic coordinates **u** and **v**. We can make the coordinate substitution {**r,t**} -> {**u == r(1+Cos[t]),v == r(1-Cos[t])**} directly in **phiCG[{3/2,-1/2,3/2},r,t,p]** or use the parabolic solutions **chi[{2,0,1},u,v,p]** (== **phi[{3/2,-1/2,3/2}]**) from Problem 20.5-1.

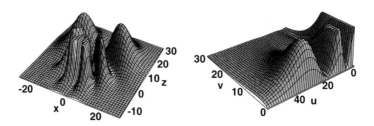

Figure 21.6-1. Hydrogenic (**Z = 1**) {**n,K,mz**} = {**4,-2,1**} probability distribution. Plotted on the left as a function of **x** and **z**, as in Figure 21.5-2, and on the right as a function of the parabolic coordinates **u** and **v**.

The resulting distribution is shown on the right in Figure 21.6-1. Here **u** and **v** act like *Cartesian* coordinates (except that both **u** and **v** are defined for {0,Infinity}) and the nodal patterns are rectangular. We thus clearly see that there are **n1 = 2** nodes along the **u** direction, and **n1 = 0** nodes along the **v** direction, as we found in Problem 20.5-1. We show in Exercise 21.6-1 that the mapping of these rectangular nodal lines onto the *x-z* plane gives the parabolic patterns evident in the **x-z** plot on the left.

The Clebsch-Gordan expansion is often used in applications to form directly linear combinations of the parabolic or Stark-state wavefunctions **chi[{n1,n2,mz},u,v,p]** which are also eigenfunctions of the *orbital* angular momentum operators **lsq** and **lz** *expressed in parabolic coordinates* (see Problem 19.0-1). One thus defines in this way **psiCGuvp[{n,l,mz},u,v,p]** in analogy with the superposition **psiCG[{n,l,mz},r,t,p]** but with the basis functions **phi[{j,m1,m2},r,t,p]** replaced by the **chi[{n1,n2, mz},u,v,p]** (see Problem 21.6-1).

Exercise 21.6-1

(a) Calculate **phiCG[{3/2,-1/2,3/2}]** as a function of **x = rSin[t]** and **z = rCos[t]** and simplify. Generate the left-hand side of Figure 21.6-1.

(b) Transform **phiCG[{3/2,-1/2,3/2}]** to parabolic coordinates **u** and **v**, simplify and generate the right-hand side of Figure 21.6-1. Both plots should take less than 5 minutes. Verify that **phiCG[{3/2,-1/2,3/2}]** transformed to parabolic coordinates equals the wavefunction **chi[{3/2,-1/2,3/2},u,v,p]** from Problem 20.5-1.

(c) Make parametric plots (**ParametricPlot**) of the coordinate pair {x,z} with **x** and **z** expressed *parametrically* as functions of **u** and **v** for fixed values of **u**. Choose **u** from the zeros of **chi[{3/2,-1/2,3/2},u,v,p]** along the **u** direction and compare with a **ContourPlot** of **phiCG[{3/2,-1/2,3/2}]** as a function of **x** and **z**. Refer to Problem 8, Appendix V.

Problem 21.6-1

(a) Define the Clebsch-Gordan expansion **psiCGuvp[{n,l,mz},u,v,p]** in analogy with the superposition **psiCG[{n,l,mz},r,t,p]** we set up in the text but using the parabolic wavefunctions **chi[{n1,n2,mz},u,v,p]** as basis functions instead. You will need to express **n1** and **n2** in terms of **j**, **m1** and **m2** as in Exercise 21.4-2. You should find for example that **psiCGuvp[{2,1,0},u,v,p] ==**

$$\frac{\text{chi}[\{0, 1, 0\}, u, v, p]}{\text{Sqrt}[2]} + \frac{\text{chi}[\{1, 0, 0\}, u, v, p]}{\text{Sqrt}[2]}$$

(b) Verify by explicit calculation for **n = 3** that the wavefunctions **psiCGuvp** diagonalize the orbital angular momentum operators **lsq** and **lz** expressed in parabolic coordinates from Problem 19.0-1.

Appendix I
Mathematica
Quick View

Some one-line examples of *Mathematica* are given for the beginner. To make *Mathematica* work, place the pointer in an *Input* cell (point and click) and hit **Enter** (*not* **Return**). *Mathematica* remembers everything you enter, and you can obtain unpredictable results if you start using a quantity in a new context different from the one you originally defined it in. If you suspect this is happening, enter **Clear**[*quantity*] and redefine it. As you work through these examples, you should modifiy them and experiment on your own.

Mathematica will describe to you any of its own built-in functions by entering **?***anything*, for example, **?%** or **?NIntegrate**. Appendices II and III advance the ideas introduced here.

```
48 + 23/576 - 16/99^2          (* exact integer arithmetic *)

% //N                          (* "history" with %;
                                  numerical approximations *)

N[ Sqrt[Pi], 100 ]             (* adjust precision *)

N[ (2143/22)^(1/4), 12 ]       (* number close to Pi *)

N[ % - Pi, 12 ]                (* previous result minus Pi *)

6 Pi^5  //N                    (* approx. proton/electron
                                  mass ratio *)
```

```
N[ Sqrt[Pi + I], 50 ]            (* complex arithmetic *)

Abs[%]

list = {a,b}            (* list => vector *)

list.list               (* dot product *)

matrix = {{m11,m12},{m21,m22}}
                        (* list of lists => matrix, tensor *)

MatrixForm[matrix]      (* formatted output *)

r = Sqrt[x^2 + y^2]     (* symbolic assignments *)

D[1/r, x]               (* symbolic derivatives *)

{ FortranForm[%], TeXForm[%] }
                        (* formatted output *)

D[1/r, x] /. x^2 + y^2 -> rsq
                        (* pattern matching and replacement *)

Plot3D[ Sin[x y], {x,-2,2}, {y,-2,2} ];

Integrate[ x^4 E^x, x ]
                        (* symbolic integration *)

D[%, x] //Expand        (* check results by differentiating *)

NIntegrate[ Sin[Sin[x]], {x,0,1} ]
                        (* numerical integration *)

Plot[ Sin[Sin[x]], {x,0,1} ];
                        (* syntax similar to Integrate *)

(1+x+y)^4 //Expand      (* algebraic manipulation *)

%/(1+x+y) //Factor

Solve[ 1 + 3x + x^3 == 0, x ] //Simplify
                        (* cubic equation *)

% //N                   (* conjugate-pairs *)

NSolve[ 1 + 3x + x^3 == 0, x ]
                        (* numerical solution faster *)

Plot[ 1 + 3x + x^3, {x,-2,2} ];
                        (* locate real roots graphically *)

BesselJ[1, 1.1 + I]     (* evaluate special functions *)

Plot3D[ BesselJ[1, 3. r], {x,-3,3}, {y,-3,3} ];

BesselJ[1,z] + O[z]^6 (* Taylor series up to order z^6 *)

BesselJ[3/2, z]         (* polynomial *)
```

Appendix II
Notebooks and
Basic Tools

As a brief introduction to *Mathematica* and physics notebooks, we investigate the motion of a *classical* projectile. First we neglect air resistance and then include it with a discussion of numerical integration.

Key points about *Mathematica* usage are included in a series of footnotes. Refer to *Appendices I* and *III* and to the *Mathematica* manual by Wolfram [62].

Projectile Motion Ignoring   AII.1
Air Resistance

Trajectory   AII.1.1

We enter the projectile's position coordinates almost as we would write them down on paper:

```
x = vo t Cos[a] Cos[p]
y = vo t Cos[a] Sin[p]
z = vo t Sin[a] - g/2 t^2

  t vo Cos[a] Cos[p]

  t vo Cos[a] Sin[p]

       2
   -(g t )
   ─────── + t vo Sin[a]
      2
```

where **g** is the acceleration due to gravity and **vo** the projectile's initial speed with elevation angle **a** above the horizontal x-y plane and azimuthal angle **p** about the vertical z axis. We can check for example the vertical acceleration by calculating the second derivative with respect to time

```
D[z,{t,2}]
```

```
 -g
```

We can also check the projectile's (constant) speed in the x-y plane with

```
Sqrt[D[x,t]^2 + D[y,t]^2] //Simplify
          2      2
 Sqrt[vo  Cos[a] ]
```

The square root of the squares can be simplified with

```
%  //PowerExpand
```

```
 vo Cos[a]
```

We easily calculate the projectile's distance to the origin as

```
r = Sqrt[ x^2 + y^2 + z^2 ]
```

$$
\text{Sqrt}[t^2\ \text{vo}^2\ \text{Cos}[a]^2\ \text{Cos}[p]^2 + (\frac{-(g\ t^2)}{2} + t\ \text{vo}\ \text{Sin}[a])^2 +
$$

$$
t^2\ \text{vo}^2\ \text{Cos}[a]^2\ \text{Sin}[p]^2]
$$

We can put this into simpler form by expanding the squares and doing some trigonometry. (You could also use the built-in function **Simplify**, although it takes longer.)

```
r = Sqrt[x^2 + y^2 + z^2] //Expand[#,Trig->True]&
```

$$
\frac{\text{Sqrt}[g^2\ t^4 + 4\ t^2\ \text{vo}^2 - 4\ g\ t^3\ \text{vo}\ \text{Sin}[a]]}{2}
$$

Here *expr* **//Expand[#,Trig->True]&** is equivalent to **Expand[*expr*, Trig->True]** and is a convenient way to apply this function while keeping the physics in the spotlight on the left and the *Mathematica* in the wings on the right. The **#** sign is a *slot* for the expression which appears to the left of the function-application bars **//**, and the **&** sign tells *Mathematica* where to end the operator.

Footnote 1

User input and *Mathematica* output are denoted **In[n]** and **Out[n]**. These labels for the *nth* output step can be referred to at any time and even used for computation. A **%** (percent sign) can also be used to refer to the previous output, and a **%n** to the *nth* previous output.

Information on any *Mathematica* command can be obtained by entering a **?** followed by the command. Entering **??** will give the command's options as well. Try for example **?D** and **??Expand**.

Mathematica input uses spaces to denote multiplication (and also an asterisk ***** if you don't mind the extra clutter!) and a hat **^** to denote raising to a power.

Mathematica works by nesting **functions** and evaluating the inner-most function first; thus *expr* **//Simplify //PowerExpand** is equivalent to **PowerExpand[Simplify[***expr***]]**. In an algebraic expression, functions including powers are evaluated first and then evaluation proceeds left to right with multiplication and division performed before addition and subtraction.

Mathematica uses several kinds of equals signs for command control. For example, an ordinary equation like **f[x]** = *expr* (called **Set**) evaluates the right hand side immediately, generating an immediate response from *Mathematica*. An assignment like **f[x]:=** *expr* (typed colon, =, called **SetDelayed**) delays evaluation of *expr* until the function **f** is called; it also suppresses a response from *Mathematica*. Logical *equal* and *not-equal* are denoted **==** (double equal) and **!=** (**.EQ.** and **.NE.** in Fortran) respectively. Logical *and* and *or* are denoted **&&** and **||** (pair of vertical bars).

Replacements, i.e. substitutions, for patterns inside expressions are made using **/.** (typed slash, period) in conjunction with arrows **->** (hyphen, >), as for example *expr* **/.** *pattern* **->** *new*.

Mathematica ignores anything enclosed by **(*** and ***)**. For example, comments can be added anywhere inside an input cell in the form **(* comments *)**. These parentheses can also be useful when debugging code to remove selected elements temporarily.

Built-in *Mathematica* functions begin with capital letters and are generally whole words. Thus it's usually a good idea to begin your own functions and symbols with lower case to avoid collision with *Mathematica* quantities.

Trajectory Plot   AII.1.2

The equations for **x**, **y** and **z** are parametric in **t**, and we can make a plot of the trajectory using **ParametricPlot3D**. Let's look at the information *Mathematica* has on this function:

```
?ParametricPlot3D
    ParametricPlot3D[{fx, fy, fz}, {t, tmin, tmax}] produces a
       three-dimensional space curve parameterized by a variable
       t which runs from tmin to tmax. ParametricPlot3D[{fx, fy,
       fz}, {t, tmin, tmax}, {u, umin, umax}] produces a three-
       dimensional surface parametrized by t and u.
       ParametricPlot3D[{fx, fy, fz, s}, ...] shades the plot
       according to the color specification s.
       ParametricPlot3D[{{fx, fy, fz}, {gx, gy, gz}, ...}, ...]
       plots several objects together.
```

Footnote 2

ParametricPlot3D is a long word. You can have *Mathematica* write out most of it for you by typing, say, the first two letters **Pa** and then **command-k**. You can then complete the selection from the pop-up menu. If you immediately type **command-i** *Mathematica* will give a template for the function. The template is also given by **?ParametricPlot3D**.

We need to give values for **vo**, **g** and **a** to specify a trajectory. A good way to do this is to use replacement (**/.**). We thus set up a vector *list* of the coordinates **x**, **y**, **z** and substitute another *list* of values for all the parameters except **t** (here **Degree** is a built-in conversion factor with the value **Pi/180**):

```
trajectory = {x,y,z} /.
        {g->9.8, vo->10, a->15 Degree, p->45 Degree} //N
                                                        2
    {6.83013 t, 6.83013 t, 2.58819 t - 4.9 t}
```

We can now plot **trajectory**. We see that **t = 1** is longer than the time of flight, if **z = 0** is the ground.

```
trajectoryPlot =
    ParametricPlot3D[
        trajectory, {t, 0, 1},
        PlotRange -> {-.75,.75}, BoxRatios -> {1,1,.6},
        FaceGrids -> {{0,0,-1}},
        ViewPoint -> {2.176, -2.401, 0.975},
        AxesLabel -> {"x","y","z"}
    ];
```

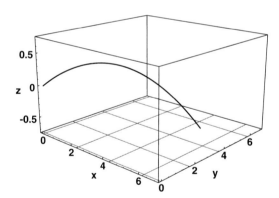

Here the **ViewPoint** was chosen using the *3D* viewpoint selector from *Prepare Input* in the *Action* menu.

AII.1.3 Range

We determine the time of flight and the *range* exactly by setting **z = 0** and solving for **t = T**, the time of flight:

```
T = Solve[ z == 0, t ]        (* logical equal *)
            2 vo Sin[a]
    {{t -> ───────────}, {t -> 0}}
                g
```

The range **R** is then obtained by substituting the nontrivial first component **T[[1]]** of the solution *list* **T** into the expression for **x** (see Exercise 1.3, Appendix III). Using the replacement command **/.**, we obtain

```
R[a_] = Sqrt[x^2 + y^2] /. T[[1]] /.
            f_ Cos[p]^2 + f_ Sin[p]^2 -> f //
            PowerExpand
```

$$\frac{2 \text{ vo}^2 \text{ Cos[a] Sin[a]}}{g}$$

We have made a pattern replacement to introduce a trigonometric identity (cf. Exercise 2.2, Appendix III) and then applied the built-in function **PowerExpand** to simplify **Sqrt**. You should examine each of these steps by repeating the calculation starting with just the first step **/. T[[1]]** and seeing what happens as each step is added in turn. You can also compare the result with that obtained using the built-in function **Simplify**.

Footnote 3

The quantity **a_** (read a-blank) is a dummy variable which stands for *any* expression, for example the value **45 Degree** or **a + 30**, without the need to declare **a** beforehand (see Exercise 1.4, Appendix III).

A list, enclosed with curly brackets **{...}**, collects together a sequence of symbols or expressions such as the position vector **{x,y,z}**, or the above solution set **T**.

Functions like **Solve[...]** require square brackets, while parentheses are reserved for algebraic expressions like **(x^2 + y^2 + z^2)^(1/2)**. Double square brackets denote components of lists and expressions, such as **{x,y,z}[[1]] == x** or **T[[1]]**. This somewhat strict syntax makes *Mathematica* input more readable, while removing ambiguities in conventional notation. For example, **c[x]** denotes a function of **x**, while **c(x)** means **c x**.

Let's now evaluate the time of flight for the above example and plot the trajectory using it. Including **//N** on the end gives a numerical result.

```
t /. T[[1]] /.{g -> 9.8, vo -> 10, a -> 15 Degree} //N

0.528202
```

We can use the result to make a *2D* parametric plot of the motion in the plane of the trajectory.

```
ParametricPlot[
    {Sqrt[x^2 + y^2], z}  /.
    {g->9.8, vo->10, a->15 Degree, p->45 Degree},
    {t, 0, %},                      (* tmax = % *)
    PlotRange -> {0,0.5},
    AxesLabel -> {" Sqrt[x^2+y^2]","z"}
];
```

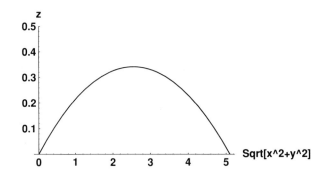

Footnote 4

We can estimate the range from the graph. With the mouse, select the plot cell by clicking on it, and move the pointer around while holding down the command key. The coordinates of the pointer are then displayed in the lower frame of the notebook. (This feature isn't supported on *3D* plots.)

You can mark a point and copy its coordinates by positioning the pointer, clicking and copying. You can then paste these coordinates back into the notebook for further use. In this way, we find the range in the above example to be about 5.1 meters. For more accuracy, you can make the plot bigger by dragging (click and hold the mouse) one of the small black handles on the frame of the plot.

AII.1.4 Units

The range **R** is easily evaluated numerically. Because of its symbolic nature, *Mathematica* naturally keeps track of units. For example, substituting values for **a**, **vo** and **g** with units attached, we obtain

```
R[15 Degree] /.
    {vo->10 meter/sec, g->9.8 meter/sec^2} //N
```

```
5.10204 meter
```

which is close to the graphical estimate (see Footnote 4) .

AII.1.5 Maximum Range

Without air resistance, **R** has a maximum for an inclination angle **a = 45 Degree**, which we prove by setting the derivative **D[R, a]** equal to zero. First, we note that using **Solve** directly doesn't give us a solution:

```
dR = D[R[a], a]
```

$$\frac{2 \ vo^2 \ Cos[a]^2}{g} - \frac{2 \ vo^2 \ Sin[a]^2}{g}$$

```
Solve[ dR == 0, a ]
```

> Solve::ifun: Warning: Inverse functions are being used by
>   Solve, so some solutions may not be found.

```
{}
```

The empty braces indicate that no solution was found, although as the message tells us *Mathematica* did try *inverse* trigonometric functions. Let's try some trig simplification first.

```
dR = D[R[a], a] //Expand[#,Trig->True]&
```

$$\frac{2\ vo^2\ Cos[2\ a]}{g}$$

And give this back to **Solve**:

```
amax = Solve[ dR == 0, a ][[1]]
```

> Solve::ifun: Warning: Inverse functions are being used by
>   Solve, so some solutions may not be found.

$$\{a \to \frac{Pi}{4}\}$$

We can easily check this result by back substitution:

```
dR /. amax
```

```
0
```

The maximum range is therefore

```
R[a] /. amax
```

$$\frac{vo^2}{g}$$

```
R[45 Degree] /.
    {vo -> 10 meter/sec, g -> 9.8 meter/sec^2} //N
```

```
10.2041 meter
```

The best way to see the effects of the inclination angle on the range is to plot several trajectories together. We simply make a table and plot it. (You might examine the effects of each plot option separately.)

```
ParametricPlot3D[
    Evaluate[
        Table[ {x,y,z} /.
            {g -> 9.8, vo -> 10., p -> 45 Degree},
            {a, 15 Degree, 75 Degree, 15 Degree}
        ]
    ],
```

```
        {t, 0, 2},
        PlotRange -> {0,5}, BoxRatios -> {1,1,.5},
        FaceGrids -> {{0,0,-1}},
        ViewPoint -> {2.176, -2.401, 0.975},
        AxesLabel -> {"x","y","z"}
    ];
```

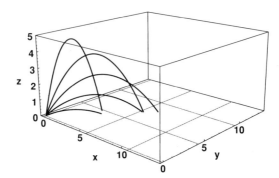

Footnote 5

Here the command **Evaluate** has been wrapped around **Table** for computational efficiency. It tells *Mathematica* to produce the table once and for all, rather than for each new value of **t**. In fact, **Plot** and its variations, like **ParametricPlot3D**, will generally run faster when **Evaluate** is wrapped in this way around the quantity being plotted.

Also, the options available for **ParametricPlot3D** can be viewed by entering **??ParametricPlot3D**.

AII.2   Including Air Resistance

AII.2.1   Numerical Integration

We can generate trajectories numerically using the built-in function **NDSolve** to integrate Newton's equation **F = m a** as a vector differential equation. Let's look at *Mathematica*'s information file on this function.

```
?NDSolve
    NDSolve[eqns, y, {x, xmin, xmax}] finds a numerical solution
        to the differential equations eqns for the function y with
        the independent variable x in the range xmin to xmax.
    NDSolve[eqns, {y1, y2, ...}, {x, xmin, xmax}] finds numerical
        solutions for the functions yi.
```

AII.2.2   Implementation

To keep this simple, we will assume that air resistance is proportional to the velocity, **Fres = -k v**, and work in the *x-z* plane of the trajectory. If we use a scaled resistance constant **k**, the mass drops out of the equations of motion. The analytic solution is, nevertheless, nontrivial (refer for example to Marion and Thornton [44]).

We write **F** = **m a** as a pair of second-order differential equations in each cartesian direction x and z, **x''[t]=Fx[t]**, **z''[t]=Fz[t]**, and integrate with respect to time **t**.

As usual when using a numerical package, we need to assign control parameters and initial conditions. To be clear, let's do this in a separate cell rather than in the argument list of **NDSolve**. Then, we can first echo-check our assignments before letting the program run. We will define the resistance parameter **k** below.

Define constants and initial conditions in the plane of the trajectory:

```
g = 9.8;                        (* gravity *)
tmax = 0.5282;
maximum time *)

vo = 10;                        (* initial speed *)
a = 15 Degree;                  (* initial angle *)
xo = 0;   vxo = vo Cos[a] //N;  (* initial position/speed *)
zo = 0;   vzo = vo Sin[a] //N;
```

Echo checks:

```
{g, tmax}

  {9.8, 0.5282}


{xo, zo, vxo, vzo, Sqrt[vxo^2 + vzo^2]}

  {0, 0, 9.65926, 2.58819, 10.}
```

Computation AII.2.3

We're now ready to compute a trajectory. We only need to define **k**, the resistance parameter. We do it here, so we can change **k** and at the same time tag the new trajectory. Then, we can easily compare with previous results and plot different trajectories together. Let's prefix variables with an **n** to indicate numerical quantities and avoid collision with variables already assigned values.

Footnote 6

You must remember that in any given session *Mathematica* keeps track of all quantities assigned. Although this history feature greatly facilitates interactive use, it can sometimes be a nuisance if your session has been a long one. For example, you might start using a symbol **x** in a context different from the one you originally defined it in. Your results will of course look strange and be unpredictable. In such cases, you can usually evaluate **Clear[x]** and reset **x**. If things are really mixed up, you might be better off saving your session, quitting and restarting *Mathematica*. Using replacements in the form *expr* **/.** *symbol* -> *new* will help avoid this problem.

```
k = 0.8;
ntrajectory[k] =
    NDSolve[
        {nx''[t] == -k nx'[t],
         nz''[t] == -k nz'[t] - g,
         nx[0] == xo, nx'[0] == vxo,
         nz[0] == zo, nz'[0] == vzo},
        {nx, nz}, {t, 0.0, 1.0}
        ][[1]]

{nx -> InterpolatingFunction[{0., 1.}, <>],

  nz -> InterpolatingFunction[{0., 1.}, <>]}
```

The result of a numerical integration is, of course, a list of discrete data representing points on the trajectory. **NDSolve**, however, automatically interpolates the resulting data points and returns an **InterpolatingFunction** (see Exercise 3.5, Appendix III). For example, we obtain the last point on the trajectory by entering

```
{nx[tmax], nz[tmax]} /. ntrajectory[k]

{4.16113, -0.0782835}
```

Thus, we can also estimate trajectory points not computed directly by the numerical integration. For example, we easily obtain the velocity by differentiating the interpolated trajectory. At **t = 0** we find that

```
D[{nx[t],nz[t]}, t] /. ntrajectory[k] /. t->0

{9.65846, 2.58697}
```

We get an idea of the accuracy of the interpolation by comparing these values with the initial velocity we inputed:

```
{vxo, vzo}

{9.65926, 2.58819}
```

The computed results can be improved by changing the options on **NDSolve**. (Take a look at **??NDSolve**.) We will check the acceleration in the case of no air resistance **k = 0** in a moment.

We plot our computed trajectory using **ParametricPlot** as before:

```
ntrajectoryPlot[k] =
    ParametricPlot[
        Evaluate[{nx[t],nz[t]} /. ntrajectory[k]],
        {t, 0, tmax},
        AxesLabel -> { " x", " z"}
    ];
```

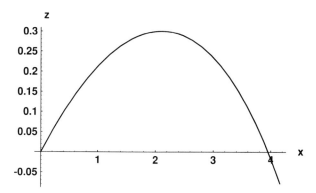

Now let's change **k** and re-enter the above input cells to recompute other trajectories. The label **[k]** on **ntrajectory** and **ntrajectoryPlot** will keep track of our results. In addition to **k = 0.8**, let's take **k = 0.0** and **k = 10.0**. We can then plot these trajectories together using the **Show** command. The comparison makes the effect of air resistance clear.

```
Show[
    ntrajectoryPlot[0.], ntrajectoryPlot[0.8],
    ntrajectoryPlot[10.]
];
```

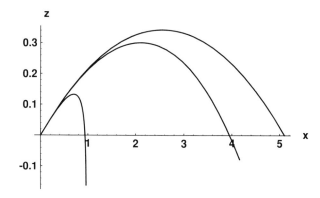

Finally, we check that in the case of no air resistance, **k = 0.0**, the vertical acceleration equals **-g** while the horizontal acceleration remains zero. At an arbitrary time, **tmax/3** for example,

```
D[{nx[t],nz[t]}, {t,2}] /.
    ntrajectory[0.0] /. t->tmax/3 //Chop
```

```
{0, -9.8}
```

Here **Chop** replaces the small x component (numbers less than **10^-10**) by exact integer **0** (see **?Chop**).

Appendix III
Home Improvement

We generally use only a small subset of all the functions built into *Mathematica*, since we do much the same thing over and over, just in different contexts. The exercises in this appendix introduce and summarize most of these basic functions and present a few tricks which may be helpful in using this book. This appendix is intended for leisurely practice and reference and shouldn't require much more than entering the examples and studying the results. You are encouraged, of course, to modify anything and experiment on your own. Don't hesitate to skip around, although if you are new to *Mathematica* you might study the previous two appendices first and then Exercises 1.1 - 1.9.

Section II on Algebra provides background for some of the packages developed for this book and listed in Appendix IV. In particular, Exercise 2.2 discusses trigonometric simplification rules and an enhanced version of the function **TrigReduce** defined in the package **Quantum`Trigonometry`**. Exercise 2.3 develops ideas for reversing the built-in function **PowerExpand** and thus introduces a new function **PowerContract** defined in the package **Quantum`PowerTools`**. Exercise 2.4 extends the built-in function **Conjugate** to symbolic expressions and thus forms the basis of the package **Quantum`QuickReIm`**. Exercise 2.5 develops rules for the built-in function **Quantum`NonCommutativeMultiply`** and the package **NonCommutativeMultiply** appropriate for defining quantum operators. Finally, Exercise 2.7 discusses integration rules for gaussian and exponential functions and the packages **Quantum`integGauss`** and **Quantum`integExp`**.

Refer to the previous two appendices and to the *Mathematica* manual by Wolfram [62]. Many of the ideas introduced here are developed in detail in Maeder's book [43], *Programming in Mathematica*. Several very good books and self-help manuals on *Mathematica* have recently appeared, but it's perhaps best simply to begin using *Mathematica*. Remember you can enter **?***anything* to obtain information on *anything*, for example, **?D**.

Contents

AIII.1   Functions

Exercise 1.1 Basic Operations

Compare the order of operations in the following examples. We use spaces and extra lines freely to keep our equations readable. Don't confuse compact programming with crowded input.

```
{2/3 5, 2/(3 5), 2/3/5}
```

$$\left\{\frac{10}{3}, \frac{2}{15}, \frac{2}{15}\right\}$$

```
{5 f/g + 2 a/h g, 5(f/g + 2 a/(h g))}
```

$$\{\frac{5\ f}{g} + \frac{2\ a\ g}{h},\ 5\ (\frac{f}{g} + \frac{2\ a}{g\ h})\}$$

Exercise 1.2 Replacements

In general, we should make temporary substitutions to assign values to variables rather than make global, permanent assignments with **=**. The permanent variety causes problems if we forget and use the variables in another context.

Here's an example of a global assignment and a gaussian function:

```
ko = 2;    xo = 1;                (* Global assignment. *)
E^(-ko (x - xo)^2)
```

$$E^{-2\ (-1\ +\ x)^2}$$

These may not, however, be the values we want later on. Consider, for example, a plane wave:

```
E^(I ko (x - xo))
```

$$E^{2\ I\ (-1\ +\ x)}$$

A better approach is to make a temporary replacement with **/.** (typed slash, period), such as

```
Sqrt[x^2 + y^2] /.{x -> 3, y -> 4}
```

```
5
```

We can also stack replacements and apply functions afterwards with **//** (typed double slash)

```
{x, y} /.
   {x -> r Cos[t], y -> r Sin[t]} /.
   {r -> 5, t -> Pi/8} //
   N
```

```
{4.6194, 1.91342}
```

Exercise 1.3 Lists, Expressions and Subscripts

Consider the following examples of extracting subelements of lists and expressions.

```
list = {aa,bb,cc,dd,ee}              (* A list.*)
```

```
{aa, bb, cc, dd, ee}
```

```
{list[[1]], list[[{1,2}]]}                    (* List subelements.*)

   {aa, {aa, bb}}


list[[2]] = xx                                (* Element re-Set.*)

   xx


list                                          (* New list.*)

   {aa, xx, cc, dd, ee}


expr = Sin[k x] E^(I w t)/Sqrt[x^2 + y^2]
                                              (* An expression.*)

    I t w
   E      Sin[k x]
   -----------------
       2    2
    Sqrt[x  + y ]


{expr[[2]], expr[[2,1]], expr[[2,1,1]]}
                                              (* List of subelements.*)

        1          2    2    2
   {-----------,  x  + y ,  x}
       2    2
    Sqrt[x  + y ]


expr[[2]] = r^-1                              (* Element re-Set.*)

   1
   -
   r


expr                                          (* New expression.*)

    I t w
   E      Sin[k x]
   -----------------
          r
```

Exercise 1.4 Dummy Variables

Let's take a look at "blank" symbols such as **x_** to define a function **f[x_]** of an arbitrary variable **x**. The symbol **_** (typed underscore), labelled here with an **x**, is a dummy variable which can represent almost anything you might want to substitute in later on. Consider for example a function which takes cube roots, analogous to the built-in function **Sqrt**.

If we try and define this without **x_**, as for example

```
cubeRoot[x] = x^(1/3)
```

```
   1/3
  x
```

we find it won't evaluate for any other argument besides **x**

```
cubeRoot[27]
```

```
  cubeRoot[27]
```

We fix this however by introducing **x_** as

```
cubeRoot[x_] = x^(1/3)
```

```
   1/3
  x
```

Thus, for example,

```
cubeRoot[-27]
```

```
        1/3
  3 (-1)
```

```
% //N                                    (* Principal value.*)
```

```
  1.5 + 2.59808 I
```

We use **f[x_,y_]** to define a function of two variables **x** and **y**, and any number of variables can be included in an argument sequence. Alternatively, we can stack on extra square brackets for two or more arguments to give certain variables special distinction. For example, we might define a Fourier-series partial sum with **nmax** terms as

```
partialSum[z_][nmax_] := Sum[c[n] Sin[n z], {n,1,nmax}]
```

```
partialSum[w t][4]
```

```
  c[1] Sin[t w] + c[2] Sin[2 t w] + c[3] Sin[3 t w] +

      c[4] Sin[4 t w]
```

The assignment := (called **SetDelayed**) is discussed in the next exercise.

Exercise 1.5 Defining Functions

An ordinary assignment like **f[x_]** = *expr* (called **Set**) evaluates *expr* immediately, generating an immediate response from *Mathematica*. An assignment like **f[x_]** := *expr* (typed **colon, =,** called **SetDelayed**) delays evaluation of *expr* on the right until **f[x]** on the left is entered or called; it also suppresses a response from *Mathematica*.

Often the difference isn't important, and either := or = can be used. For example, we could assign **Sin[x]** to **psi[x_]** equally well with either **psi[x_]** = **Sin[x]** or **psi[x_]** := **Sin[x]**.

Situations arise, however, which require, or are better handled with, one or the other. For example, we defined **partialSum** in the previous Exercise 1.4 for an arbitrary number of terms **nmax** using **SetDelayed**. We might want however just a 6-term version of this sum for plotting different values of **w**. We can evaluate this once and for all using **Set** as

```
pS6[w_] = partialSum[w t][6] /. c[n_] -> 1/n^2
```

$$
\text{Sin}[t\ w] + \frac{\text{Sin}[2\ t\ w]}{4} + \frac{\text{Sin}[3\ t\ w]}{9} + \frac{\text{Sin}[4\ t\ w]}{16} +
$$

$$
\frac{\text{Sin}[5\ t\ w]}{25} + \frac{\text{Sin}[6\ t\ w]}{36}
$$

Here we've introduced a particular set of expansion coefficients **c[n]** with a simple pattern replacement (refer to Exercise 2.2 below.)

In the same way, we often use **Set** to define a result once and for all which is the outcome of a long calculation. For example, suppose we compute the derivative of the following function **f** and simplify it, and then save the result as **df**:

```
f = Sin[x]/(Sin[x]+Cos[x])    (* Sample function. *)
```

$$
\frac{\text{Sin}[x]}{\text{Cos}[x] + \text{Sin}[x]}
$$

```
D[f,x]                        (* Its derivative. *)
```

$$
-\left(\frac{(\text{Cos}[x] - \text{Sin}[x])\ \text{Sin}[x]}{(\text{Cos}[x] + \text{Sin}[x])^2}\right) + \frac{\text{Cos}[x]}{\text{Cos}[x] + \text{Sin}[x]}
$$

```
ExpandAll[Together[%,Trig->True],Trig->True]
```

$$
\frac{1}{1 + \text{Sin}[2\ x]}
$$

```
df[x_] = %                    (* Save the simplified result;
                                 SetDelayed won't work here. *)
```

$$
\frac{1}{1 + \text{Sin}[2\ x]}
$$

```
D[f,x] - df[x] /. x -> Pi/8 //N
                              (* Numerical check.*)
```

$$
2.71051\ 10^{-19}
$$

Here's an example which requires **SetDelayed**. Suppose we want to define an operator **d[fnc]** that takes the derivative of **fnc** with respect to **x**. Let's first try this with **Set** and then with **SetDelayed**.

```
d[fnc_] = D[fnc,x]          (* Output hints this won't work.*)

   0
```

Since the dummy variable **fnc** nowhere contains **x**, the derivative on the right side evaluates immediately to zero, which is the value assigned to **d[fnc]**. We thus obtain zero for any expression. For example,

```
d[Tan[x]]

   0
```

SetDelayed prevents the derivative from evaluating until we call **d[fnc]** and specify a function of **x** for **fnc**. Hence, a proper definition is

```
d[fnc_] := D[fnc,x]         (* SetDelayed *)
```

```
d[Tan[x]]                   (* What we want.*)
        2
   Sec[x]
```

Finally, compare the following. In our first attempt to define **plot[n]**, the use of **Set** forces **Plot** to evaluate immediately, generating an error message, since a value for **n_** hasn't been specified.

```
plot[n_] = Plot[E^(-x) Sin[n x], {x,0,5}]
                                  Sin[n x]
      Plot::plnr: CompiledFunction[{x}, --------, -CompiledCode-][x]
                                     x
                                    E
            is not a machine-size real number at x = 0..
```

The use of **SetDelayed** eliminates the error in our second attempt, since **Plot** isn't asked to evaluate until **n** is specified with a command such as **plot[3]**.

```
plot[n_] := Plot[ E^(-x) Sin[n x], {x,0,5} ]
```

```
plot[3];
```

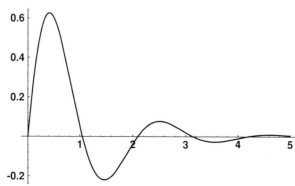

Exercise 1.6 Multiple Rules

In many situations we want to give a function more than one definition, which we generally do using **SetDelayed**. For example, we might enter a time-independent function such as

```
psi[n_,x_] := Sin[n Pi x]
```

Then we can easily include the definition of a time-dependent counterpart upon entering

```
psi[n_,x_,t_] := E^(-I n^2 Pi^2 t) psi[n,x]
```

Mathematica keeps track of both definitions and distinguishes them essentially with the function argument list, **n**, **x** in the first case and **n**, **x**, **t** in the second. We can check all the definitions we've entered with

```
?psi
```

```
Global`psi

psi[n_, x_] := Sin[n*Pi*x]
psi[n_, x_, t_] := psi[n, x]/E^(I*n^2*Pi^2*t)
```

Note that *Mathematica* won't change these definitions or delete one unless we reset or **Clear** it ourselves. Therefore, when debugging definitions, it's a good idea to collect them together into one cell and introduce a **Clear** at the top of the cell to avoid using previous definitions inadvertently. Hence, for example, we might redefine **psi** as

```
Clear[psi]
psi[x_]     := E^(-x^2/2)
psi[x_,t_] := E^(-I t/2) psi[x]
```

Exercise 1.7 Comparing Patterns

We can check if two expressions are identical patterns with logical equals **==** (typed = twice, called **Equal**). For example,

```
{I == Sqrt[-1], a == b, 1 - x^2 == 1 - x^2}
```

```
{True, a == b, True}
```

Note this check simply returns the equation if the patterns are not identical. We can use **===** (typed triple =, called **SameQ**) to obtain either **True** or **False**:

```
{a === a, a === b}
```

```
{True, False}
```

We can also check patterns with logical *not* equals by introducing an exclamation point **!** as

```
{a != a, a != b, a =!= a, a =!= b}
```

```
{False, a != b, False, True}
```

It's important to recognize that these logical operators compare patterns and don't check algebraic equality, which is in general difficult to do. For example, the two patterns `1 + 2x + x^ 2` and `(1 + x)^2` are different (the first contains three terms, the second two terms raised to a power), but are of course equivalent algebraically. That is,

```
1 + 2x + x^2 == (1 + x)^2
```

$$1 + 2 x + x^2 == (1 + x)^2$$

```
% //ExpandAll
```

```
True
```

The function **ExpandAll** rather than **Expand** is needed here to penetrate both sides of the equation (see Exercise 2.1 below).

Exercise 1.8 Function Wrapping

Standard-Wrap. Functions in *Mathematica* are applied by "wrapping" them around an expression or variable. For example,

```
Factor[1-x^6]
```

$$(-1 + x) (1 + x) (-1 + x - x^2) (1 + x + x^2)$$

```
Expand[%/(1+x)]
```

$$1 - x + x^2 - x^3 + x^4 - x^5$$

We can also do this in a single step as

```
Expand[Factor[1-x^6]/(1+x)]
```

$$1 - x + x^2 - x^3 + x^4 - x^5$$

Postfix Form. We can also apply functions from the right side of an expression. For example,

```
1-x^6 //Factor
```

$$(-1 + x) (1 + x) (-1 + x - x^2) (1 + x + x^2)$$

This has the advantage that the object of principal interest, here the expression `1-x^6`, is given the stage on the left, while the *Mathematica* operation, which has the lessor role of performing a task, stays in the wings on the right.

Again, we can perform multiple wraps by stacking up functions on the right. We can also move the *Mathematica* to the next line to separate it further from the physics, as for example,

```
-1/(1-x) + 1/(1+x) //
    Together //ExpandAll
```

$$\frac{2\ x}{-1 + x^2}$$

Also, if we want to include a function's options, we simply add brackets [] and a "slot" # for the expression being operated on along with the option, and signal the end of the function with &. (In effect, we introduce a *pure function*. See Exercise 1.9.) For example,

```
Sqrt[Pi] //N[#,50]&
```

```
1.7724538509055160272981674833411451827975494561224
```

Note the slot goes in the operator's argument sequence where the expression normally goes. Here's another example.

```
Sin[x]^2 //Expand[#,Trig -> True]&
```

$$\frac{1}{2} - \frac{Cos[2\ x]}{2}$$

Prefix Form. Finally, we can also apply functions from the left, as for example (note parentheses are needed here)

```
Factor @ (1-x^6)
```

$$(-1 + x)\ (1 + x)\ (-1 + x - x^2)\ (1 + x + x^2)$$

And we can include options as we did with the postfix form:

```
N[#,50]& @ Sqrt[Pi]
```

```
1.7724538509055160272981674833411451827975494561224
```

The syntax `Q @ psi` is particularly useful when we want to think of `Q` as an operator which operates on the function `psi`. For example, we might apply the derivative operator defined in Exercise 1.5 as

```
{d @ Tan[x], d @ (1/(1+x^2))}
```

$$\{Sec[x]^2,\ \frac{-2\ x}{(1 + x^2)^2}\}$$

In any case, all three forms are fully equivalent:

```
{Sqrt @ 8, Sqrt[8], 8 //Sqrt}
```

$$\{2^{3/2},\ 2^{3/2},\ 2^{3/2}\}$$

Exercise 1.9 Pure Functions

Pure functions are a powerful feature not available with most traditional computer languages. The idea is to have a compact way of representing functions without having to give them a name. With pure functions we just indicate what operations are to be performed and where the arguments are placed. *Mathematica* uses # to stand for the argument or *"slot"* of a pure function (or #1, #2, etc. for two or more arguments) and & to signal its end (actually its **Head**). Pure functions are thus also referred to as *anonymous* functions. Consider the following examples.

```
cube := #^3 &                    (* Pure function that cubes.*)

{cube @ x, cube[x], x //cube}     (* Operate on x.*)

    3   3   3
  {x , x , x}
```

Pure functions provide a handy way to define quantum operators with respect to an unspecified wavefunction, represented by the slot. Here's an example of a momentum operator:

```
p := -I D[#,x]&

p @ (E^(I k x))                  (* Operate on E^(I k x).*)

    I k x
   E      k
```

Here we separate out powers of **x** in a polynomial,

```
(a+x+b)^3 //Collect[#,x]&

   3     2        2    3        2              2
  a + 3 a b + 3 a b + b + (3 a  + 6 a b + 3 b ) x +

              2    3
    (3 a + 3 b) x  + x
```

and extract their coefficients:

```
% //Coefficient[#,x^2]&

  3 a + 3 b
```

Here's a convenient way to separate variables (note the %% here).

```
{Select[%%,FreeQ[#,x]&], Select[%%,!FreeQ[#,x]&]}

    3     2        2    3         2              2
  {a + 3 a b + 3 a b + b , (3 a  + 6 a b + 3 b ) x +

              2    3
    (3 a + 3 b) x  + x}
```

Refer to ?**Select** and ?**FreeQ** and note that !**FreeQ** means logical *not* **FreeQ**.

Exercise 1.10 Dummy Expressions

We use "single blanks" _ to represent *single* expressions in a function definition or a pattern-matching rule. In some situations, however, we need to represent a *sequence* of expressions. For this we use "double blank" __ (double underscore) to stand for a sequence of one or more expressions and "triple blank" ___ (triple underscore) for a sequence of *zero* or more expressions. (Triple blanks require some caution as they can lead to infinite loops. See Wolfram [62].)

Double Blanks. Let's define a function that computes the distance from the origin to a point in an n-dimensional space. The following function **distance** accepts an arbitrary number of coordinates and thus works for arbitrary dimension.

```
distance[point__] := Sqrt[{point}.{point}]
                         (* double blank *)
```

```
distance[x,y]              (* Distance to origin in x-y plane.*)
        2    2
  Sqrt[x  + y ]
```

```
distance[x,y,z,I t]
        2    2    2    2
  Sqrt[-t  + x  + y  + z ]
```

Triple Blanks. These are often used for passing options on to another function. For example, let's define a function that plots the derivative of a quantity and has all the options of **Plot** available to it.

```
plotD[f_,{x_,xo_,x1_},opts___] :=
         Plot[ Evaluate[D[f,x]], {x,xo,x1}, opts ]
                    (* opts = any Plot option.*)
                    (* Evaluate derivative before plotting.*)
```

```
plotD[
   E^-x^2,
   {x,-3,3}, PlotRange ->{-1.1,1.1},
   AxesLabel->{"  x","D[E^-x^2,x]"}
];
```

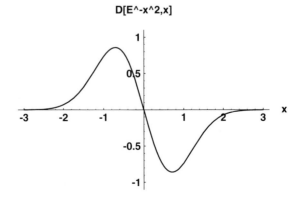

The triple blanks allow us to omit the options altogether and enter for example just `plotD[E^-x^2,{x,-3,3}]`.

Exercise 1.11 Restricting Variables

You can include constraints on dummy variables with **/; *constraint*** and enter function definitions almost as you would write them down on paper. Consider for example a step function

```
step[x_,xo_] := 0 /; x  < xo   (* /; => "whenever" or "if".*)
step[x_,xo_] := 1 /; x >= xo
```

Note that `SetDelayed :=` is required when defining functions with `Condition` `/;` (refer to Exercise 1.5). Of course, *Mathematica* automatically keeps track of our definitions:

```
?step
```

```
  Global'step

  step[x_, xo_] := 0 /; x < xo

  step[x_, xo_] := 1 /; x >= xo
```

We can't do too much with this symbolically (try taking a derivative), but numerically it works fine. For example, here's a good test of our definition:

```
Plot[ step[x,1], {x,-5,5}, PlotRange -> {0,2.1} ];
```

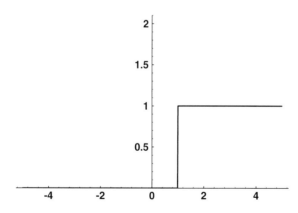

It's sometimes convenient, particularly for pattern matching, to attach the constraints directly to the dummy variables with the syntax **x_?(*constraint*)** with *constraint* expressed as a pure function (cf. Exercise 1.9). Consider, for example, another version of the step function:

```
step[x_?(# <  -1 &)] := 0
step[x_?(# >= -1 &)] := 1
```

```
Plot[ step[x], {x,-2,2}, PlotRange -> {0,2.1} ];
```

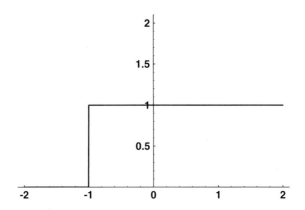

Exercise 1.12 Default Values

For many functions you define, parameters in the argument list will require certain values that you rarely will want to change. For such parameters *Mathematica* allows you to set up default values, so that the parameters need not even appear in the argument list unless the user wants to change their values. Consider the following two examples.

Define a function **root** which computes the nth root of a quantity but defaults to the square root if **n** isn't specified.

```
root[x_,n_:2] := x^(1/n)
                        (* Colon designates default value.*)
root[x]

   Sqrt[x]


root[-32,5]
          1/5
    2 (-1)
```

We can simplify the argument list of the **step** function defined in excercise 1.11. Note we introduce **Clear[step]** at the top of the cell to avoid collision with the previous rules (see Exercise 1.6).

```
Clear[step]
step[x_,xo_:0] := 0 /; x  < xo        (* Default xo = 0.*)
step[x_,xo_:0] := 1 /; x >= xo
```

Here are two examples for **xo** = 0 and 2.

```
Show[
    GraphicsArray[
        Plot[
            step[x]                     (* Default xo = 0 used.*)
            {x,-5,5}, PlotRange -> {0,2.1},
            DisplayFunction -> Identity
        ],
```

```
            Plot[
              step[x,2],                        (* xo = 2.*)
              {x,-5,5}, PlotRange -> {0,2.1},
              DisplayFunction -> Identity
            ]}
        ],
        DisplayFunction -> $DisplayFunction
    ];
```

Note the option `DisplayFunction -> Identity` prevents display of the results generated by `Plot`, while `DisplayFunction -> $DisplayFunction` releases the results collected by `GraphicsArray` for `Show`.

Exercise 1.13 Tagging Functions and Variables

Suppose we want to declare two variables **x** and **y** to be independent such that, for example, their total derivatives with respect to one another vanish. Thus, consider

```
Dt[x,y] = 0

  Set::write: Tag Dt in Dt[x, y] is Protected.

  0
```

The difficulty is that `Dt` is at the top level or **Head** of the left-hand side of this expression, and being a built-in function cannot be equated to zero (or anything else for that matter) since it is protected by the system. Because our declaration is really about **x** and **y**, we need to by pass the top level in this case. We can eliminate the difficulty by attaching the definitions to the variables **x** and **y** with **/:** (**TagSet**) in the following way:

```
 x/: Dt[x,y] = 0;          y/: Dt[y,x] = 0;
```

Then, for example,

```
{Dt[x,y], Dt[y,x], Dt[Sqrt[x^2 + y^2], y]}

                   y
   {0, 0, ───────────────}
                2   2
          Sqrt[x + y ]
```

As an aside, note that we can define global constants for total derivatives independently of all other variables. For example,

```
SetAttributes[c,Constant]
```

```
{Dt[Sqrt[x^2 + a^2],x], Dt[Sqrt[x^2 + c^2],x]}
```

$$\left\{ \frac{2\ x\ +\ 2\ a\ Dt[a,\ x]}{2\ Sqrt[a^2\ +\ x^2]},\ \frac{x}{Sqrt[c^2\ +\ x^2]} \right\}$$

where **a** is still possibly a function of **x**.

We can also make other kinds of declarations such as

```
k/: IntegerQ[k] = True;          a/: Sign[a] = -1;
```

Thus, consider

```
e[n_?IntegerQ] := h w (n + 1/2)
```

```
{e[0], e[k], e[2.1]}
```

$$\left\{ \frac{h\ w}{2},\ h\ (\frac{1}{2}\ +\ k)\ w,\ e[2.1] \right\}$$

and

```
{Integrate[1/Sqrt[a x^2 + x], x],
 Integrate[1/Sqrt[b x^2 + x], x]}
```

$$\left\{ \frac{ArcSinh\left[\dfrac{1\ +\ 2\ a\ x}{2\ Sqrt[\frac{-1}{4\ a}]\ Sqrt[a]}\right]}{Sqrt[a]}, \right.$$

$$\left. \frac{Log[1\ +\ 2\ b\ x\ +\ 2\ Sqrt[b]\ Sqrt[x\ +\ b\ x^2]]}{Sqrt[b]} \right\}$$

AIII.2 Algebra

Exercise 2.1 Algebraic Simplification

Our supervision is crucial for guiding *Mathematica* through algebra, both for speed and for achieving desired results. We can use the built-in function **Simplify**, but it might take us in a direction we don't want to go. Often we can accomplish the same thing but with more control and usually faster by applying combinations of **Factor**, **Expand**, and **Together**, and also **Map[Factor,expr]** and **Collect**. It takes practice, however, just like algebra by hand does.

Consider the expression

```
expr1 =
    (a^2 + 2 a b + b^2)/(a - I b) +
    (a^2 + 2 a b + b^2)/(a + I b)
```

$$\frac{a^2 + 2\,a\,b + b^2}{a - I\,b} + \frac{a^2 + 2\,a\,b + b^2}{a + I\,b}$$

This is *Mathematica*'s simplification:

```
expr1 //Simplify
```

$$\frac{2\,a\,(a + b)^2}{a^2 + b^2}$$

We obtain the same result starting with **Factor**, which first finds a common denominator (by applying **Together**) and then factors the result:

```
expr1 //Factor
```

$$\frac{2\,a\,(a + b)^2}{(-I\,a + b)\,(I\,a + b)}$$

Applying **Expand** to just the denominator with **ExpandDenominator** then gives

```
% //ExpandDenominator
```

$$\frac{2\,a\,(a + b)^2}{a^2 + b^2}$$

which agrees with **Simplify**.

We can prevent **Factor** from finding a common denominator and force it to factor each term separately by mapping **Factor** onto each term as

```
expr1 //Map[Factor,#]&
```

$$\frac{-I\,(a + b)^2}{-I\,a + b} + \frac{I\,(a + b)^2}{I\,a + b}$$

Finally, the function **Together** finds a common denominator but doesn't factor the result. Thus,

```
expr1 //Together //ExpandDenominator
```

$$\frac{2\,(a^3 + 2\,a^2\,b + a\,b^2)}{a^2 + b^2}$$

Here's another expression

```
expr2 = (d + (-n - 1)/a)^2
```

$$(d + \frac{-1 - n}{a})^2$$

that *Mathematica* "simplifies" as

```
expr2 //Simplify
```

$$\frac{(1 - a\ d + n)^2}{a^2}$$

Now suppose we want to express it in inverse powers of **a**. We can do this with **Collect** as

```
expr2 //Expand //Collect[#,a]&
```

$$d^2 + \frac{-2\ d - 2\ d\ n}{a} + \frac{1 + 2\ n + n^2}{2\ a}$$

and clean up the numerators with **Map[Factor,expr]** as

```
% //Map[Factor,#]&
```

$$d^2 - \frac{2\ d\ (1 + n)}{a} + \frac{(1 + n)^2}{a^2}$$

In this case, we can also achieve almost the same thing with **Expand[expr2,a]**, which expands terms in **expr2** containing **a** and avoids terms independent of **a**. Thus,

```
expr2 //Expand[#,a]&
```

$$d^2 + \frac{2\ d\ (-1 - n)}{a} + \frac{(-1 - n)^2}{a^2}$$

We also need to note that most functions like **Factor** don't reach all subelements of an expression. Although **Expand** can be applied to all levels with **ExpandAll**, we have to use **MapAll** to apply a function like **Factor** to all levels.

As an example, let's wrap **Sqrt** around **expr1** and apply **Factor** as

```
Sqrt[expr1] //Factor
```

$$\mathrm{Sqrt}[2]\ \mathrm{Sqrt}\!\left[\frac{a^3 + 2\ a^2\ b + a\ b^2}{(a - I\ b)\ (a + I\ b)}\right]$$

This didn't do anything but find a common denominator (with **Together**). Now let's apply **Factor** to all subelements with

```
Sqrt[expr1] //MapAll[Factor,#]&
```

$$\mathrm{Sqrt}[2]\ \mathrm{Sqrt}\!\left[\frac{a\ (a + b)^2}{(a - I\ b)\ (a + I\ b)}\right]$$

We can expand the denominator with **ExpandAll**, but this also expands the numerator:

```
% //ExpandAll
```

$$\text{Sqrt}[2]\ \text{Sqrt}[\frac{a^3 + 2\ a^2\ b + a\ b^2}{a^2 + b^2}]$$

We expand just the denominator by mapping **ExpandDenominator** to all subelements (note the %% here) with

```
%% //MapAll[ExpandDenominator,#]&
```

$$\text{Sqrt}[2]\ \text{Sqrt}[\frac{a\ (a + b)^2}{a^2 + b^2}]$$

This is also the result *Mathematica* obtains:

```
Sqrt[expr1] //Simplify
```

$$\text{Sqrt}[2]\ \text{Sqrt}[\frac{a\ (a + b)^2}{a^2 + b^2}]$$

MapAll also comes in handy with logical equals to help demonstrate algebraic equivalence of two patterns. For example, we can compare the two versions of an expression we generated in Exercise 1.5 with

```
D[f,x] == df[x]
```

$$-(\frac{(Cos[x] - Sin[x])\ Sin[x]}{(Cos[x] + Sin[x])^2}) + \frac{Cos[x]}{Cos[x] + Sin[x]} == \frac{1}{1 + Sin[2\ x]}$$

We prove their equivalence by applying **Together** to the left side with **MapAll** as (note the nested pure functions)

```
% //MapAll[ Together[#,Trig->True]&, #]& //
    ExpandAll[#,Trig->True]&
```

```
True
```

You might note that **Simplify** by itself doesn't work here.

Exercise 2.2 Simplification with Pattern Matching

With pattern matching and replacement, we have precise control over how *Mathematica* does algebra. Here's a simple function

```
expr3 = E^(I k Sqrt[x^2 + y^2])/Sqrt[x^2 + y^2]
```

$$\frac{E^{I\ k\ Sqrt[x^2 + y^2]}}{Sqrt[x^2 + y^2]}$$

We simplify the arguments of **Sqrt** with the replacement

```
% /.x^2 + y^2 -> r^2
```

$$\frac{E^{I\ k\ Sqrt[r^2]}}{Sqrt[r^2]}$$

Note that replacements penetrate all levels of the expression. We can further simplify with **PowerExpand** as

```
% //PowerExpand
```

$$\frac{E^{I\ k\ r}}{r}$$

Pattern matching provides a powerful and elegant way to apply functions to subelements of an expression. Consider for example

```
expr4 = E^(-k^2/w + k/w + ko)/Sqrt[1/(a + I b) + 1/(a - I b)]
```

$$\frac{E^{ko + k/w - k^2/w}}{Sqrt\left[\dfrac{1}{a - I\ b} + \dfrac{1}{a + I\ b}\right]}$$

Suppose we want to simplify the denominator without changing the numerator. We can do this with

```
expr4 /. 1/Sqrt[a_] :> 1/Sqrt[Simplify[a]]
```

$$\frac{E^{ko + k/w - k^2/w}}{Sqrt[2]\ Sqrt\left[\dfrac{a}{a^2 + b^2}\right]}$$

We can also use **MapAt**, although the necessary positions of the subelements can be tedious to determine (see **?MapAt** and **?Position** and refer to Wolfram [62]).

Note that we've introduced **:>** (**RuleDelayed**) here rather than **->** (**Rule**). With **Rule**, the right side of our replacement rule would evaluate immediately before trying to match the left side with something in **expr4**; hence, nothing would happen, since **Simplify** would operate on simply the dummy variable **a**. With **RuleDelayed**, the right side is held unevaluated until the left side matches something in **expr4**, in which case a value for **a** is assigned and the right side evaluated. In short, we need **RuleDelayed** whenever we want to wrap functions in the right side of a replacement rule around something generated by the pattern matching. This is completely analogous to the action of **Set** and **SetDelayed** (see Exercise 1.5).

Here's a rule for completing the square:

```
expr4 /. a_ x_^2 + b_ x_ + c_ :>
              a(x + b/(2a))^2 + Together[-b^2/(4a) + c]
```

$$\frac{-((-(1/2) + k)^2/w) + (1 + 4 \text{ ko } w)/(4 w)}{\text{Sqrt}[\dfrac{1}{a - I \ b} + \dfrac{1}{a + I \ b}]}$$

which we can check (and much more quickly than with **ExpandAll**) with

```
expr4 == % /. E^q_ :> E^Expand[q]
```

    True

Again, we see that replacement rules penetrate all levels of an expression.

Replacements are also handy for applying trigonometric identities, and generally faster than the built-in function-option **Trig -> True**. Thus, consider

```
expr5 = expr3 /.{x -> r Cos[t], y -> r Sin[t]}
```

$$\frac{I \ k \ \text{Sqrt}[r^2 \ \text{Cos}[t]^2 + r^2 \ \text{Sin}[t]^2]}{\text{Sqrt}[r^2 \ \text{Cos}[t]^2 + r^2 \ \text{Sin}[t]^2]}$$

We can apply the fundamental identity **Sin[t]^2 + Cos[t]^2 == 1** with

```
expr5 /. c_. Sin[t_]^2 + c_. Cos[t_]^2 -> c //PowerExpand
```

$$E^{\frac{I \ k \ r}{r}}$$

where **c_** stands for common factors while **c_.** works the same way but has the default value unity. Thus, the rule still works even if **r = 1**:

```
expr5 /. r -> 1 /. c_. Sin[t_]^2 + c_. Cos[t_]^2 -> c
```

$$E^{I \ k}$$

Note, in an isolated example, it's generally easier to match just a single term in the identity, as for example

```
expr5 /. Sin[t_]^2 -> 1 - Cos[t]^2 //ExpandAll //PowerExpand
```

$$E^{\frac{I \ k \ r}{r}}$$

There are several variations of the fundamental trig identity which can be helpful in simplifying algebra in many applications. For example,

```
Trigrules := {
    a_. Sin[x_]^2 + a_. Cos[x_]^2 :> a,
    a_. Sec[x_]^2 + b_ Tan[x_]^2 :> a /; b == -a,
    a_. Csc[x_]^2 + b_ Cot[x_]^2 :> a /; b == -a,
    a_. Sin[x_] Tan[x_] + a_. Cos[x_] :>  a Sec[x],
    a_. Cos[x_] Cot[x_] + a_. Sin[x_] :>  a Csc[x]}
```

If you don't have much trig to do, you can make replacements as above on a case-by-case basis, or use built-in functions with the option **Trig -> True**. Thus,

```
Sin[x] Tan[x] + Cos[x] /. Trigrules
```

```
  Sec[x]
```

```
Sin[x] Tan[x] + Cos[x] //Factor[#,Trig -> True]&
```

```
  Sec[x]
```

However, if your calculation is trig intensive, the function **TrigReduce** from the package **Algebra`Trigonometry`** can be helpful and faster. For increased efficiency, an enhanced version of **Trigrules** has been appended to **TrigReduce** and included in the package **Quantum`Trigonometry`**, which otherwise has all of the functionality of **Algebra`Trigonometry`** (see Appendix IV). Thus, for example

```
Needs["Quantum`Trigonometry`"]
```

```
{expr5, Sec[x] - Cos[x]} //TrigReduce //PowerExpand
```

$$\left\{\frac{E^{I\ k\ r}}{r}, \ Sin[x]\ Tan[x]\right\}$$

Exercise 2.3 Reversing PowerExpand

We have seen in some of the previous examples that a product raised to a power is not automatically expanded as a product of powers. Thus, for example,

```
Sqrt[a b c]
```

```
  Sqrt[a b c]
```

Instead, one has to apply **PowerExpand** as

```
% //PowerExpand
```

```
  Sqrt[a] Sqrt[b] Sqrt[c]
```

The reason for this is that the two forms may not be equivalent for certain expressions (see **?PowerExpand**). When applicable, however, **PowerExpand** can be a very powerful function for simplifying algebra. Compare for example forms of **Sqrt[expr1]** from Exercise 2.1 with and without **PowerExpand**:

```
{Sqrt[expr1] //Simplify, Sqrt[expr1] //Simplify //PowerExpand}
```

$$\left\{\text{Sqrt}[2] \; \text{Sqrt}\left[\frac{a \; (a + b)^2}{a^2 + b^2}\right], \; \frac{\text{Sqrt}[2] \; \text{Sqrt}[a] \; (a + b)}{\text{Sqrt}[a^2 + b^2]}\right\}$$

Note these two are equivalent only if **a + b** is positive or zero.

In addition, it's easy to see that we could greatly enhance the utility of **PowerExpand** if we had a way to reverse it. For example, we could further simplify the second (right-hand) form of **Sqrt[expr1]** by recombining the **Sqrt** powers. Since there is no built-in function which performs this operation, let's set up a rule that does. Consider therefore the following pattern replacement:

```
Sqrt[a] Sqrt[b] Sqrt[c] //. m_^q_ n_^q_ -> (m n)^q
```

```
Sqrt[a b c]
```

Here we've introduced a common exponent **q_** on the factors **m_** and **n_** to collect like powers. The command **//.** (**ReplaceRepeated**) is also required for expressions involving three or more factors in order to apply the rule until the result no longer changes.

As it stands, our rule only cleans up numerators and denominators. For example,

```
Sqrt[a b c/d/e] //PowerExpand
```

$$\frac{\text{Sqrt}[a] \; \text{Sqrt}[b] \; \text{Sqrt}[c]}{\text{Sqrt}[d] \; \text{Sqrt}[e]}$$

```
% //. m_^q_ n_^q_ -> (m n)^q
```

$$\frac{\text{Sqrt}[a b c]}{\text{Sqrt}[d e]}$$

We need another rule for combining positive and negative powers. Consider, therefore, the pattern **m_^q_ n_^p_** with **q** constrained to be positive or zero and **p == -q** (recall that **&&** is logical *and*):

```
% //.{m_^q_ n_^q_  :> (m n)^q,
      m_^q_ n_^p_  :> (m/n)^q /; q >= 0 && p == -q}
```

$$\text{Sqrt}\left[\frac{a b c}{d e}\right]$$

Note that **RuleDelayed** (**:>**) arrows are required when constraints are attached to rules with **/;** (refer to Exercise 2.2 and to Exercise 1.11). These rules are in fact analogous to those used to define **PowerExpand** (see the package **StartUp`Expand`**).

Powers of integers are handled separately by *Mathematica*, and the above rules don't work when integers are involved:

```
PowerExpand[ Sqrt[2 a b d/c/e] ] //.
   { m_^q_ n_^q_ :> (m n)^q,
     m_^q_ n_^p_ :> (m/n)^q /; q >= 0 && p == -q}
```

$$\frac{\text{Sqrt}[2]\ \text{Sqrt}[a]\ \text{Sqrt}[b]\ \text{Sqrt}[d]}{\text{Sqrt}[c]\ \text{Sqrt}[e]}$$

We can repair them, however, by adding constraints with !**Integer** to check that the factors **m** and **n** are not integers. Thus, consider

```
PowerExpand[ Sqrt[2 a b d/c/e] ] //.
   { m_^q_ n_^q_ :> (m n)^q /; !IntegerQ[m] && !IntegerQ[n],
     m_^q_ n_^p_ :> (m/n)^q /; q >= 0 && p == -q &&
                                    !IntegerQ[m] && !IntegerQ[n]}
```

$$\text{Sqrt}[2]\ \text{Sqrt}[\frac{a\ b\ d}{c\ e}]$$

The result also now works on our original expression:

```
PowerExpand[Simplify[Sqrt[expr1]]] //.
   { m_^q_ n_^q_ :> (m n)^q /; !IntegerQ[m] && !IntegerQ[n],
     m_^q_ n_^p_ :> (m/n)^q /; q >= 0 && p == -q &&
                                    !IntegerQ[m] && !IntegerQ[n]}
```

$$\text{Sqrt}[2]\ (a + b)\ \text{Sqrt}[\frac{a}{a^2 + b^2}]$$

These rules are so handy it's useful to define a function for them, which can be considered the inverse of **PowerExpand**. We shall call it **PowerContract**,

```
PowerContract[expr_] := expr //.
   { m_^q_ n_^q_ :> (m n)^q /; !IntegerQ[m] && !IntegerQ[n],
     m_^q_ n_^p_ :> (m/n)^q /; q >= 0 && p == -q &&
                                    !IntegerQ[m] && !IntegerQ[n]}
```

and use it throughout this book. It's thus included in the package **Quantum `PowerTools `** (see Appendix IV). You can check it by entering

```
Needs["Quantum`PowerTools`"]
```

```
Sqrt[2 a b d/c/e] //PowerExpand //PowerContract
```

$$\text{Sqrt}[2]\ \text{Sqrt}[\frac{a\ b\ d}{c\ e}]$$

Exercise 2.4 Complex Conjugate

In quantum mechanics it is important that we're able to calculate the complex conjugate of any expression, whether numeric or symbolic. Since the built-in functions `Conjugate`, `Re`, and `Im` are defined for numbers only, we shall develop some simple rules for extending their definitions to *symbolic* expressions.

This is relatively straightforward in physics because our expressions are generally combinations of real variables and `I == Complex[0,1]`. Hence, if `z` is complex, we write `z = x + I y` explicitly and take `x` and `y` to be real. We can thus obtain the complex conjugate of any expression in *Mathematica* simply by replacing all occurrences of `I` by `-I`. With pattern matching, this works quickly and at all levels of the expression.

Consider therefore the rule

```
conjugate[expr_] := expr /. Complex[x_,y_] -> x - I y
```

and two sample expressions

```
{z = x + I y, 1/z}
```

$$\left\{ x + I\ y,\ \frac{1}{x + I\ y} \right\}$$

Their absolute magnitudes squared can be calculated as

```
conjugate[%] % //ExpandAll
```

$$\left\{ x^2 + y^2,\ \frac{1}{x^2 + y^2} \right\}$$

With this definition, we also easily extract the real and imaginary parts of an expression by combining the expression and its conjugate according to

```
re[expr_] := (expr + conjugate[expr])/2     //Expand
im[expr_] := (expr - conjugate[expr])/(2I) //Expand
```

Then, for example,

```
{re[1/z], im[1/z]} //Together //ExpandDenominator
```

$$\left\{ \frac{x}{x^2 + y^2},\ -\left(\frac{y}{x^2 + y^2}\right) \right\}$$

We can often obtain additional algebraic simplification directly by applying the built-in quantity `ComplexExpand` which attempts to transform functions of `I` into the general form `a + I b` with `a` and `b` real. It also assumes by default that all variables are real (see `?ComplexExpand`), although in general we have to include the option `TargetFunctions` to get the simplest results. For example,

```
conjugate[1/z] 1/z //
    ComplexExpand[#,TargetFunctions -> {Re,Im}]& //
    Together
```

$$\frac{1}{x^2 + y^2}$$

(Other **TargetFunctions** are also supported, e.g. **{Abs,Arg}**. See Wolfram [62], Section 3.3.7.) In addition, we can declare **y** for example to be complex according to

```
E^z //ComplexExpand[#,y]&
```

$$E^{x \,-\, Im[y]} \; Cos[Re[y]] \; + \; I \; E^{x \,-\, Im[y]} \; Sin[Re[y]]$$

When working with trig functions and complex-valued exponentials, it's sometimes convenient to load the package **Algebra`Trigonometry`** in order to introduce Euler's formula **E^(I a) -> Cos[a] + I Sin[a]**.

```
Needs["Algebra`Trigonometry`"]
```

Refer to **?TrigToComplex** and **?ComplexToTrig**, and consider for example the superposition of two simple signals with phases **p1** and **p2**:

```
s = E^(I p1) + E^(I p2)
```

$$E^{I \; p1} \; + \; E^{I \; p2}$$

We calculate the signal strength and simplify with

```
conjugate[s] s //ComplexToTrig //Simplify
```

$$4 \; Cos[\frac{p1 \,-\, p2}{2}]^2$$

(Try this without **ComplexToTrig**.) We can do the same thing in this case with **ComplexExpand**, though not as quickly:

```
conjugate[s] s //ComplexExpand //Simplify
```

$$4 \; Cos[\frac{p1 \,-\, p2}{2}]^2$$

However, we can return to our original expression by dividing by its conjugate and applying **TrigToComplex**:

```
%/conjugate[s] //TrigToComplex //Together
```

$$E^{I \; p1} \; + \; E^{I \; p2}$$

We shall thus find it useful to include our conjugate rule in a package which can be conveniently loaded when needed. We therefore define **Quantum`QuickReIm`** to extend the built-in functions **Conjugate**, **Re** and **Im** to symbolic arguments (see Appendix IV).

```
Needs["Quantum`QuickReIm`"]
```

Then, for example,

```
Conjugate[1/z] 1/z //ExpandDenominator
```

$$\frac{1}{x^2 \,+\, y^2}$$

This package is very short and therefore a much faster version of the *Mathematica* package **Algebra `ReIm`**. We have simply provided a rule for **Conjugate** and passed the burden of algebraic simplification on to the user, who must supply the appropriate built-in functions and pattern matching. The result is thus quite general. The package **Algebra `ReIm`** on the other hand defines rules for simplifying various functions involving **Conjugate**, **Re** and **Im** directly and enhances **ComplexExpand**. However, its format is just the opposite of **Quantum `QuickReIm`** in that it assumes all variables are complex unless declared real explicitly, not the usual situation encountered in physics applications.

We note in passing that our package **Quantum `PowerTools`** also defines a function **PowerExpandComplex** which further enhances **ComplexExpand** and expresses roots of complex-valued polynomials in Cartesian form. Thus, for example,

```
Needs["Quantum `PowerTools`"]
```

```
Sqrt[q + I r] //PowerExpandComplex
```

$$\frac{I \; \text{Sqrt}[q^2 + r^2 - q \; \text{Sqrt}[q^2 + r^2]]}{\text{Sqrt}[2] \; (q^2 + r^2)^{1/4}} +$$

$$\frac{\text{Sqrt}[q^2 + r^2 + q \; \text{Sqrt}[q^2 + r^2]]}{\text{Sqrt}[2] \; (q^2 + r^2)^{1/4}}$$

which is useful in determining quantities such as the attenuation of a wave described by a complex wavenumber.

Exercise 2.5 NonCommutativeMultiply

It's useful to be able to formally define noncommuting operators whose effect depends on their order of application, analogous to matrices and matrix multiplication. We can represent the product of two such quantities **A** and **B** in *Mathematica* by the special operation **A ** B == NonCommutativeMultiply[A, B]**, which preserves the order of application of **A** and **B**. For example, the commutator

```
A ** B - B ** A
```

```
A ** B - B ** A
```

doesn't vanish automatically, as it does with ordinary multiplication

```
A B - B A
```

```
0
```

For generality, *Mathematica* attaches few rules to ****** and essentially defines it merely to be associative, so that for example.

```
(A**B) **C == A** (B**C)
```

```
True
```

In particular, the distributive law of multiplication is not invoked

```
(A + B) ** C
```

```
(A + B)  ** C
```

although in this book we would like this to simplify to **A ** C + B ** C**, since operators in quantum mechanics are linear. Consider therefore two rules for attaching the distributive law to **:

```
NCMrules := {
    A_ ** (B_ + C_) :> A ** B + A ** C,
    (B_ + C_) ** A_ :> B ** A + C ** A}
```

Then, for example,

```
(A + B) ** (C + D) //. NCMrules
```

```
A ** C + A ** D + B ** C + B ** D
```

as desired. We find, however, that multiplication and division by ordinary numbers still do not automatically simplify, as for example

```
(2 A + B/3) ** C //. NCMrules
```

$$(2\ A)\ **\ C + (\tfrac{B}{3})\ **\ C$$

Thus, consider additional rules to ensure that ** commutes with ordinary numbers, as quantum operators do (except antilinear operators involving complex conjugation, like time reversal):

```
NCMrules := {
    A_ ** (B_ + C_)           :> A ** B + A ** C,
    (B_ + C_) ** A_           :> B ** A + C ** A,

    A_ ** c_?NumberQ          :> c A,
    c_?NumberQ ** A_          :> c A,
    A_ ** (B_ c_?NumberQ)     :> c A ** B,
    (A_ c_?NumberQ) ** B_     :> c A ** B,

    A_ ** (B_ c_Rational)     :> c A ** B,
    (A_ c_Rational) ** B_     :> c A ** B}
```

The last two rules here are needed because a factor like **1/3** is expressed internally as **Rational[1,3]** (enter for example **A/3 //FullForm**). We then obtain

```
(2 A + B/3) ** C //. NCMrules
```

$$2\ A\ **\ C + \frac{B\ **\ C}{3}$$

Note that generally we need **//.** (**ReplaceRepeated**) to apply the rules until the expression no longer changes (cf. Exercise 2.3).

Our rules also have the desirable feature that multiplication by zero vanishes and that multiplication by one is the identity operation. For example,

```
{0 ** A, 1 ** A} //. NCMrules
```

> {0, A}

Also, we can declare quantities to be numbers with **/:** (**TagSet**, see Exercise 1.13), as for example

```
h /: NumberQ[h] = True;
```

```
(h A + B) ** C //. NCMrules
```

> h A ** C + B ** C

Finally, we note that factors like **Sqrt[3]**, expressed internally with **Power**, need separate rules in order to commute with ******. We thus include two additional rules

```
NCMrules := {
    A_ ** (B_ + C_)          :> A ** B + A ** C,
    (B_ + C_) ** A_          :> B ** A + C ** A,
    A_ ** c_?NumberQ         :> c A,
    c_?NumberQ ** A_         :> c A,
    A_ ** (B_ c_?NumberQ)    :> c A ** B,
    (A_ c_?NumberQ) ** B_    :> c A ** B,
    A_ ** (B_ c_Rational)    :> c A ** B,
    (A_ c_Rational) ** B_    :> c A ** B,

    A_ ** (B_ c_Power)       :> c A ** B,
    (A_ c_Power) ** B_       :> c A ** B}
```

such that

```
(Sqrt[2] A) ** (B ** C/Sqrt[3]) //. NCMrules
```

$$\mathrm{Sqrt}[\frac{2}{3}] \; A \; ** \; B \; ** \; C$$

We can also define a function to apply the rules. Here's an example which uses the built-in operation **FixedPoint** to apply the rules and transform an expression repeatedly with **Expand** until the result no longer changes (cf. Maeder [43], Chapter 6).

```
ExpandNCM[expr_] := FixedPoint[Expand[(#//. NCMrules)]&,expr]
```

Thus, (of course **//** is now needed rather than **//.**)

```
{(2 A + B/3) ** C, h ** A, (Sqrt[2] A) ** (B ** C/Sqrt[3])} //
    ExpandNCM
```

$$\{2 \; A \; ** \; C + \frac{B \; ** \; C}{3}, \; h \; ** \; A, \; \mathrm{Sqrt}[\frac{2}{3}] \; A \; ** \; B \; ** \; C\}$$

In setting up a quantum operator algebra in this book with ******, we shall find it convenient to convert these rules to assignments with **:=** (**SetDelayed**), as for example **A_ ** c_?NumberQ := c A**, so that *Mathematica* will automatically use them on any expression it evaluates involving ******. At the same time, it's appropriate to include these assignments in a package which can be loaded when needed. Thus, entering (see Appendix IV)

```
Needs["Quantum`NonCommutativeMultiply`"]
```

we obtain for example

```
{(2 A + B/3) ** C, h ** A, (Sqrt[2] A) ** (B ** C/Sqrt[3])}
```

$$\{2 \; A \; ** \; C \; + \; \frac{B \; ** \; C}{3}, \; h \; ** \; A, \; Sqrt[\frac{2}{3}] \; A \; ** \; B \; ** \; C\}$$

without the need for **//. NCMrules** or **//ExpandNCM**.

Exercise 2.6 Kronecker Delta Function

In many applications we use the symbol **KD[n,k]** to represent the *Kronecker delta function*, which equals one if **n = k** and zero if **n ≠ k**. Although we can let **KD[n,k]** evaluate in certain situations, we need a symbol for formal expressions and manipulations when **n** and **k** are left unspecified.

For example, we can let **KD[n,k]** evaluate fully when **n** and **k** are numbers. We can also let it evaluate to unity when **n** and **k** are identical patterns. Thus, consider

```
KD[n_,n_] := 1
KD[n_?NumberQ, m_?NumberQ] := 0
```

so that for example

```
{KD[1,1], KD[1,2], KD[1+n,1+n], KD[n,k], KD[1+n,k]}
```

```
    {1, 0, 1, KD[n, k], KD[1 + n, k]}
```

We can nevertheless substitute values for **n** and **k** later on:

```
% /. k -> n+1
```

```
    {1, 0, 1, KD[n, 1 + n], 1}
```

```
% /. n -> 1
```

```
    {1, 0, 1, 0, 1}
```

Finally, it's often useful to let the following unlike patterns simplify

```
KD[n_ + k_,n_] := 0
KD[n_,n_ + k_] := 0
```

such that for example

```
KD[m+n,m+n+k]
```

```
    0
```

Exercise 2.7 Integration Rules

A variety of applications in quantum mechanics involve gaussian and exponential wavefunctions, and it's convenient to be able to integrate these quantities efficiently. Although the built-in command **Integrate** will generally evaluate the required integrals, it can be sluggish if the integrand is complicated.

We can construct a specialized set of rules which will perform the desired integrals quickly by pattern matching. In particular, the request **Integrate[integrand,x]** can be thought of simply as a replacement rule of the form **/. integrand -> integral**. Even with a short list of rules, we obtain a fairly powerful integrator which will apply to a wide class of functions because of *Mathematica*'s abilities to transform algebraic expressions.

To see how this works, consider gaussian functions and powers of the form **x^n E^(-a x^2)** integrated over **{x,-Infinity,Infinity}**. We obtain expressions for the basic integrals from almost any table of integrals, as for example Gradshteyn and Ryzhik [28], p. 337, or even the built-in function **Integrate**. Thus, consider

```
integGaussRules := {
         E^(p_ x_^2 + c_.) :> E^c Sqrt[Pi/-p],
   x_^n_. E^(p_ x_^2 + c_.) :> 0 /; OddQ[n],
   x_^n_  E^(p_ x_^2 + c_.) :>
          E^c (n-1)!!/(-2p)^(n/2) Sqrt[Pi/-p] /; EvenQ[n]}
```

For example, we can integrate **x^4 E^(-a x^2)** on **{x,-Infinity,Infinity}** simply by the replacement

```
x^4 E^(-a x^2) /. integGaussRules //PowerExpand

 3 Sqrt[Pi]
 ──────────
     5/2
  4 a
```

Note this integral is defined only if **Re[a] > 0**, and hence our rules hold only if **Re[p] < 0**. We could protect ourselves from misuse by attaching a condition to each rule to ensure that **Re[p] < 0**, but then our rules wouldn't apply to symbolic exponential arguments, which is undesirable. The built-in function **Integrate** works similarly.

As long as we integrate over all **x** and therefore can shift the origin arbitrarily, we can evaluate somewhat more general functions of the form **E^(-a x^2 + b x)** by completing the square in the exponent. Thus, introducing for example our rule from Exercise 2.2, we find that

```
ib = E^(-a x^2 + b x + c) /.
         a_. x^2 + b_. x + c_. :>
         a(x + b/(2a))^2 + Together[b^2/(-4a) + c] /.
         integGaussRules

      2
  (b  + 4 a c)/(4 a)         Pi
 E                    Sqrt[──]
                            a
```

We can evaluate integrals involving powers of **x** and **E^(-a x^ 2 + b x)** by parametric differentiation of **ib** with respect to **b** to pull down factors of **x** into the integrand from the exponent. For example, the integral of $x^4 E^{(-a x^2 + b x + c)}$ is given by

```
D[ib,{b,4}] //ExpandAll //PowerExpand //Factor
```

$$\frac{(12\ a^2 + 12\ a^2\ b^2 + b^4)\ E^{b^2/(4\ a)\ +\ c}\ \text{Sqrt}[Pi]}{16\ a^{9/2}}$$

We can partially check this expression by comparing with our previous result in the limit $b \to 0, c \to 0$:

```
% /. {b -> 0, c -> 0}
```

$$\frac{3\ \text{Sqrt}[Pi]}{4\ a^{5/2}}$$

Again, it's useful to include these rules in a package which can be loaded when needed. At the same time, we can define a function **integGauss** with a syntax identical to **Integrate** in order to apply the rules to any gaussian integrand. We thus have for example (see Appendix IV)

```
Needs["Quantum`integGauss`"]
```

```
integGauss[x^4 E^(-a x^2 + b x + c), {x,-Infinity,Infinity}] //
    PowerExpand //Factor
```

$$\frac{(12\ a^2 + 12\ a^2\ b^2 + b^4)\ E^{b^2/(4\ a)\ +\ c}\ \text{Sqrt}[Pi]}{16\ a^{9/2}}$$

In a similar fashion we can set up a function **integExp** to evaluate integrals involving powers of r and $E^{(-a\ r)}$ with respect to the radial coordinate $\{r, 0, \text{Infinity}\}$ for $\text{Re}[a] > 0$ (see Appendix IV). We thus load

```
Needs["Quantum`integExp`"]
```

and find

```
integExp[r^n E^(-a r + c), {r,0,Infinity}]
```

$$a^{-1 - n}\ E^{c}\ n!$$

Since the built-in function **n!** evaluates as **Gamma[1+n]** for noninteger and complex arguments **n**, this result holds for all $\text{Re}[n] > -1$ (see Gradshteyn and Ryzhik [28], p. 317). Comparing with the built-in integrator, we verify our expression with

```
Integrate[r^n E^(-a r + c), {r,0,Infinity}]
```

$$a^{-1 - n}\ E^{c}\ \text{Gamma}[1 + n]$$

```
% == %% /. n -> 1.1 + I //N
```

```
True
```

Computing   AIII.3

Exercise 3.1 Dynamic Programming

The **SetDelayed** := assignment facilitates very compact programming, especially since functions can call themselves in *Mathematica*. To illustrate this concept of dynamic programming, let's examine a procedure to compute factorials. (This is for fun, since *Mathematica* already has factorial functions, **n!** and **n!!** built-in. See **?!** and **?!!**.) Thus, consider

```
fact[n_] := n fact[n-1]
fact[1]  := 1                    (* Boundary conditions.*)
fact[0]  := 1
```

We must be sure to include the boundary conditions for **fact[0]** and **fact[1]**, so that **fact** doesn't "chase its tail" forever. First, let's check our definitions

```
?fact
   Global'fact

   fact[0] := 1

   fact[1] := 1

   fact[n_] := n*fact[n - 1]
```

and then test them out and compare the result with the built-in factorial function:

```
fact[30]

   265252859812191058636308480000000
```

```
% - 30!

   0
```

When you want *Mathematica* to remember values, there is a handy idiom for telling a dynamic program to store all the quantities it computes. One simply uses := in conjuction with = as **f[x_]:=f[x]=rhs**. For example,

```
fact[n_] := fact[n] = n fact[n-1]
```

When we compute say **fact[30]** with this definition and then check **?fact**, we find that all intermediate values have been saved. This construct can thus speed up certain computations, since results already computed need only be pulled from memory and not be re-computed.

We can test the speed increase by loading a utility for keeping track of and displaying machine CPU time.

```
Needs["Utilities'ShowTime'"]
```

Compare how long it takes to compute 100 values and then 10 more:

```
fact[100];
```

    2.71667 Second

```
fact[110];
```

    0.466667 Second

(You can use **Off[ShowTime]** and **On[ShowTime]** to toggle **ShowTime** off and on. See **?ShowTime**.)

Exercise 3.2 Procedural vs. Structured Iteration

Good *Mathematica* programming avoids **Do** loops when at all possible and uses instead *structured* iteration on lists. *Mathematica* code is thus more compact and elegant than conventional programs and will minimize departure from the physics problem at hand.

Consider the following examples to compare conventional procedural iteration with good *Mathematica*.

```
list = {aa,bb,cc,dd,ee}          (* Sample list or "array."*)
```

    {aa, bb, cc, dd, ee}

Procedural iteration. This is the conventional method, which *Mathematica* nevertheless supports.

```
squarelist = {};                 (* Initialize empty list.*)
Do[ AppendTo[ squarelist, list[[n]]^2 ], {n,1,5} ]
                                 (* Square each element in turn
                                       and add to list.*)
squarelist                       (* Print result.*)
```

           2   2   2   2   2
        {aa , bb , cc , dd , ee}

Structured iteration. This is better *Mathematica*, since built-in iteration primitives are used.

```
Table[ list[[n]]^2, {n,1,5} ]
```

           2   2   2   2   2
        {aa , bb , cc , dd , ee}

```
#^2 & /@ list                    (* Map pure function at list.*)
```

           2   2   2   2   2
        {aa , bb , cc , dd , ee}

```
list^2                           (* Listability.*)
```

           2   2   2   2   2
        {aa , bb , cc , dd , ee}

Here's a handy way to apply a function **g** to just the **y** values of a set of data:

```
data = {{x1,y1,z1},{x2,y2,z2},{x3,y3,z3}};
```

```
{#[[1]],g[#[[2]]],#[[3]]} & /@ data
```

```
    {{x1, g[y1], z1}, {x2, g[y2], z2}, {x3, g[y3], z3}}
```

By merely including specific pure-function slots, we can also extract and transform just the **y** values, or say pairs of data to pass on for example to **ListPlot**.

```
g[#[[2]]]& /@ data
```

```
    {g[y1], g[y2], g[y3]}
```

```
{#[[1]],g[#[[3]]]} & /@ data
```

```
    {{x1, g[z1]}, {x2, g[z2]}, {x3, g[z3]}}
```

Exercise 3.3 Nesting Operations

Let's take another look at structured iteration, introduced in the previous exercise, with pure functions. Again the idea is to use built-in *Mathematica* functions and perform structured iteration when possible. The functions **Nest**, **NestList** and **FixedPoint** are often used in this way to write code which is free of **Do** loops and independent of the number of equations and variables. Refer for example to the development of the package **RungeKutta.m** in Maeder [43]. Here's an example taken from that book, Sec. 4.3.3.

We consider Newton's iteration formula **(xold + r/xold)/2 -> xnew** for approximating the square root of a number **r** given an initial guess **xi**. Let's take **r = 2**:

```
r = 2
```

```
    2
```

Procedural iteration.

```
x = N[1,60]                      (* Initialize to x = 1 and
                                    60 places of accuracy.*)
```

```
    1.
```

```
Do[x = (x+r/x)/2; Print[x], {6}] (* Iterate 6 times.*)
    1.5
    1.41666666666666666666666666666666666666666666666666666666667
    1.41421568627450980392156862745098039215686274509803921568627
    1.41421356237468991062629557889013491011655962211574404458491
    1.41421356237309504880168962350253024361498192577619742849829
    1.41421356237309504880168872420969807856967187537723400156101
```

```
x - N[Sqrt[r], 60]               (* Compare with Sqrt.*)
                    -49
    2.8592838433 10
```

Structured iteration. Now let's iterate using the built-in functions **Nest** and **NestList** with Newton's formula expressed as a pure function. First, with **Nest**:

```
?Nest
    Nest[f, expr, n] gives an expression with f applied
    n times to expr.
```

```
Nest[(# + r/#)/2 &, N[1,60], 6]        (* Initialize to x=1,
                                          iterate 6 times.*)
```

```
1.4142135623730950488016887242096980785696718753772340015610l
```

Now with **NestList**.

```
?NestList
```

> NestList[f, expr, n] gives a list of the results of applying
>   f to expr 0 through n times.

```
NestList[(# + r/#)/2 &, N[1,60], 6]   (* List of results.*)
```

```
{1., 1.5,
```

```
   1.4166666666666666666666666666666666666666666666666666666667,
```

```
   1.4142156862745098039215686274509803921568627450980392l568627,
```

```
   1.414213562374689910626295578890134910116559622115744044584g1,
```

```
   1.4142135623730950488016896235025302436149819257761974284g829,
```

```
   1.4142135623730950488016887242096980785696718753772340015610l}
```

Finally, we use **FixedPoint** to do essentially the same thing. The difference is **FixedPoint** computes until the result no longer changes.

```
?FixedPoint
```

> FixedPoint[f, expr] starts with expr, then applies f repeatedly
>   until the result no longer changes. FixedPoint[f, expr, n]
>   stops after at most n steps.

```
FixedPoint[(# + r/#)/2 & , N[1,60] ]
```

```
1.4142135623730950488016887242096980785696718753769480731766g
```

We can still print out each iteration by including **Print[#]** followed by a semicolon **;** as part of the pure function definition, all enclosed by **(...) &**.

```
FixedPoint[ (Print[#]; (# + r/#)/2 )&, N[1,60] ]
```

```
1.
1.5
1.4166666666666666666666666666666666666666666666666666666667
1.4142156862745098039215686274509803921568627450980392156g627
1.41421356237468991062629557889013491011655962211574404458491
1.4142135623730950488016896235025302436149819257761974284g829
1.4142135623730950488016887242096980785696718753772340015610l
1.4142135623730950488016887242096980785696718753769480731766g
1.4142135623730950488016887242096980785696718753769480731766g
```

Exercise 3.4 Data Fitting

Let's take a brief look at simple data fitting with synthetic data generated using lorentzian line shapes with small random numbers added in. Such data might simulate photon or particle spectra. (Refer also to the *Mathematica* packages **Statistics-'LinearRegression'** and **Statistics 'NonlinearFit'**.)

Consider, for example, three lorentzian functions all with the same widths but with peaks located at **y = -1, 1** and **3**. Summing these three functions with various amplitudes, we thus take

```
data =
    Table[
        {y,
         0.5/(1+(y+1)^2) + 0.9/(1+(y-1)^2) + 0.2/(1+(y-3)^2) +
         0.09 (Random[]-0.5)},
        {y,-5.,5.,0.2}
    ];
```

where we have suppressed the output (with ;) to conserve space. Instead, we plot it as

```
plotdata = ListPlot[data, AxesLabel -> {" y","data"}];
```

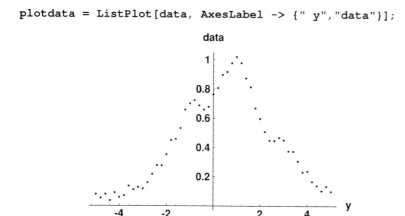

We perform a least-squares fit of the data with a list of functions of **y** of our choosing which we pass to the built-in function **Fit**. The fit is linear in that **Fit** determines only the best coefficients in a linear combination of the functions we give it. In particular, **Fit** doesn't calculate the relative positions or widths of our functions; we have to adjust these ourselves (cf. however **Statistics 'NonlinearFit '**).

To keep this simple, let's just use our three lorentzians, and pretend we didn't know how the data was produced. We thus obtain the best linear combination of the three lorentzians as a function of **y** with (refer to **?Fit**)

```
fitdata =
    Fit[
        data,
        {1/(1+(y+1)^2), 1/(1+(y-1)^2), 1/(1+(y-3)^2)},
        y
    ]
```

$$\frac{0.212267}{1 + (-3 + y)^2} + \frac{0.880855}{1 + (-1 + y)^2} + \frac{0.501506}{1 + (1 + y)^2}$$

We note that the coefficients, determined here by **Fit**, are close to the values we used to set up **data**. Of course, we can plot this result as a continuous function of **y** as

```
plotfit = Plot[fitdata,{y,-5,5}, AxesLabel -> {" y","data fit"}];
```

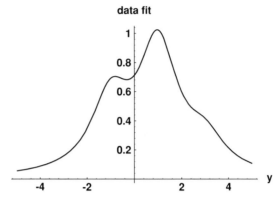

and compare with the original data:

```
Show[plotfit, plotdata];
```

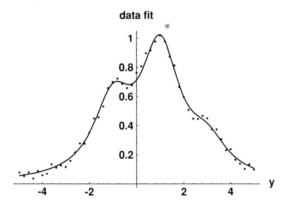

Exercise 3.5 Data Interpolation

We can also interpolate our data in the previous exercise with the built-in function **Interpolation**, which fits polynomials to overlapping subsets of **data**. The result is a smooth function of **y** which, unlike a least-squares fit, passes through all of the original data points and also interpolates points in between. Since third-order polynomials are used by default, the result is the simplest kind of cubic spline.

We thus interpolate **data** simply by wrapping it with **Interpolation**:

```
spline = Interpolation[data]
```

```
InterpolatingFunction[{-5., 5.}, <>]
```

The resulting **InterpolatingFunction** object is a pure function which needs to be given an argument in order to return a value (refer to **??Interpolation** and Exercise 1.9). The range **{-5.,5.}** indicates the values of **y** for which the interpolation was performed and for which the pure function is defined. For example, at the peaks of the lorentzians

```
{spline[-1.0], spline[1.0], spline[3.0]}
```

```
{0.699465, 1.01924, 0.450112}
```

Likewise, we can plot **spline[y]** as an ordinary continuous function of **y**. Here **Evaluate** is included to speed up the plotting.

```
plotspline =
    Plot[
        Evaluate[spline[y]], {y,-4.99,4.99},
        AxesLabel -> {" y","splined data"}
    ];
```

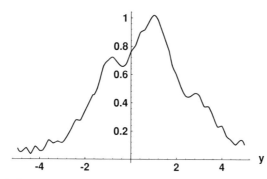

We can also differentiate **spline[y]** symbolically. However, since the interpolation was performed with a third-order polynomial, fourth-order and higher derivatives are identically zero. Therefore, for example

```
D[spline[y],{y,4}] /. y -> 1
```

```
0
```

Although cubic splines are generally the most efficient choice, the order can also be changed by changing the option **InterpolationOrder** (see **??Interpolation**). Refer in addition to the package **Graphics 'Spline'**.

Finally, we can plot the interpolation on top of the original data, as before.

```
Show[ plotspline, plotdata ];
```

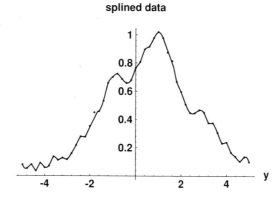

Appendix IV
Quantum Packages

The packages developed for this book and located in the **Quantum** folder (or subdirectory) on the disk that comes with the book are listed. Several of the underlying rule sets are developed and discussed in the exercises in Appendix III, Section 2. In order for the calls to these packages (**Get** and **Needs** commands) to work properly, the **Quantum** folder must be placed in the *Mathematica* **Packages** folder (or directory).

The format of these packages is standard and described in detail in Maeder [43], Section 2.4. Maeder also gives examples of defining rule sets and setting up packages.

Contents

Quantum `Clebsch`   AIV.1

Racah's formula is implemented to compute the Clebsch-Gordan coefficient and its related *3j* symbol. Definitions follow those in Brink and Satchler

[12], p. 34, eq. 2.34, and App. I, p. 136. This package was adapted from one written by S. Wolfram for version 1.0 of *Mathematica*. Refer also to the *Mathematica* package **StartUp `ClebschGordan `**, written and developed by Paul Abbott, which implements the built-in functions **ClebschGordan**, **ThreeJSymbol** and **SixJSymbol**.

```
BeginPackage["Quantum`Clebsch`"]

Clebsch::usage = "Clebsch[{A,a},{B,b},{C,c}] computes the
Clebsch-Gordan coefficient for nonsymbolic arguments using
Racah's formula. Refer to Brink&Satchler, ``Angular
Momentum'', 2nd ed.(Oxford), p.34."

Threej::usage = "Threej[{A,a},{B,b},{C,c}] computes the
Wigner 3-j symbol for nonsymbolic arguments using the
proportionality with the Clebsch-Gordan coefficient
Clebsch and hence Racah's formula. Refer to Brink & Satchler,
``Angular Momentum'', 2nd ed. (Oxford), p.34, and App. I, p.136."

Begin[" `private`"]

(* see Brink&Satchler, p.140 *)
Clebsch[{A_,a_},{A_,b_},{0,0}] :=
                     (-1)^(A-a)/Sqrt[2A+1] /; b == -a
Clebsch[{A_,a_},{B_,b_},{0,0}] := 0

f1[C_,A_,B_,a_,b_] :=
            Sum[
                (-1)^k (k! (A+B-C-k)! (A-a-k)! (B+b-k)! *
                (C-B+a+k)!(C-A-b+k)!)^-1,
                { k, Max[0,B-C-a,A-C+b], Min[A+B-C,A-a,B+b]}
            ]  (* These Max and Min prevent the arguments of
            Factorial (!) from becoming negative when the sum
            over k is performed. *)

f2[C_,A_,B_] := (-C+A+B)! (C-A+B)! (C+A-B)! /(1+C+A+B)!

f3[C_,A_,B_,c_,a_,b_] :=
            (1+2C) (C-c)! (C+c)! (A-a)! (A+a)! (B-b)! (B+b)!

(* see Brink&Satchler, p.34, eq. 2.34 *)
Clebsch[ {A_, a_}, {B_, b_}, {C_, c_} ] :=
    If[
        c!=a+b || C < Abs[A-B] || C > A+B ||
        Abs[a] > A || Abs[b] > B || Abs[c] > C,
        0,
        Block[{},
            f1[C,A,B,a,b] Sqrt[ f2[C,A,B] f3[C,A,B,c,a,b] ]
        ]
    ]
```

```
(* see Brink&Satchler, App. I, p.136 *)
Threej[ {A_,a_},{B_,b_},{C_,c_} ] :=
         (-1)^(B-A+c)/Sqrt[2C+1] Clebsch[{A,a},{B,b},{C,-c}]
End[]


EndPackage[]
```

Quantum `integExp `   AIV.2

Rules for integrating patterns involving `r^n E^(-a r)` for `{r,0,Infinity}` are given. Refer to Exercise 2.7, Appendix III.

```
BeginPackage["integExp`"]


integExp::usage = "integExp[integrand,{r,0,Infinity}] integrates
linear combinations of patterns of the form r^n E^(-a r) for
Re[a] > 0. WARNING: The requirement Re[a] > 0 must be enforced
by the user."


Begin[" `Private`"]


rules :=
{
     E^(a_. r + b_. r + c_.) :> E^c E^((a+b) r),
r^n_. E^(a_ r + c_.) :> E^c n! (-a)^(-n-1) /; FreeQ[c,r],
     E^(a_ r + c_.) :> E^c (-a)^(-1) /; FreeQ[c,r]
}


integExp[ seq_, {x_,0,Infinity} ] :=
        Block[ {r = x}, Expand[seq] //. rules ]
End[]


EndPackage[]
```

Quantum `integGauss `   AIV.3

Rules for integrating patterns involving `x^n E^(-a x^2 + b x + c)` for `{x,-Infinity,Infinity}` are given. This short list works faster and usually gives simpler expressions than the built-in function **Integrate**. Refer to Exercise 2.7, Appendix III and to Gradshteyn and Ryzhik [28], p. 337.

```
BeginPackage["integGauss`"]


integGauss::usage = "integGauss[integrand,{x,-Infinity,Infinity}]
integrates linear combinations of patterns of the form
x^n E^(-a x^2 + b x + c) for Re[a] > 0 and integer n >= 0.
WARNING: The requirements Re[a] > 0 and integer n >= 0
must be enforced by the user."
```

```
Begin[" `Private`"]

rules =
{
      E^(a_. x^n_. + b_. x^n_. + c_.) :> E^((a+b) x^n + c),
      E^(p_ x^2 + c_.) :> E^c Sqrt[Pi/-p] /; FreeQ[c,x],
x^n_. E^(p_ x^2 + c_.) :> 0 /; OddQ[n] && FreeQ[c,x],
x^n_  E^(p_ x^2 + c_.) :> E^c (n-1)!!/(-2p)^(n/2) Sqrt[Pi/-p] /;
                                    EvenQ[n] && FreeQ[c,x],
      E^(p_ x^2 + q_. x + c_.) :>
              E^c Sqrt[Pi/(-p)] E^(q^2/(-4 p)) /; FreeQ[c,x],
x^n_. E^(p_ x^2 + q_. x + c_.) :>
          ( E^c Sqrt[Pi/(-p)]/(-p) *
              D[ w/2 E^(w^2/(-4 p)), {w,n-1} ] /. w->q ) /;
                                    FreeQ[c,x]
}

integGauss[ seq_, {y_,-Infinity,Infinity} ] :=
        Block[ {x = y}, ExpandAll[seq] //. rules ]
End[]

EndPackage[]
```

AIV.4   Quantum `NonCommutativeMultiply`

Rules for making the built-in function **NonCommutativeMultiply** distributive as well as associative (which is built-in) and for making it commute with multiplication and division by ordinary numbers and powers are given. Note that the rules are implemented globally and applied by *Mathematica* each time ****** is used. Refer to Exercise 2.5, Appendix III.

```
BeginPackage["NonCommutativeMultiply`"]

Begin[" `Private`"]

protected = Unprotect[NonCommutativeMultiply]

A_ ** (B_ + C_)         := A ** B + A ** C
(B_ + C_) ** A_         := B ** A + C ** A

A_ ** c_?NumberQ        := c A
c_?NumberQ ** A_        := c A
A_ ** (B_ c_?NumberQ)   := c A ** B
(A_ c_?NumberQ) ** B_   := c A ** B

A_ ** (B_ c_Rational)   := c A ** B
(A_ c_Rational) ** B_   := c A ** B
```

```
A_ ** (B_ c_Power)        := c A ** B
(A_ c_Power) ** B_        := c A ** B

Protect[ Release[protected] ]

End[]

EndPackage[]
```

Quantum 'PowerTools '   AIV.5

A function **PowerContract** for reversing the built-in command **Power-Expand** is defined.   Refer to Exercise 2.3, Appendix III. Also, functions for simplifying roots of polynomials involving complex **I** as well as for simplifying exponents are given.

```
BeginPackage["PowerTools '"]

PowerContract::usage ="PowerContract[expr] nests expanded
powers. Intended as an inverse to PowerExpand."

PowerExpandComplex::usage = "PowerExpandComplex[expr]
expresses roots of a complex-valued polynomial in cartesian
form. Variables are assumed to be real otherwise. Results
may need Algebra 'Trigonometry ' to simplify further."

FactorExponents::usage ="FactorExponents[expr] applys Factor
to exponents which appear in expr."

TogetherExponents::usage ="TogetherExponents[expr] applys
Together to exponents which appear in expr."

Begin[" 'Private '"]

PowerContract[expr_] := expr //.
    { m_^q_ n_^q_ :> (m n)^q /;
            !IntegerQ[m] && !IntegerQ[n],
      m_^q_ n_^p_ :> (m/n)^q /;
              q >= 0 && p == -q &&
              !IntegerQ[m] && !IntegerQ[n]}

PECRules =
    Dispatch[
      {(a_ . + b_ . Complex[l_,m_] )^(n_Rational) :>
              ((a + b l)^2 + (b m)^2)^(n/2) *
              ( Cos[n ArcTan[a + b l, b m]] +
                      I Sin[n ArcTan[a + b l,b m]] ),
```

```
(* Avoid non-delayed equals and number substitution after
          using following two rules *)

        Cos[ n_Rational ArcTan[x_,y_] ] :>
            Sqrt[(1 + Cos[2n ArcTan[x,y]])]/Sqrt[2] /;
                            EvenQ[Denominator[n]],
        Sin[ n_Rational ArcTan[x_,m_?Negative y_.] ] :>
            -Sqrt[(1 - Cos[2n ArcTan[x,-m y]])]/Sqrt[2] /;
                            EvenQ[Denominator[n]],
        Sin[ n_Rational ArcTan[x_,y_] ] :>
            Sqrt[(1 - Cos[2n ArcTan[x,y]])]/Sqrt[2] /;
                            EvenQ[Denominator[n]],
        Cos[ArcTan[x_,y_]] :> x/Sqrt[x^2 + y^2],
        Sin[ArcTan[x_,y_]] :> y/Sqrt[x^2 + y^2]
    }
    ]
PowerExpandComplex[expr_] :=
    FixedPoint[Expand[PowerExpand[# //.PECRules]]&,expr];
FactorExponents[expr_] :=
    expr //. e_^m_ :> e^Factor[m]
TogetherExponents[expr_] :=
    expr //. e_^m_ :> e^Together[m]
SetAttributes[{PowerExpandComplex,PowerContract,
    FactorExponents,TogetherExponents}, Listable]
End[]

EndPackage[]
```

AIV.6   Quantum `QuantumRotations`

```
BeginPackage["QuantumRotations`"]

ReducedMatrix::usage = "ReducedMatrix[j,m,n,b]
computes the m-nth matrix element of the reduced rotation
matrix as a function of the Euler angle
b (= beta) using Wigner's sum from Brink and Satchler [12],
p.22. Result is a polynomial in Sin[b/2] and Cos[b/2].
May need TrigReduce to simplify.  Indeterminate
results of the form 0^0 may arise for some values
b = p Pi, p = integer. Such results can be avoided
by first evaluating ReducedMatrix for arbitrary b
then using /. b -> p Pi."

QuantumRotationMatrix::usage =
"QuantumRotationMatrix[j,m,n,a,b,g]
computes the m-nth matrix element of the rotation matrix
as a function of the Euler angles a (= alpha), b (= beta),
g (= gamma) using the function ReducedMatrix[j,m,n,b].
```

May need TrigReduce to simplify. Indeterminate results
of the form 0^0 may arise for some values b = p Pi,
p = integer. Such results can be avoided by first
evaluating QuantumRotationMatrix for arbitrary
b then using /. b -> p Pi."

Begin[" 'private'"]

(* Wigner's formula from Brink & Satchler [12], p.22.
Result is always a polynomial in Sin[b/2] and Cos[b/2]. *)

(* No rotation generates identity matrix *)
ReducedMatrix[j_, m_, m_, 0] := 1
ReducedMatrix[j_, mp_, m_, 0]:= 0

(* 2Pi rotation changes sign of identity matrix
   for half-integral j *)
ReducedMatrix[j_, m_, m_, 2Pi] := (-1)^(2j)
ReducedMatrix[j_, mp_, m_, 2Pi]:= 0

(* Periodic with period {4Pi} *)
ReducedMatrix[j_, m_, m_, 4Pi] := 1
ReducedMatrix[j_, mp_, m_, 4Pi]:= 0

(*
These rules help avoid indeterminate results of the
form 0^0 when b=0,2Pi,4Pi. We need in principle similar
rules for all b = p Pi, p = integer, since such values
also give indeterminate results.

The indeterminate results can be avoided by first evaluating
ReducedMatrix then using /. b -> p Pi. Introducing a dummy
variable bp along with /. bp -> b inside the function
definition ReducedMatrix complicates other results
when the argument b is passed specific values. *)

```
ReducedMatrix[ j_, m_, n_ , b_] :=
    Block[{kk},
            Expand[
                    Sqrt[ (j+m)! (j-m)! (j+n)! (j-n)! ] *
                    Sum[
                        (-1)^kk ( (j+m-kk)! (j-n-kk)! kk! *
                        (kk+n-m)!)^-1 *
                        Cos[b/2]^(2j+m-n-2kk) *
                        Sin[b/2]^(2kk+n-m),
                        {kk, Max[0,m-n], Min[j-n, j+m]}
                    ]
            ]
    ]
```

```
(*
A better method might be to use a formula for the reduced
rotation matrices based on Jacobi polynomials, which are
built-in. See Biedenharn and Louck [7], p. 49.

This formula, however, generates negative powers of Sin
and Cos and therefore gives the false impression that the
reduced matrices are sometimes singular.  Biedenharn and
Louck on p. 50 give another formula which will always
return the reduced matrices as nonsingular polynomials.
This improved formula involves a somewhat more complicated
algorithm, although it is very similar to the Wigner
formula above.

It is based on a solution of a differential equation,
discussed in Gottfried [27], p. 278.

The d. e. solution is related to a hypergeometric
function 2F1 and thus appears to be similar to the
formulas Paul Abbott uses to define the built-in
function ClebschGordan. Refer to Abbott's package
StartUp`ClebschGordan`.
*)

QuantumRotationMatrix[j_,m_,n_,a_,b_,g_] :=
    E^(-I m a) ReducedMatrix[j,m,n,b] E^(-I n g)

End[]

EndPackage[]
```

AIV.7   Quantum `QuickReIm`

The built-in functions **Conjugate**, **Re** and **Im** are extended to symbolic expressions. Refer to Exercise 2.4 in Appendix III. In order for these definitions to work properly, all symbols must be assumed to be real and all complex-valued expressions entered explicitly in terms of complex **I**. For example, a complex variable z must be of the form $z = x + I\,y$ with x and y real.

```
BeginPackage["QuickReIm`"]

Begin[" `Private`"]

protected = Unprotect[ Re, Im, Conjugate ]

Conjugate[expr_] := expr /. Complex[x_,y_] -> x - I y
```

```
Re[expr_] := Expand[ (expr + Conjugate[expr])/2    ]
Im[expr_] := Expand[ (expr - Conjugate[expr])/(2I) ]

Protect[ Release[protected] ]

End[]

EndPackage[]
```

Quantum `Trigonometry`   AIV.8

This package is identical with the package **Algebra`Trigonometry`** shipped with *Mathematica* Version 2.0, except that rules have been added to enhance the functionality of **TrigReduce**. In particular, variations of the identity **Sin^2 + Cos^2 == 1** have been included to help simplify expressions more directly. Refer to Exercise 2.2, Appendix III. The package was originally developed by R. Maeder.

Only the rule set **`TrigReduceRel** in the package is affected, and the new rules have been included at the top of the set as follows:

```
`TrigReduceRel = {

              (* begin Feagin's additions *)
         (* Variations on Sin^2 + Cos^2 == 1 *)

    a_. Sin[x_]^2 + a_. Cos[x_]^2 :> a,

    a_. + b_. Sin[t_]^2 + c_ :> a - b Cos[t]^2 /; c == -b,
    a_. + b_. Cos[t_]^2 + c_ :> a - b Sin[t]^2 /; c == -b,

  a_. Sec[x_]^2 + b_ Tan[x_]^2 :> a /; b == -a,
  a_. Csc[x_]^2 + b_ Cot[x_]^2 :> a /; b == -a,

    a_. + b_. Tan[t_]^2 + b_. :> a + b Sec[t]^2,
    a_. + b_. Cot[t_]^2 + b_. :> a + b Csc[t]^2,
    a_. + b_. Sec[t_]^2 + c_  :> a + b Tan[t]^2 /; c == -b,
    a_. + b_. Csc[t_]^2 + c_  :> a + b Cot[t]^2 /; c == -b,
    a_. Sin[x_] Tan[x_] + a_. Cos[x_] :>  a Sec[x],

  a_. Sin[x_] Tan[x_] + b_  Sec[x_] :> -a Cos[x]       /; b == -a,
  a_. Sec[x_]         + b_  Cos[x_] :>  a Sin[x] Tan[x] /; b == -a,

  a_. Cos[x_] Cot[x_] + a_. Sin[x_] :>  a Csc[x],
  a_. Cos[x_] Cot[x_] + b_  Csc[x_] :> -a Sin[x]       /; b == -a,
  a_. Csc[x_]         + b_  Sin[x_] :>  a Cos[x] Cot[x] /; b == -a,
              (* end Feagin's additions *)
```

```
(* the following two formulas are chosen to allow easy
   reconstruction of TrigExpand[Sin[x]^n] or
   TrigExpand[Cos[x]^n].  In these cases, Sin[n x] with
   even n does not occur.  There we use another
   formula. *)

Cos[n_Integer x_] :> 2^(n-1) Cos[x]^n +
Sum[ Binomial[n-i-1, i-1] (-1)î n/i 2^(n-2i-1)
     Cos[x]^(n-2i), {i, 1, n/2} ]     /; n > 0,

Sin[m_Integer?OddQ x_] :>
    Block[{`p = -(m^2-1)/6, `s = Sin[x], `k},
      Do[s += p Sin[x]^k;
            p *= -(m^2 - k^2)/(k+2)/(k+1),
         {k, 3, m, 2}];
      m s]                            /; m > 0,

Sin[n_Integer?EvenQ x_] :>
Sum[ Binomial[n, i] (-1)^((i-1)/2) Sin[x]î Cos[x]^(n-i),
     {i, 1, n, 2} ]                   /; n > 0,

Tan[n_Integer x_] :> Sin[n x]/Cos[n x],

Sin[x_ + y_] :> Sin[x] Cos[y] + Sin[y] Cos[x],
Cos[x_ + y_] :> Cos[x] Cos[y] - Sin[x] Sin[y],
Tan[x_ + y_] :> (Tan[x] + Tan[y])/(1 - Tan[x] Tan[y]),

(* rational factors, "symb" does not have a value *)
Sin[r_Rational x_] :> (Sin[Numerator[r] `symb] /.
    TrigReduceRel /. `symb -> x/Denominator[r])
/; Numerator[r] != 1,
Cos[r_Rational x_] :> (Cos[Numerator[r] `symb] /.
    TrigReduceRel /. `symb -> x/Denominator[r])
/; Numerator[r] != 1,

(* half angle args *)
Tan[x_/2] :> (1 - Cos[x])/Sin[x],
Cos[x_/2]^(n_Integer?EvenQ) :>
          ((1 + Cos[x])/2)^(n/2),
Sin[x_/2]^(n_Integer?EvenQ) :>
          ((1 - Cos[x])/2)^(n/2),
Sin[x_/2]^n_. Cos[x_/2]^m_. :> Tan[x/2]^n /; m == -n,
Sin[r_ x_.] Cos[r_ x_.]     :> Sin[2 r x]/2
                                    /; IntegerQ[2r]
}
```

Appendix V
Grad, Div, Curl

Throughout theoretical physics much use is made of derivative and vector derivative operators. Examples in quantum mechanics include the momentum and the kinetic energy of a particle proportional to the gradient (a vector) and the laplacian (the "square" of a vector), respectively, of the wavefunction in the coordinate representation. In order to represent best the physical symmetries of a system in a particular application, we want to be able to transform such quantities to an appropriate set of coordinates.

This appendix provides an introduction to vector calculus on the computer and, in particular, the transformation of vector derivative operators from Cartesian to curvilinear coordinates. It developed out of the need to be able to transform more complicated operators such as angular momentum from Cartesian to curvilinear coordinates.

Many of the results we derive here for *grad*, *div* and *curl* can also be found in the *Mathematica* package **Calculus `VectorAnalysis`**. However, our emphasis here is on the method and in obtaining compact results, since derived quantities such as angular momentum can be rather complicated. To this end, we express everything in terms of the total derivative **Dt** rather than the partial derivative **D** (see Section AV.2.1 below).

Vector Products   AV.1

Our principle task will be to apply the idea of ordinary vector *dot* and *cross* products to vector derivative operators. Consider two vectors **avec** and **bvec** referred to Cartesian axes with their **x**, **y** and **z** components labeled for convenience as **1**, **2** and **3**. For calculations, we collect vector components into *lists*

```
avec = {a[1],a[2],a[3]};      bvec = {b[1],b[2],b[3]};
```

which we explicitly label with the suffix **vec**. The dot-product function **Dot** is built-in, while the cross-product function **Cross** is defined in the package

```
<<LinearAlgebra`CrossProduct`
```

Thus, we can compute the dot and cross products **avec** • **bvec** and **avec** × **bvec** as

```
{avec.bvec, Cross[avec,bvec]}
   {a[1] b[1] + a[2] b[2] + a[3] b[3],
     {-(a[3] b[2]) + a[2] b[3], a[3] b[1] - a[1] b[3],
     -(a[2] b[1]) + a[1] b[2]}}
```

(A more general version of **Cross** is provided by the function **CrossProduct** from the package **Calculus`VectorAnalysis`**.)

Now for many applications, it is useful to think of these vector products as sums over particular components of a matrix. Although we won't pursue it, the general notion is that of *tensor contraction* in which vectors are tensors of first *rank* which contract via the dot product to produce scalars or tensors of zero rank. The physical idea which is especially useful is to define mathematical objects which behave in a particular way under coordinate transformation. Consider the 3×3 matrix

```
Table[a[j] b[k], {j,1,3},{k,1,3}] //MatrixForm
   a[1] b[1]    a[1] b[2]    a[1] b[3]
   a[2] b[1]    a[2] b[2]    a[2] b[3]
   a[3] b[1]    a[3] b[2]    a[3] b[3]
```

It's easy to see that the dot product **avec** • **bvec** is just the sum over the diagonal components of this matrix:

```
avec.bvec == Sum[a[i] b[i], {i,1,3}]
   True
```

The cross product clearly involves the off-diagonal elements, although with a somewhat subtle choice of signs. A close look shows that the *k*th component is **a[i] b[j] - a[j] b[i]**, if the sequence **i, j, k** is an even permutation of **1, 2, 3**, and **a[j] b[i] - a[i] b[j]**, if it's an odd one. Although this sequencing is often defined by introducing unit vectors and expanding the cross product as a determinant, we can also generate it by introducing the built-in function **Signature**, which tags the evenness or oddness of a permutation. For example,

```
{Signature[{1,2,3}],  Signature[{3,2,1}],  Signature[{1,1,2}]}
```

```
{1, -1, 0}
```

Note especially that if any two elements of a sequence are the same then its **Signature** vanishes. This function is thus also known as the *permutation symbol* or *Levi-Civita density*. Hence, the cross product **avec** × **bvec** is given by

```
Cross[avec, bvec] ==
    Table[
        Sum[Signature[{i,j,k}] a[i] b[j],{i,1,3},{j,1,3}],
        {k,1,3}
    ]
```

```
True
```

It's clear for example from this definition that **avec** × **bvec** = − **bvec** × **avec** and therefore that **avec** × **avec** = **0**. Although this definition is useful when considering coordinate transformations and for *deriving* vector identities, its primary utility for us will be in defining vector derivative operators.

Exercise 1

(a) Show that `Signature[{i,j,k}]a[i]b[j]c[k]` summed over **i, j, k** is equivalent to **avec** • (**bvec** × **cvec**). It's easy to see from this definition that **avec** • (**bvec** × **cvec**) == **bvec** • (**cvec** × **avec**) , etc. for **avec, bvec, cvec** in cyclic order.

(b) Show that `Signature[{i,j,k}]` `KD[i,j]` summed over **i, j** *vanishes* for any **k** with `KD[i,j]` the Kronecker delta function (see Exercise 2.6, Appendix III). Show that `Signature[{i,j,k}]^2` summed over **i, j, k** equals **6**.

Exercise 2

It is straightforward to prove vector identities by "brute force," *if they're already given,* by expanding the left- and right-hand sides of an identity and comparing the results.

(a) Verify that **avec** • (**bvec** × **cvec**) == **bvec** • (**cvec** × **avec**) , etc. for **avec, bvec, cvec** in cyclic order. Thus, **avec** • (**avec** × **bvec**) == **bvec** • (**avec** × **bvec**) = **0**.

(b) Verify that **avec** × **bvec** × **cvec** == (**avec** • **cvec**) **bvec** − (**avec** • **bvec**) **cvec**, and finally that (**avec** × **bvec**) • (**cvec** × **dvec**) == (**avec** • **cvec**)(**bvec** • **dvec**) − (**avec** • **dvec**)(**bvec** • **cvec**).

Exercise 3

If **a** and **b** are the magnitudes of **avec** and **bvec** (i.e. a^2 = **avec** • **avec**), show that **a b** Sin[phi] is the magnitude of **avec** × **bvec**, where **phi** is the angle between **avec** and **bvec**. (Use **avec** • **bvec** == **a b** Cos[phi].)

Problem 1

Consider a cube with one corner at the origin and sides oriented along the x, y, and z axes.

(a) Show that the angle between the body diagonals is `ArcCos[1/3]`.

(b) Determine the components of the unit vector perpendicular to the plane defined by the body diagonals. Find a way to check your result.

AV.2 Cartesian Coordinates

AV.2.1 Derivative Format

For compactness, we will use the *total* derivative function `Dt` to define derivative operators. Whereas `Dt[f,x]` is left unevaluated for arbitrary `f`, the *partial* derivative `D[f,x]` evaluates to zero. Thus,

```
{D[f,x], Dt[f,x]}
```

```
{0, Dt[f, x]}
```

When `D` computes, `f` is assumed by default to be independent of `x`, otherwise the coordinate dependence of `f` has to be given explicitly using for example `D[f[x, ...],x]` or `D[f,x, NonConstants -> {f}]`. This can become rather cumbersome, however, for what we need to do. In most applications of interest, the coordinate dependence of `f` will be clear and will not need to be written out all the time.

A related inconvenience occurs when we want to use a result derived with `D` to define a new operator or expression. Consider the following simple operator defined using both `D` and `Dt`:

```
{dx[f_] = D[f[x],x], dtx[f_] = Dt[f,x]}
```

```
{f'[x], Dt[f, x]}
```

It's natural to think of these quantities as operations on the original function `f`. However, a definition for `f[x]` does not immediately apply to `f'[x]`. Note for example what happens if we give `dx` the argument `Sin[x]`:

```
dx[Sin[x]]
```

```
(Sin[x])'[x]
```

Instead of being differentiated, `f` is literally replaced by `Sin[x]`. The reason is the pattern `f'[x]` has the internal form `Derivative[1][f][x]`, which nowhere explicitly contains the pattern `f[x]` (see `//FullForm` and also Exercise 4 below). The use of `Dt` eliminates the difficulty and gives us what we want:

```
dtx[Sin[x]]
```

> Cos[x]

Another way around this problem would be to use **D** in conjunction with **SetDelayed** (`:=`) to define operators, but this prevents us from viewing their form directly, which can be the result of a lengthy calculation (see Exercise 1.5, Appendix III). Thus, **Dt** allows us to derive and examine rather complicated expressions and still conveniently use them later on to define other operators.

The function **Dt** is mathematically equivalent to **D** if the coordinates on which **f** depends are declared to be independent and constants declared to be constant. This can be done once and for all using global declarations like **x/: Dt[x,y] = 0**, etc. and **SetAttributes[{c1,c2, ... }, Constant]**. (Note that setting the global attribute **NonConstants** is not possible.)

It is convenient to set up a function which makes the necessary tags (see Exercise 1.13, Appendix III) and declares *three* variables to be independent with respect to **Dt**. We thus define

```
IndependentVariables[x_,y_,z_] :=
    Module[{},
        x/: Dt[x,y] := 0;    x/: Dt[x,z] := 0;
        y/: Dt[y,x] := 0;    y/: Dt[y,z] := 0;
        z/: Dt[z,x] := 0;    z/: Dt[z,y] := 0
    ]
```

such that for example

```
IndependentVariables[x,y,z];

{Dt[x,y], Dt[x,z], Dt[y,z]}
```

> {0, 0, 0}

Gradient   AV.2.2

To introduce the *gradient* of a scalar function **f [x,y,z]**, we first compute its total differential

```
d[f_] = Dt[f[x,y,z]]
```

> $$\text{Dt}[z] \ f^{(0,0,1)}[x, y, z] + \text{Dt}[y] \ f^{(0,1,0)}[x, y, z] + \text{Dt}[x] \ f^{(1,0,0)}[x, y, z]$$

This is a fundamental theorem of multivariable calculus, which is built into **Dt**. We can put it into a compact and more useful form (see also Exercise 4) by introducing the definitions

```
ToDt = {Derivative[1,0,0][f_][x,y,z] -> Dt[f,x],
        Derivative[0,1,0][f_][x,y,z] -> Dt[f,y],
        Derivative[0,0,1][f_][x,y,z] -> Dt[f,z]}
```

$$\{(f\_)^{(1,0,0)}[x, y, z] \rightarrow Dt[f, x], (f\_)^{(0,1,0)}[x, y, z] \rightarrow$$
$$Dt[f, y], (f\_)^{(0,0,1)}[x, y, z] \rightarrow Dt[f, z]\}$$

Thus,

```
d[f_] = Dt[f[x,y,z]] /.ToDt
```

Dt[x] Dt[f, x] + Dt[y] Dt[f, y] + Dt[z] Dt[f, z]

Exercise 4

Without the **ToDt** rules, **d[f_] = Dt[f[x,y,z]]** leads to somewhat unexpected results. Compare for example **d[Tan[x]]** before and after application of the rules. (Use **//FullForm**.) Note that without the rules, **d[Tan[x]]** simply substitutes the *entire* expression **Tan[z]** for the *head* **f** of the **function f[x,y,z]**. Thus, try also **d[Tan[#1]&]** and **d[Tan[#2]&]**.

This expression can be interpreted as the *dot* product of an *infinitesimal displacement* vector (or differential of the position vector),

```
Dt[rvec == {x,y,z}]
```

Dt[rvec] == {Dt[x], Dt[y], Dt[z]}

with the vector

```
grad[f_] = {Dt[f,x], Dt[f,y], Dt[f,z]}
```

{Dt[f, x], Dt[f, y], Dt[f, z]}

which defines the Cartesian *gradient* of a scalar function **f**, often written $\nabla\mathbf{f}$ and read "del **f**". Hence,

```
d[f] == grad[f].Dt[rvec] /.rvec -> {x,y,z}
```

True

Note how **Dt**, unlike **D**, allows us to view our results without having to be explicit about arguments of **f**. Nevertheless, for some applications, we shall want derivatives evaluated at points other than **x**, **y** and **z**. In such cases, we can define the gradient as **grad[x_,y_,z_][f_]** (see Exercise 6).

Although the gradient is defined for scalar arguments, our rule for **grad** works just as well with *vector* arguments. This is because **Dt** is automatically threaded over lists, i.e. **Listable**.

Exercise 5

(a) Show that the gradient of $r = \text{Sqrt}[x\hat{}\,2 + y\hat{}2 + z\hat{}2]$ is the *unit* vector **rvec/r**.

(b) Show that the **grad[V[r]] = V'[r]grad[r]** for an arbitrary function **V[r]**.

(c) If **avec** is a constant vector, show that the gradient of **avec • rvec** equals **avec**.

Exercise 6

Define the relative coordinate vector **Rvec = rvec[1] - rvec[2]**, where **rvec[i] =
{x[i],y[i],z[i]}**, and extend the definition of the gradient such that **grad[x_,
y_,z_][f_]** evaluates derivatives with respect to any point **x**, **y** and **z**. (Declare the vector components to be independent using a set of rules of the form
x/: Dt[x[i_],y[j_]]= 0, etc.)

(a) Show that the gradient of **R = Sqrt[Rvec • Rvec]** with respect to **rvec[i]** is the
unit vector **+Rvec/R** for **i = 1** and **-Rvec/R** for **i = 2**. Also, show that **grad[V[R]] =
V'[R]grad[R]** for an arbitrary function **V[R]**. (Compare with the previous exercise.)

(b) Evaluate the gradient of **avec • Rvec** with respect to **rvec[i = 1,2]** for a constant
vector **avec**. (Compare with the previous exercise.)

Taylor Expansion   AV.2.3

As an important example, consider the Taylor series expansion of a *scalar*
function **c[rvec + avec]** about the point **rvec**. Here are the first few
terms obtained with **Series** and expanding about **avec = 0**:

```
Series[c[x + ax,y + ay,z + az], {ax,0,1},{ay,0,1},{az,0,1}] /.
   ToDt //Normal //Expand
 c[x, y, z] + ax Dt[c, x] + ay Dt[c, y] + az Dt[c, z] +
            (0,1,1)                       (1,0,1)
    ay az c       [x, y, z] + ax az c       [x, y, z] +
            (1,1,0)                       (1,1,1)
    ax ay c       [x, y, z] + ax ay az c       [x, y, z]
```

The built-in function **Normal** converts the result to an ordinary algebraic
expression. Thus, we conclude from the first line of this result that to first
order in **avec**

$$c[rvec + avec] \;==\; c[rvec] + avec \bullet grad[c[rvec]] + \ldots \quad (2.3\text{-r1})$$

(In fact, one can show that this is an *exponential* series in **avec • grad**.)
Hence, the change **c[rvec + avec] - c[rvec]** in c is maximized if **avec**
is parallel to **grad[c]** (since then the cosine of the angle between **avec**
and **grad[c]** is unity). Therefore, the gradient determines the slope of the
function along its direction of *maximum* increase from the point **rvec**.

Problem 2

Consider the function (a two-center Coulomb potential) $v = -1/\text{Sqrt}[x\hat{}2 + y\hat{}2 +
(z+R/2)\hat{}2] - 1/\text{Sqrt}[x\hat{}2 + y\hat{}2 + (z-R/2)\hat{}2]$ plotted in Figure 2.3-1 for $y = 0$
and $R = 10$.

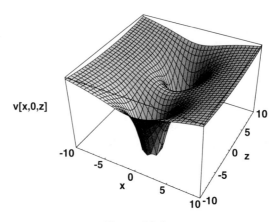

v[x,0,z]

Figure 2.3-1

(a) Make a `ContourPlot` of **v** for **y** = 0 and **R** = 10 and compare it with a plot of the function's gradient generated with `PlotVectorField` from the package `Graphics`PlotField`. (For best results, don't get too close to the singularities at **z** = ±**R/2**.) Notice how the gradient arrows point along the directions of maximum increase, i.e. are perpendicular to the *equipotential lines* in the contour plot. (Hint: Align the two plots and animate.)

(b) As is evident from Figure 2.3-1, the point **rvec** = 0 is a *saddle point*. Show that **v** is hyperbolic in **x**, **y** and **z** with negative curvature parallel to the *z* axis and positive perpendicular. Thus, expand **v** to second order in **x**, **y** and **z** about **rvec** = 0 and note the consequence of the rotational or *cylindrical* symmetry about the *z* axis. (Expand first about **z** in order to evaluate the `Sqrt` properly.)

Compare the coefficients of **x**, **y** and **z** with **avec** • **grad[v]** and **avec** • **grad[avec** • **grad[v]]**, whose **avec** = {**ax,ay,az**} is a *constant* displacement vector.

AV.2.4 Divergence and Curl

It is natural to consider the "dot" and "cross" products of the vector operator **grad** with a vector field **fvec** = {**fx,fy,fz**} whose each component is a separate function of **x**, **y** and **z**. In this way, we define the *divergence* and *curl* of **fvec**, respectively, i.e. ∇ • **fvec** and $\nabla \times$ **fvec** (read "del-dot" and "del-cross").

In analogy with the dot and cross products defined in the previous section for ordinary vectors, thus consider the (Jacobian) matrix formed from the derivatives of the components of **fvec**, which we conveniently generate using **grad** with a *vector* argument:

```
grad[{fx,fy,fz}] //MatrixForm
    Dt[fx, x]    Dt[fy, x]    Dt[fz, x]

    Dt[fx, y]    Dt[fy, y]    Dt[fz, y]

    Dt[fx, z]    Dt[fy, z]    Dt[fz, z]
```

Exercise 7

Show that this matrix is also given by a (built-in) `Outer` product of `{fx,fy,fz}` and `rvec` with differentiation replacing multiplication.

We thus define the divergence and curl as the following sums over elements of the above matrix:

```
div[{fx_,fy_,fz_}] = Sum[ grad[{fx,fy,fz}][[i,i]], {i,1,3} ]

   Dt[fx, x] + Dt[fy, y] + Dt[fz, z]

curl[{fx_,fy_,fz_}] =
    Table[
        Sum[
            Signature[{i,j,k}] grad[{fx,fy,fz}][[i,j]],
              {i,1,3},{j,1,3}
        ],
        {k,1,3}
    ]
   {-Dt[fy, z] + Dt[fz, y], Dt[fx, z] - Dt[fz, x],

      -Dt[fx, y] + Dt[fy, x]}
```

The simplicity of these results derives from the fact that in Cartesian coordinates the coordinate axes are fixed in space, and thus the corresponding unit vectors are independent of the coordinates, i.e.

```
eRep = {ex -> {1,0,0}, ey -> {0,1,0}, ez -> {0,0,1}};
```

Derivatives of these vectors therefore vanish. The unit vectors are depicted along with the position vector in Figure 2.4-1.

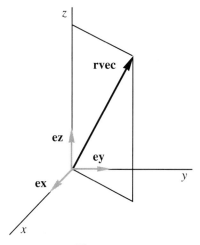

Figure 2.4-1

In general, however, one works with coordinate axes and unit vectors which conform to the *local* shape of a *curvilinear* space, usually chosen on the basis of a physically relevant symmetry. Thus, in general, the unit vectors will have nonvanishing derivatives. In this connection, we note that $\nabla \cdot$ **fvec** and $\nabla \times$ **fvec** can also be computed by first expanding **fvec** in terms of the unit vectors and taking derivatives and then forming the "dot" and "cross" products:

```
div[{fx,fy,fz}] ==
    ex.Dt[ex fx + ey fy + ez fz, x] +
    ey.Dt[ex fx + ey fy + ez fz, y] +
    ez.Dt[ex fx + ey fy + ez fz, z] /. eRep

    True


curl[{fx,fy,fz}] ==
    Cross[ ex, Dt[ex fx + ey fy + ez fz, x] ] +
    Cross[ ey, Dt[ex fx + ey fy + ez fz, y] ] +
    Cross[ ez, Dt[ex fx + ey fy + ez fz, z] ] /. eRep

    True
```

As we shall see, this idea is inherent to the calculation of the divergence and curl in curvilinear coordinates.

As the name implies, **div[fvec]** is a measure of how much the vector field **fvec[rvec]** diverges or spreads out at the point **rvec**. Likewise, **curl[fvec]** is a measure of how much **fvec[rvec]** curls or wraps around the point **rvec**. The direction of the **curl** is given by the right-hand rule (fingers along **fvec** and thumb along the **curl**).

Exercise 8

Using the function **PlotVectorField** from the package **Graphics`PlotField`** make plots of the three vector fields defined by {**x,y,0**} = **rvec[z=0]**, {**1,1,0**} = *constant vector* and {**-y,x,0**} = **Cross[{0,0,1},rvec]**. Interpret your results by computing **div** and **curl** for these vectors.

AV.2.5 Laplacian

Divergence and curl are all the *first-order* derviatives we can calculate with ∇. By applying ∇ twice, we can construct five species of *second-order* derivatives. By far, the most important is the *laplacian* defined to be the divergence of the gradient of a *scalar* field **f**, i.e. $\nabla \cdot \nabla$**f** (read "del-dot-del or del-squared")

```
laplacian[f_] = div[ grad[ f ] ]

    Dt[f, {x, 2}] + Dt[f, {y, 2}] + Dt[f, {z, 2}]
```

Now the curl of the gradient $\nabla \times \nabla$**f** and the divergence of the curl $\nabla \cdot \nabla \times$ **fvec** are both *always* zero, important facts which we easily demonstrate:

```
{curl[grad[f]], div[curl[{fx,fy,fz}]]}
```

```
   {{0, 0, 0}, 0}
```

The gradient of the divergence $\nabla\,(\nabla \cdot \mathbf{fvec})$ does not have a special name because it seldom occurs in physical applications. Finally, the curl of the curl $\nabla \times \nabla \times \mathbf{fvec}$ gives nothing new since (in analogy with the triple vector product) it's equal to $\nabla\,(\nabla \cdot \mathbf{fvec}) - \nabla \cdot \nabla\,\mathbf{fvec}$:

```
curl[curl[{fx,fy,fz}]] ==
    grad[div[{fx,fy,fz}]] - laplacian[{fx,fy,fz}] //
        ExpandAll
```

```
   True
```

This example also demonstrates how straightforward it is to prove vector calculus identities, albeit by brute force, *when the identities are given*. One should note, however, that more direct and elegant methods exist, as they do for establishing vector identities.

Although we could go on and work out *third* derivatives, one finds that first and second derivatives suffice for nearly all physical applications.

Exercise 9

(a) Show that `div[rvec]` = 3 and that `curl[rvec]` = 0. Show also that `laplacian[1/r]` = `div[rvec/r^3]` = 0, where **r** is the magnitude of **rvec**.

(b) Show that `curl[Cross[nvec, rvec]/2]` = **nvec**, if **nvec** is a constant vector.

Exercise 10

Redo the previous exercise using the relative coordinate vector `Rvec = rvec[1] -rvec[2]` in place of **rvec**. As in Exercise 6, you'll have to extend the definition of the divergence and the curl such that `div[x,y,z][fvec]` and `curl[x,y,z][fvec]` evaluate derivatives with respect to a point **x**, **y** and **z**.

Exercise 11

Let `gvec = g[r] rvec`, where **r** is the magnitude of **rvec**. Show that `g[r]` = *constant*/r^3 if we require `div[gvec]` to vanish. Show that `curl[gvec]` = 0 for any **g**.

Exercise 12

Verify the following identities:

(a) $\nabla\,(\mathbf{avec} \cdot \mathbf{bvec}) = \mathbf{avec} \times (\nabla \times \mathbf{bvec}) + \mathbf{bvec} \times (\nabla \times \mathbf{avec}) + (\mathbf{avec} \cdot \nabla)\,\mathbf{bvec} + (\mathbf{bvec} \cdot \nabla)\,\mathbf{avec}$;

(b) $\nabla \cdot (\mathbf{avec} \times \mathbf{bvec}) = \mathbf{bvec} \cdot (\nabla \times \mathbf{avec}) - \mathbf{avec} \cdot (\nabla \times \mathbf{bvec})$;

(c) $\nabla \times (\mathbf{avec} \times \mathbf{bvec}) = (\mathbf{bvec} \cdot \nabla)\,\mathbf{avec} - (\mathbf{avec} \cdot \nabla)\,\mathbf{bvec} + \mathbf{avec}\,(\nabla \cdot \mathbf{bvec}) - \mathbf{bvec}\,(\nabla \cdot \mathbf{avec})$.

AV.2.6 Notation

In what follows, particularly when using curvilinear coordinates, we need to be able to distinguish vectors and operators in various representations. For example, the position vector **rvec = {x,y,z}** in Cartesian coordinates (see Figure 2.4-1) is represented in spherical polar coordinates and along spherical polar axes by **{r,0,0}**, where **r = Sqrt[x^ 2 + y^2 + z^2]**.

Note in particular we have to indicate both the coordinates and the set of axes being used to express the components of a vector. For example, we can also refer **{r,0,0}** to Cartesian axes as

> {r,0,0} -> {r Sin[t] Cos[p], r Sin[t] Sin[p], r Cos[t]}
>
> (2.6-r1)

where **t** (= θ) and **p** (= ϕ) are spherical polar angles (see Figure 4.0-1 below). The general idea, of course, is to expand a vector in a new set of unit vectors, in this case spherical-polar unit vectors referred to Cartesian axes. However, when computing vectors as lists, it's not always apparent which set of axes or unit vectors is being used.

Exercise 13

Show that **r** is the magnitude of **{r Sin[t]Cos[p], r Sin[t]Sin[p], r Cos[t]}**.

Although the context will make this distinction clear most of the time, it's convenient to have a labeling scheme when working with more than one representation. Thus, we shall label the position vector referred to Cartesian and to spherical axes as

> {rvecC = {x,y,z}, rvecS = {r,0,0}};

where **C** and **S** indicate Cartesian and spherical components, respectively. When in addition we want to be explicit about the coordinate dependence, we shall append a suffix **xyz** or **rtp** and write, for example, **rvecCxyz** or **rvecCrtp**.

This scheme distinguishes various possibilities in a relatively straightforward way, as for example

> {rvecSxyz = {Sqrt[x^2 + y^2 + z^2], 0, 0},
> rvecCrtp = {r Sin[t] Cos[p], r Sin[t] Sin[p], r Cos[t]}};

Note that the label **vec** is dropped when denoting *magnitudes*,

> {rxyz = Sqrt[x^2 + y^2 + z^2], rrtp = r};

and that the labels **C** and **S** are inappropriate for scalars.

Curvilinear Coordinates   AV.3

We want to consider now a coordinate transformation from Cartesian to some other system of coordinates. In this section, we sketch how this is done for a general set of *curvilinear* coordinates, which we denote as **u**, **v** and **w**. Since everything here will depend only on **u**, **v** and **w**, we drop the suffix **uvw** to distinguish the coordinate dependence.

The transformation is defined by the Cartesian components of the position vector **rvec** written as functions of **u**, **v** and **w**:

```
rvecC = {x[u,v,w], y[u,v,w], z[u,v,w]};
```

Although we won't do much actual calculation in this section, it's convenient to declare the new coordinates to be independent (see Section AV.2.1):

```
IndependentVariables[u,v,w]
```

Again, it's useful to introduce a few rules to simplify derivatives when invoking the chain rule. Thus, we define

```
ToDt = {Derivative[1,0,0][f_][u,v,w] -> Dt[f,u],
        Derivative[0,1,0][f_][u,v,w] -> Dt[f,v],
        Derivative[0,0,1][f_][u,v,w] -> Dt[f,w]};
```

Then we can compute and simplify, for example, the total differential of a function of **u**, **v** and **w**:

```
d[f_] = Dt[f[u,v,w]] /.ToDt

  Dt[u] Dt[f, u] + Dt[v] Dt[f, v] + Dt[w] Dt[f, w]
```

Unit Vectors   AV.3.1

In order to represent a general vector in the these coordinates, we need a set of unit vectors along the directions of *increasing* **u**, **v** and **w**. Such vectors are easily generated by calculating the change in **rvec** as a function of each coordinate in turn, i.e by taking the derivatives of **rvecC** with respect to **u**, **v** and **w**. Here **rvecC** is used since it defines the connection with **x**, **y** and **z**:

```
{su = Dt[rvecC, u] /.ToDt, sv = Dt[rvecC, v] /.ToDt,
 sw = Dt[rvecC, w] /.ToDt}

  {{Dt[x, u], Dt[y, u], Dt[z, u]}, {Dt[x, v], Dt[y, v], Dt[z, v]},
   {Dt[x, w], Dt[y, w], Dt[z, w]}}
```

These vectors are not yet normalized, rather their magnitudes define *scale factors* along the directions **u**, **v** and **w**. It's convenient to introduce these here as a set of replacement rules (the suffix **rep** on **h** refers to replacement rules):

```
hrep =
    {hu -> Sqrt[su.su], hv -> Sqrt[sv.sv], hw -> Sqrt[sw.sw]}
```

$$\{hu \to Sqrt[Dt[x, u]^2 + Dt[y, u]^2 + Dt[z, u]^2\ ],$$

$$hv \to Sqrt[Dt[x, v]^2 + Dt[y, v]^2 + Dt[z, v]^2\ ],$$

$$hw \to Sqrt[Dt[x, w]^2 + Dt[y, w]^2 + Dt[z, w]^2\ ]\}$$

In general, these factors depend on the coordinates. Their squares are also known as the diagonal components of the *metric* (or *transformation*) *tensor*. We thus define the Cartesian components of a set of curvilinear unit vectors *referred* to the Cartesian axes:

```
eCrep = {eu -> su/hu, ev -> sv/hv, ew -> sw/hw}
```

$$\{eu \to \{\frac{Dt[x, u]}{hu}, \frac{Dt[y, u]}{hu}, \frac{Dt[z, u]}{hu}\},$$

$$ev \to \{\frac{Dt[x, v]}{hv}, \frac{Dt[y, v]}{hv}, \frac{Dt[z, v]}{hv}\},$$

$$ew \to \{\frac{Dt[x, w]}{hw}, \frac{Dt[y, w]}{hw}, \frac{Dt[z, w]}{hw}\}\}$$

We easily check for example that these vectors are properly normalized:

```
{eu.eu, ev.ev, ew.ew} /.eCrep /.hrep //Together
```

```
{1, 1, 1}
```

For a general coordinate transformation, however, these unit vectors will not be orthogonal, although we shall restrict our attention to coordinate systems for which they are. Such systems are called, not surprisingly, *orthogonal* and are characterized by a *diagonal* metric tensor.

The dependence of the unit vectors on **u**, **v** and **w** means that unlike their Cartesian counterparts, they move and change directions as **rvecC** moves. The idea is depicted in Figure 3.1-1. The unit vectors thus define a set of curvilinear axes which moves and rotates with **rvecC** and describes the local shape of the coordinate system. It's often useful to refer the unit vectors to the curvilinear axes they define. In that case, we clearly have

```
eGrep = {eu -> {1,0,0}, ev -> {0,1,0}, ew -> {0,0,1}};
```

Exercise 14

Consider the trivial *identity* transformation for which the curvilinear coordinates are just the Cartesian coordinates. Derive the scale factors **hx**, **hy** and **hz** and the unit vectors **ex**, **ey** and **ez** referred to the Cartesian axes.

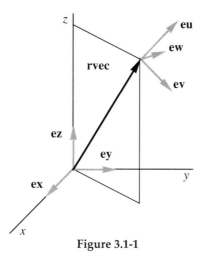

Figure 3.1-1

Curvilinear Gradient   AV.3.2

The scale factors determine the lengths of infinitesimal displacements along the directions **u**, **v** and **w**. Therefore, the element of volume is given by

```
d[vol] = hu Dt[u]   hv Dt[v]   hw Dt[w]
```

```
  hu hv hw Dt[u] Dt[v] Dt[w]
```

whereas the general infinitesimal displacement *vector* is defined as

```
d[rvec] = hu eu Dt[u] + hv ev Dt[v] + hw ew Dt[w]
```

```
  eu hu Dt[u] + ev hv Dt[v] + ew hw Dt[w]
```

As in Cartesian coordinates, this is still the differential of the position vector:

```
Dt[rvecC] == d[rvec] /.ToDt /.eCrep
```

```
  True
```

Thus, we define the gradient as

```
grad[f_] = eu/hu Dt[f,u] + ev/hv Dt[f,v] + ew/hw Dt[f,w]
```

$$\frac{eu\ Dt[f,\ u]}{hu} + \frac{ev\ Dt[f,\ v]}{hv} + \frac{ew\ Dt[f,\ w]}{hw}$$

such that its dot product with **d[rvec]** gives the differential of a function **f** of the curvilinear coordinates:

```
d[f_] = grad[f].d[rvec] /.eGrep

 Dt[u] Dt[f, u] + Dt[v] Dt[f, v] + Dt[w] Dt[f, w]
```

Here **eGrep** has been used to ensure that the coordinate system is orthogonal. Finally, we can refer the components of the gradient to either Cartesian or curvilinear axes. For example, along the curvilinear axes we define

```
gradG[f_] = grad[f] /.eGrep

   Dt[f, u]   Dt[f, v]   Dt[f, w]
 {————————, ————————, ————————}
     hu         hv         hw
```

To complete the transformation to curvilinear coordinates, we need to calculate the divergence and curl. The basic idea is to account for the change in the unit vectors as a function of the coordinates **u**, **v** and **w**, i.e. the unit vectors have to be differentiated. (By comparision, recall that the unit vectors in Cartesian coordinates are fixed in space and have vanishing derivatives.) This is straightforward once the Cartesian components of the unit vectors have been calculated. Hence, we shall postpone the calculation of the divergence and curl until a specific choice of curvilinear coordinates has been made, as in the next section.

It should be pointed out in this connection that general expressions for **div** and **curl** can be derived rather elegantly in terms of the scale factors from the *divergence* theorem and *Stoke's* theorem of vector calculus. Such expressions are used in the *Mathematica* package **Calculus`VectorAnalysis`**, which you might want to examine.

AV.4 **Spherical Coordinates**

As an example of transforming to curvilinear coordinates, let's calculate **grad**, **div**, **curl** in spherical polar coordinates **r**, **t** and **p**. Here **r** = **Sqrt[x^2 + y^2 + z^2]** is the magnitude of **rvec**, **t** (= θ) is the *polar* angle between **rvec** and the z axis, **and p** (= ϕ) is the *azimuthal* angle between the x-z plane and the plane formed by **rvec** and the z axis. Thus, the ranges **{t,0,Pi}** and **{p,0,2Pi}** cover the entire unit sphere **r** = 1. These coordinates are depicted in Figure 4.0-1. By convention, **t** = 0 along the positive z axis, and **p** = 0 along the positive x axis.

To distinguish vector components referred to Cartesian and spherical coordinate axes, we shall use the labels **C** and **S**, respectively, as introduced in Section 2.6. Since everything in this section will depend only on the spherical coordinates, we drop the suffix **rtp** to label coordinate dependence.

Also, since we're going to be doing a good deal of trigonometry here, it'll be helpful and faster when simplifying results if we introduce our enhanced version of the function **TrigReduce** from the package **Quantum`Trigonom-**

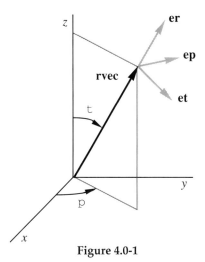

Figure 4.0-1

etry` (see Exercise 2.2, Appendix III). At the same time, we can check to make sure that the function **Cross** is defined.

```
Needs["Quantum`Trigonometry`"]
Needs["LinearAlgebra`CrossProduct`"]
```

Spherical Unit Vectors   AV.4.1

We introduce the transformation to spherical coordinates by writing the *Cartesian* components of the position vector **rvec** in terms of **r**, **t** and **p**. These can be read off Figure 4.0-1 and thus define the vector **rvecC**:

```
rvecC = {r Sin[t] Cos[p], r Sin[t] Sin[p], r Cos[t]};
```

In order to compute partial derivatives using **Dt**, we need to declare **r**, **t** and **p** to be independent coordinates. We make the declarations with the function we defined in Section AV.2.1:

```
IndependentVariables[r,t,p]
```

Following the general prescription developed in the previous section, we then easily generate vectors along the directions of *increasing* **r**, **t** and **p**:

```
{srC = Dt[rvecC,r], stC = Dt[rvecC,t], spC = Dt[rvecC,p]}

   {{Cos[p] Sin[t], Sin[p] Sin[t], Cos[t]},

     {r Cos[p] Cos[t], r Cos[t] Sin[p], -(r Sin[t])},

     {-(r Sin[p] Sin[t]), r Cos[p] Sin[t], 0}}
```

We need to compute the magnitudes of these vectors in order to define spherical-polar *unit* vectors. We thus compute the *scale factors* or squares of the elements of a diagonal *metric tensor*

```
{hr, ht, hp} =
     {Sqrt[srC.srC], Sqrt[stC.stC], Sqrt[spC.spC]} //
          TrigReduce //PowerExpand

  {1, r, r Sin[t]}
```

Exercise 15

Show that **srC**, **stC** and **spC** are orthogonal. (Thus prove that the off-diagonal elements of the metric tensor vanish.)

Exercise 16

Show that the element of volume in spherical coordinates is that of the parallelpiped **Dt[r]Dt[t]Dt[p]srC • (stC × spC)**.

We can now define the spherical-polar unit vectors **erC**, **etC** and **epC** referred to Cartesian axes. It's convenient to do this as a set of replacement rules:

```
eCrep = {er -> srC/hr, et -> stC/ht, ep -> spC/hp}
  {er -> {Cos[p] Sin[t], Sin[p] Sin[t], Cos[t]},

   et -> {Cos[p] Cos[t], Cos[t] Sin[p], -Sin[t]},

   ep -> {-Sin[p], Cos[p], 0}}
```

Note that **er** is just **rvec/r**. We shall also need these vectors referred to the spherical axes. Clearly, when this is done, we have

```
eSrep = {er -> {1,0,0}, et -> {0,1,0}, ep -> {0,0,1}}
  {er -> {1, 0, 0}, et -> {0, 1, 0}, ep -> {0, 0, 1}}
```

Exercise 17

Show that **er**, **et** and **ep** form an *orthogonal triad*, i.e. that **er** × **et** = **ep**, etc. in cyclic order.

As an example, let's expand an arbitrary vector **avec** in terms of the spherical unit vectors given its components **ax**, **ay** and **az** in Cartesian coordinates. The coefficients in the expansion are of course just the projections of **avec** onto the spherical axes: **avec • er**, **avec • et**, and **avec • ep**. Since the Cartesian components of **avec** are given, we compute the projections by introducing the Cartesian components of the spherical unit vectors:

```
avecC = {ax,ay,az};
avec  = avecC.(er /.eCrep) er + avecC.(et /.eCrep) et +
            avecC.(ep /.eCrep) ep
```

```
  ep (ay Cos[p] - ax Sin[p]) +
    et (ax Cos[p] Cos[t] + ay Cos[t] Sin[p] - az Sin[t]) +
      er (az Cos[t] + ax Cos[p] Sin[t] + ay Sin[p] Sin[t])
```

We then generate the spherical components **avecS** by referring the spherical unit vectors in this result to the spherical axes:

```
avecS = avec /.eSrep
  {az Cos[t] + ax Cos[p] Sin[t] + ay Sin[p] Sin[t],
    ax Cos[p] Cos[t] + ay Cos[t] Sin[p] - az Sin[t],
    ay Cos[p] - ax Sin[p]}
```

We can check this vector by computing its magnitude:

```
Sqrt[avecS.avecS] //Expand //TrigReduce
       2    2    2
  Sqrt[ax + ay + az ]
```

Exercise 18

As an additional check, refer the unit vectors to the Cartesian axes and get back **avecC**.

3D Parametric Plots   AV.4.2

As an example of a useful tool, let's make a *3D* parametric plot of a vector field **fvec** as a function of **t** and **p**. We create an interesting plot by taking a field as simple as **fvecS = {fr,0,0}** with **fr = fr[t,p]** restricted to the unit sphere with **r = 1**. For example, we can generate a family of self-intersecting bowls using **fr = Sin[t] (2 + Cos[p/n])** when **n** is an integer such that **{p,0,2Pi n}**.

We plot with **ParametricPlot3D** from the package **Graphics `ParametricPlot3D`** (it seems to be faster than the built-in function **ParametricPlot3D**). Note that this function requires as its argument the *Cartesian* components of the vector field being plotted (take a look at **?ParametricPlot3D**). However, we can easily transform from **fvecS** to **fvecC** by working backwards through our derivation of **avecS** in the previous example. Using first **eSrep** then **eCrep**, we define with **n = 4**,

```
fvecS = {Sin[t] (2 + Cos[p/4]), 0, 0};
fvecC = fvecS.(er /.eSrep) er + fvecS.(et /.eSrep) et +
           fvecS.(ep /.eSrep) ep /.eCrep

          p                    2            p                  2
  {(2 + Cos[-]) Cos[p] Sin[t] , (2 + Cos[-]) Sin[p] Sin[t] ,
          4                                 4

      p
  (2 + Cos[-]) Cos[t] Sin[t]}
      4
```

```
Needs["Graphics`ParametricPlot3D`"]

ParametricPlot3D[
    Evaluate[N[fvecC]],
    {t,Pi/2,Pi,Pi/12}, {p,0,8Pi,Pi/12},
    Axes -> None, Boxed -> False
];
```

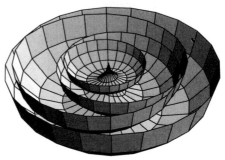

Figure 4.2-1

which can be given to **ParametricPlot3D**, as in Figure 4.2-1. In order to open the bowl, we restrict **t** to the lower hemisphere, i.e. to the range {**t,Pi/2,Pi**}. Here **Pi/12** is the plot step size, which in effect determines the number of plotpoints. (We could also use the option **PlotPoints**.)

You might note that the same plot can be obtained directly from **fr** using the plot function **SphericalPlot3D** from the same package. Our example should make clear how this and a related function **CylindricalPlot3D** work.

Exercise 19

Check that **Sqrt[fr^2 + ft^2 + fp^2]** is the magnitude of a general vector **fvecC** with spherical components **fr**, **ft** and **fp** referred to the Cartesian axes.

Exercise 20

The unit vectors **er**, **et** and **ep** change directions as **rvec** moves around the unit sphere and therefore have nonvanishing derivatives. Evaluate their derivatives with respect to **r**, **t** and **p** and derive the following set of rules:

```
deRep =
{Dt[er, r] :> 0, Dt[er, t] :>  et, Dt[er, p] :>  Sin[t] ep,
 Dt[et, r] :> 0, Dt[et, t] :> -er, Dt[et, p] :>  Cos[t] ep,
 Dt[ep, r] :> 0, Dt[ep, t] :>   0, Dt[ep, p] :>
  -Cos[t] et - Sin[t] er};
```

Hint: Use **eCrep**, and introduce replacement rules to recognize for example that (**f_** . **et** /. **eCrep**) -> **f et**.

Euler Angles AV.4.3

Our example from the end of Section AV.4.1 with transforming **avecC ->** **avecS** affords a more general point of view. Let **er** define the z' axis of a new system of coordinates $x'\ y'\ z'$ and **et** and **ep** the new x' and y' axes, respectively, as depicted in Figure 4.3-1. The components **avecS** can then be seen to be those of **avec** referred to the new axes:

```
ePrep = {er->{0,0,1}, et->{1,0,0}, ep->{0,1,0}};
avecP = avec /.ePrep

  {ax Cos[p] Cos[t] + ay Cos[t] Sin[p] - az Sin[t],

    ay Cos[p] - ax Sin[p], az Cos[t] + ax Cos[p] Sin[t] +

    ay Sin[p] Sin[t]}
```

The generalization is that the new coordinate system can be viewed as having been obtained from the old one by a rotation through the angles **t** and **p**. In fact, these angles are related to a triplet of *Euler angles* **a** $(= \alpha)$, **b** $(= \beta)$ and **g** $(= \gamma)$ which can be used to describe a general rotation of axes. One can always transform from one Cartesian coordinate system to another by performing just three independent rotations in a particular sequence. The Euler angles are defined as the three angles of rotation and in turn define a geometrical rotation matrix **R[a,b,g]** which connects the two sets of components of **avecC** and **avecP** under a general rotation.

The sequence of rotations is not unique and two choices of Euler angles are commonly used in physics, the so-called *x-* and *y-conventions* (Goldstein [25], Chapter 4). The *x-convention*, familiar in classical mechanics, is illustrated in Figure 4.3-2. One starts by rotating the initial axes *xyz* by the angle **a** counterclockwise about the *z* axis. This moves the *x* axis to an intermediate

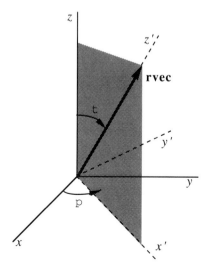

Figure 4.3-1

position in the *x-y* plane and defines the so-called *line of nodes*. The second rotation is through the angle **b** counterclockwise about the line of nodes (the intermediate *x* axis) and moves the *z* axis into its final position as the new *z'* axis. The line of nodes is thus the intersection of the original *x-y* plane with the new *x'-y'* plane. Finally, the intermediate *x* axis is moved into its final position as the new *x'* axis by a counterclockwise rotation about the *z'* axis by the angle **g**.

The rotation matrix **Rx[a,b,g]** in the *x-convention* can therefore be calculated by a product of three rotation matrices, one for each step in the sequence, in each case simply a two-dimensional rotation about the appropriate axis:

```
R1 = {{Cos[a],Sin[a],0}, {-Sin[a],Cos[a],0}, {0,0,1}};
R2 = {{1,0,0}, {0,Cos[b],Sin[b]}, {0,-Sin[b],Cos[b]}};
R3 = {{Cos[g],Sin[g],0}, {-Sin[g],Cos[g],0}, {0,0,1}};

Rx[a_,b_,g_] = R3.R2.R1 //Expand;
Rx[a,b,g] //MatrixForm
```

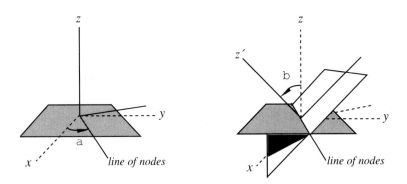

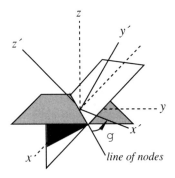

Figure 4.3-2

```
Cos[a]Cos[g] - Cos[b]Sin[a]Sin[g]      Cos[g]Sin[a] + Cos[a]Cos[b]Sin[g]   Sin[b]Sin[g]

-(Cos[b]Cos[g]Sin[a]) - Cos[a]Sin[g]   Cos[a]Cos[b]Cos[g] - Sin[a]Sin[g]   Cos[g]Sin[b]

Sin[a]Sin[b]                           -(Cos[a]Sin[b])                     Cos[b]
```

This matrix is identical to **RotationMatrix3D[a,b,g]** defined in the *Mathematica* package **Geometry`Rotations`**, which you're invited to compare with. It is easily verified that this is an *orthogonal* matrix such that its inverse equals its transpose (see Exercise 21). This ensures that scalar products (viz. lengths of vectors) are preserved by the rotation. Moreover, it is a *proper* matrix with determinant +1 so that right-handed coordinate systems are transformed into right-handed systems.

The *y convention* is obtained from the *x* convention simply by performing the second rotation instead about the intermediate *y* axis. Thus, the angle **b** retains its meaning, while **a** is now the angle between the *y* axis and the new line of nodes (the intermediate *y* axis) and **g** the angle between the line of nodes and the final y' axis. Hence, the angles in the two conventions are related as **ax = ay + Pi/2, bx = by, gx = gy - Pi/2** (see Goldstein [25], Appendix B) such that the rotation matrix in the *y* convention is given by

```
Ry[a_,b_,g_] = Rx[a+Pi/2, b, g-Pi/2] //TrigReduce;
Ry[a,b,g] //MatrixForm
```

```
Cos[a]Cos[b]Cos[g] - Sin[a]Sin[g]      Cos[b]Cos[g]Sin[a] + Cos[a]Sin[g]   -(Cos[g]Sin[b])

-(Cos[g]Sin[a]) - Cos[a]Cos[b]Sin[g]   Cos[a]Cos[g] - Cos[b]Sin[a]Sin[g]   Sin[b]Sin[g]

Cos[a]Sin[b]                           Sin[a]Sin[b]                        Cos[b]
```

This choice of Euler angles has become the standard in quantum mechanics (see for example Brink and Satchler [12] and references therein).

Returning to our example, it's convenient to use the *y* convention with the replacements **a -> p, b -> t**, and with **g = 0** to fix the new x'-z' axes in the *z*- **rvec** plane (see Figure 4.3-1):

```
Ry[p,t,0] //MatrixForm

   Cos[p] Cos[t]    Cos[t] Sin[p]    -Sin[t]

   -Sin[p]          Cos[p]           0

   Cos[p] Sin[t]    Sin[p] Sin[t]    Cos[t]
```

We can quickly check this result by verifying that the components of the unit vectors **er**, **et** and **ep** have the desired form in the new frame:

```
{Ry[p,t,0].et, Ry[p,t,0].ep, Ry[p,t,0].er} /. eCrep //
   TrigReduce

   {{1, 0, 0}, {0, 1, 0}, {0, 0, 1}}
```

Clearly, normalization has been preserved. Thus, **avecP** is computed as

```
avecP == Ry[p,t,0].avecC
```

    True

We note in passing that we have emphasized here the transformation of vectors in the *passive sense*. That is, we have considered the matrix **R[p,t,s]** to operate on the components of **avec** in the unprimed system to obtain its components in the primed system. From this point of view, the matrix acts on the coordinate system only and the vector **avec** is left unchanged. We could also emphasize, however, the *active sense* in which the matrix rotates the vector **avecC** changing it to the new vector **avecP** but *with both vectors expressed in the same coordinate system*. Both points of view are mathematically equivalent and are frequently encountered in physical applications.

Exercise 21

Show that **Rx** and therefore **Ry** are *proper, orthogonal* matrices such that their determinants equal +1 and their inverses equal their transpose. Verify that the inverse rotation is also given by **Rx[-g, -b, -a]]**, as is evident from Figure 4.3-2. Thus, verify that **avecC == Transpose[Ry[p,t,0]].avecP == Ry[0,-t,-p]].avecP**.

AV.4.4 Gradient

Given the scale factors **hr**, **ht** and **hp** it is straightforward to set up the gradient. Following the general prescription from Section AV.3.2, we define

```
grad[f_] = er/hr Dt[f,r] + et/ht Dt[f,t] + ep/hp Dt[f,p]
```

$$\frac{ep\ Csc[t]\ Dt[f,\ p]}{r} + er\ Dt[f,\ r] + \frac{et\ Dt[f,\ t]}{r}$$

and thus ensure that

```
d[f_] = grad[f].Dt[rvecC] /.eCrep //Expand //TrigReduce
```

    Dt[p] Dt[f, p] + Dt[r] Dt[f, r] + Dt[t] Dt[f, t]

where the infinitesimal displacement vector is given by

```
Dt[rvecC] == hr er Dt[r] + ht et Dt[t] + hp ep Dt[p] /.
    eCrep
```

    True

We can refer the gradient to the spherical axes simply by substituting the unit vectors referred to these axes:

```
gradS[f_] =
    er/hr Dt[f,r] + et/ht Dt[f,t] + ep/hp Dt[f,p] /. eSrep
```

$$\{Dt[f,\ r],\ \frac{Dt[f,\ t]}{r},\ \frac{Csc[t]\ Dt[f,\ p]}{r}\}$$

Although this is the familiar form of the gradient in spherical polar coordinates, we can also refer it to the Cartesian axes. In effect, we apply the *chain rule* to the Cartesian derivatives **Dt [f, x]**, etc.:

```
gradC[f_] =
    er/hr Dt[f,r] + et/ht Dt[f,t] + ep/hp Dt[f,p] /. eCrep
```

$$\{ \frac{Cos[p] \ Cos[t] \ Dt[f, \ t]}{r} - \frac{Csc[t] \ Dt[f, \ p] \ Sin[p]}{r} +$$

$$Cos[p] \ Dt[f, \ r] \ Sin[t],$$

$$\frac{Cos[p] \ Csc[t] \ Dt[f, \ p]}{r} + \frac{Cos[t] \ Dt[f, \ t] \ Sin[p]}{r} +$$

$$Dt[f, \ r] \ Sin[p] \ Sin[t],$$

$$Cos[t] \ Dt[f, \ r] \ - \ \frac{Dt[f, \ t] \ Sin[t]}{r} \}$$

Because the scale factors vanish for **r -> 0** or **t -> 0**, **gradS** and **gradC** are not defined at the origin or along the z axis. Hence, the same is also true of **div**, **curl** and **laplacian** in spherical coordinates.

Problem 3

Show that **gradC** is what one obtains by applying the chain rule to {**Dt [f, x]**, **Dt [f, y]**, **Dt [f, z]**}. Thus, compute **Dt [f [r, t, p], x]**, etc. and compare with **gradC**. Hint: The most straightforward way to obtain the derivatives **Dt [r, x]**, **Dt [t, x]**, etc. is to differentiate the transformation equations {**x, y, z**} == **rvecC** directly and solve for the derivatives.

As an example, we can check that **grad [avec • rvec] = avec** when **avec** is a constant vector (see Exercise 5). Operating with **gradS**, we obtain

```
avecC = {ax,ay,az};   SetAttributes[avecC, Constant];
```

```
gradS[avecC.rvecC] //Expand //TrigReduce
   {az Cos[t] + ax Cos[p] Sin[t] + ay Sin[p] Sin[t],

     ax Cos[p] Cos[t] + ay Cos[t] Sin[p] - az Sin[t],

     ay Cos[p] - ax Sin[p]}
```

But this is just **avecS**, which we computed at the end of Section AV.4.1:

```
% == avecS
   True
```

Our check is more convincing, however, if we operate with **gradC**:

```
gradC[avecC.rvecC] //Expand //TrigReduce
   {ax, ay, az}
```

Exercise 22

Using **gradS** and **gradC**, check that **grad[r] = rvec/r = er** and that **grad[1/r] = -er/r^2** (cf. Exercise 5).

AV.4.5   Divergence and Curl

We turn now to the calculation of the divergence and curl of a vector field **fvec** given its spherical components {**fr, ft, fp**}. We can do this in analogy with what we did in Cartesian coordinates in Section AV.2.4 without too much effort.

In transforming from Cartesian to spherical coordinates (and curvilinear coordinates in general), two new considerations arise. First, of course, we must transform derivatives to spherical coordinates using the chain rule. As we have seen in Problem 3, this is accomplished using **gradC**. Second, we must remember that the spherical unit vectors, unlike their Cartesian counterparts, depend on the coordinates (viz. the angles **t** and **p**) and therefore have nonvanishing derivatives. The simplest way to compute these derivatives and at the same time those of **fvec** is first to determine the *Cartesian* components of **fvec** and then pass them on to **gradC**. (Note, in particular, we wouldn't get the correct result using the spherical components of **fvec**.) We thus define

```
fvecC = fr er + ft et + fp ep /.eCrep
  {ft Cos[p] Cos[t] - fp Sin[p] + fr Cos[p] Sin[t],

    fp Cos[p] + ft Cos[t] Sin[p] + fr Sin[p] Sin[t],

    fr Cos[t] - ft Sin[t]}
```

Then, in analogy with Cartesian coordinates, we compute **div** and **curlC** as (**curlC** takes about 2 minutes to compute)

```
Clear[div]

div[{fr_, ft_, fp_}] =
    Sum[gradC[fvecC][[i,i]], {i,1,3}] //
        Expand //TrigReduce
```

$$\frac{2\ fr}{r} + \frac{ft\ Cot[t]}{r} + \frac{Csc[t]\ Dt[fp,\ p]}{r} + Dt[fr,\ r] + \frac{Dt[ft,\ t]}{r}$$

```
curlC[{fr_, ft_, fp_}] =
    Table[
        Sum[
            Signature[{i,j,k}] gradC[fvecC][[i,j]],
            {i,1,3},{j,1,3}
        ],
        {k,1,3}
    ] //Expand //TrigReduce
```

$$\{-(\text{Cos}[p] \ \text{Cos}[t] \ \text{Dt}[fp, \ r]) + \frac{\text{Cos}[p] \ \text{Cot}[t] \ \text{Dt}[fr, \ p]}{r} -$$

$$\frac{\text{Cos}[p] \ \text{Dt}[ft, \ p]}{r} - \frac{ft \ \text{Sin}[p]}{r} + \frac{\text{Dt}[fr, \ t] \ \text{Sin}[p]}{r} -$$

$$\text{Dt}[ft, \ r] \ \text{Sin}[p] + \frac{\text{Cos}[p] \ \text{Dt}[fp, \ t] \ \text{Sin}[t]}{r},$$

$$\frac{ft \ \text{Cos}[p]}{r} - \frac{\text{Cos}[p] \ \text{Dt}[fr, \ t]}{r} + \text{Cos}[p] \ \text{Dt}[ft, \ r] -$$

$$\text{Cos}[t] \ \text{Dt}[fp, \ r] \ \text{Sin}[p] + \frac{\text{Cot}[t] \ \text{Dt}[fr, \ p] \ \text{Sin}[p]}{r} -$$

$$\frac{\text{Dt}[ft, \ p] \ \text{Sin}[p]}{r} + \frac{\text{Dt}[fp, \ t] \ \text{Sin}[p] \ \text{Sin}[t]}{r},$$

$$\frac{fp \ \text{Csc}[t]}{r} + \frac{\text{Cos}[t] \ \text{Dt}[fp, \ t]}{r} - \frac{\text{Dt}[fr, \ p]}{r} - \frac{\text{Cot}[t] \ \text{Dt}[ft, \ p]}{r} +$$

$$\text{Dt}[fp, \ r] \ \text{Sin}[t]\}$$

We have cleared **div** first to avoid collision with our previous definition in Cartesian coordinates. (If we need both definitions, we can use the label **divrtp**. See Section 2.6.) Note, that **Set** (=), and not **SetDelayed** (:=), has been used in order to evaluate **div** and **curlC** once and for all (see Exercise 1.5, Appendix III).

Recall that **curlC** has its components referred to the *Cartesian* axes, which accounts for its relatively complicated form. We can simplify it considerably by referring it back to the spherical axes, thus defining **curlS**. This can be done in analogy with the earlier calculation of **avecS** from **avecC** at the end of Section 4.1:

```
curlCf = curlC[{fr,ft,fp}];
curlf  = curlCf.(er /.eCrep) er + curlCf.(et /.eCrep) et +
              curlCf.(ep /.eCrep) ep;

curlS[{fr_,ft_,fp_}] = curlf /.eSrep //Expand //TrigReduce
```

$$\{\frac{fp \ \text{Cot}[t]}{r} + \frac{\text{Dt}[fp, \ t]}{r} - \frac{\text{Csc}[t] \ \text{Dt}[ft, \ p]}{r},$$

$$-(\frac{fp}{r}) - \text{Dt}[fp, \ r] + \frac{\text{Csc}[t] \ \text{Dt}[fr, \ p]}{r}, \frac{ft}{r} - \frac{\text{Dt}[fr, \ t]}{r} +$$

$$\text{Dt}[ft, \ r]\}$$

Our results for **div** and **curlS** are essentially what one finds in any table of vector-calculus identities, except that here *Mathematica* automatically expands out the dervatives (cf. Exercise 27). You might thus compare our expressions with those from the *Mathematica* package **Calculus `Vector-**

Analysis ` computed directly from the scale factors **hr**, **ht** and **hp**. Note, however, that **gradCrtp** and **curlCrtp** are not provided by this package.

As an example, let's redo Exercise 9 using spherical coordinates with **rvecS = {r,0,0}**. We compute five quantities together as a single list. Here **avecS** is the *constant* vector from the previous section.

```
{div[{r,0,0}], div[{1/r^2,0,0}], curlS[{f[r],0,0}],
 curlC[{f[r],0,0}],
 curlC[Cross[avecS,{r,0,0}]/2] //Expand //TrigReduce}

  {3, 0, {0, 0, 0}, {0, 0, 0}, {ax, ay, az}}
```

Exercise 23

Using the constant vector **avecS**, show that **div[avecS]** and **curl[avecS]** vanish in spherical coordinates. (This is, of course, a trivial exercise in Cartesian coordinates.)

Exercise 24

Show that **curl[grad[f]]** and **div[curl[fvec]]** vanish identically in spherical coordinates (cf. Section AV.1.2).

Exercise 25

Compute the **div** and **curlC** of the vectors **{x,y,0}** = **rvec (z = 0)**, **{1,1,0}** and **{-y,x,0}** = **Cross[{0,0,1},rvec]** from Exercise 8.

Problem 4

Derive **div** and **curlS** for a general vector **fvecS = {fr, ft, fp}** by writing **fvecS** explicitly in terms of the spherical unit vectors, computing derivatives with respect to **r**, **t** and **p** using **deRep** from Exercise 20, and then computing the dot and cross products. Finally, derive **curlC** from **curlS**.

Problem 5

Derive **div** in spherical coordinates by applying the chain rule directly to **Dt[fx,x]** + **Dt[fy,y]** + **Dt[fz,z]**, with **fx**, **fy** and **fz** the Cartesian components of **fvecS**. (Refer to Problem 3.)

AV.4.6 Laplacian

Finally, we define the laplacian or "del-squared." Note that we use **gradS** since **div** requires its argument referred to spherical axes.

```
laplacian[f_] = div[ gradS[f] ]
```

$$\frac{2\,Dt[f, r]}{r} + \frac{Cot[t]\,Dt[f, t]}{r^2} + \frac{Csc[t]^2\,Dt[f, \{p, 2\}]}{r^2}$$

$$+ \, Dt[f, \{r, 2\}] + \frac{Dt[f, \{t, 2\}]}{r^2}$$

As a simple but often useful example, consider the laplacian of a function **f** of **r** only. Note how the first derivative can be eliminated by writing **f[r]** in terms of a new function **u[r]/r**:

```
{laplacian[f[r]], laplacian[u[r]/r] //Expand}
```

$$\{\frac{2\ f'[r]}{r} + f''[r], \ \frac{u''[r]}{r}\}$$

As an important corollary of this result, we see again that the laplacian of **1/r** vanishes for **r** $\neq$ **0** (see Exercise 9).

Exercise 26

Show that **laplacian[avec • rvecC] = 0** if **avec** is a constant vector. Explain.

Exercise 27

Show that our expression for **laplacian** is equivalent to the more usual, compact form which *Mathematica* automatically expands:

$$\frac{Dt[r^2\ Dt[f, r], r]}{r^2} + \frac{Dt[Sin[t]\ Dt[f, t], t]}{r^2\ Sin[t]} + \frac{Dt[f, \{p, 2\}]}{(r\ Sin[t])^2}$$

Problem 6

Show that the built-in functions **SphericalHarmonicY[l,m,t,p]** are *eigenfunctions* of **laplacian**. That is, show for fixed **r** that **laplacian[SphericalHarmonicY[l,m,t,p]] == l(l+1)/r^2 SphericalHarmonicY[l,m,t,p]** for l = 0, 1, 2, 3 and **Abs[m]** $\leq$ 1.

Problem 7

Derive the transformation to *cylindrical* coordinates **r**, **p** (= ϕ) and **z** defined by the relations **x == r Cos[p], y == r Sin[p], z == z**.

(a) Work out the scale factors and the unit vectors. Show that the unit vectors form an orthogonal triad. Derive **gradC[f]**.

(b) Derive the divergence and curl for a general vector field **fvecCyl = {fr, fp, fz}**. Show that **laplacian[f] ==**

$$\frac{Dt[r\ Dt[f, r], r]}{r} + \frac{Dt[f, \{p, 2\}]}{r^2} + Dt[f, \{z, 2\}]$$

Problem 8

Derive the transformation to *parabolic* coordinates **u, v** and **p** defined in terms of the spherical polar coordinates by the relations **u == r + z == r (1 + Cos[t]), v ==**

`r - z == r (1 - Cos[t])`, and `p == p`, where **u** and **v** take values from `{0, Infinity}` and **p** from `{0, 2Pi}`. (Note here **u** and **v** are the coordinates ξ and η, respectively, defined in Bethe and Salpeter [6], p. 27, or η and ξ defined in Schiff [59], p. 95.)

(a) We see that curves of constant **u** (for all **p**) are parabolas about the z or polar axis with focus at the origin that open in the direction of positive z or **t = 0**. The curves of constant **v** are similarly parabolas that open in the direction of negative z or **t = Pi**. These coordinates are the large-**R** limit of the elliptic coordinates introduced in the next problem.

Make a set of 2D parametric plots to display the curves of constant **u** and **v**. Or use **ParametricPlot3D**.

(b) Work out the scale factors and the unit vectors and simplify with **PowerContract** from the package from **Quantum`PowerTools`**. Verify that the unit vectors form an orthogonal triad and that the element of volume is **(u+v)/4 Dt[u] Dt[v] Dt[p]**. Show that **gradC[f] ==**

$$\left\{ \frac{2\ \text{Sqrt}[u\ v]\ \text{Cos}[p]\ \text{Dt}[f,\ u]}{u + v} + \frac{2\ \text{Sqrt}[u\ v]\ \text{Cos}[p]\ \text{Dt}[f,\ v]}{u + v} - \right.$$

$$\frac{\text{Dt}[f,\ p]\ \text{Sin}[p]}{\text{Sqrt}[u\ v]}, \quad \frac{\text{Cos}[p]\ \text{Dt}[f,\ p]}{\text{Sqrt}[u\ v]} + \frac{2\ \text{Sqrt}[u\ v]\ \text{Dt}[f,\ u]\ \text{Sin}[p]}{u + v} +$$

$$\left. \frac{2\ \text{Sqrt}[u\ v]\ \text{Dt}[f,\ v]\ \text{Sin}[p]}{u + v}, \quad \frac{2\ u\ \text{Dt}[f,\ u]}{u + v} - \frac{2\ v\ \text{Dt}[f,\ v]}{u + v} \right\}$$

(c) Derive the divergence and curl for a general vector field **fvecP = {fu, fv, fp}**. Show that **laplacian[f] ==**

$$\frac{4\ (\text{Dt}[u\ \text{Dt}[f,\ u],\ u] + \text{Dt}[v\ \text{Dt}[f,\ v],\ v])}{u + v} + \frac{\text{Dt}[f,\ \{p,\ 2\}]}{u\ v}$$

Problem 9

Derive the transformation to two-center *elliptical* coordinates **l** $(= \lambda)$, **m** $(= \mu)$ and **p** $(= \phi)$ defined by the relations **R l == r[1] + r[2]**, **R m == r[1] - r[2]** and Figure 4.6-1. Here, `{l, 1, Infinity}`, `{v, -1, 1}` and `{p, 0, 2Pi}`. (For checking your results, refer to the **ProlateEllipsoidal** coordinates **xi** (= **l**), **eta** (= **m**) in the package **Calculus`VectorAnalysis`**.)

(a) It's easy to see from the figure that **r[i=1, 2] == Sqrt[r^2 + (R/2 1 z)^2]** where **r == Sqrt[x^2 + y^2]**. Show that **x == r Cos[p]** and **y == r Sin[p]** with **r == R/2 Sqrt[(1-1)^2 (1-m)^2]** and that **z == R/2 l m**. (Hint: Use the built-in function **Eliminate**.)

(b) Recalling the definitions of the *conic sections* from geometry, we see that curves of constant **l** (for all **p**) are ellipses with foci at **z = ±R/2**, while curves of constant **m** are hyperbolae about the z or polar axis with foci also at **z = ±R/2** and which open away from the origin. The surfaces of constant **p** are the same as in the spherical coordinate system: planes through the polar axis.

Make a set of 2D parametric plots to display the curves of constant **l** and **m**. Or use **ParametricPlot3D**.

(c) Work out the scale factors and the unit vectors **el, em, ep** and simplify. Show that

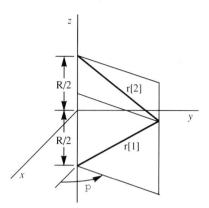

Figure 4.6-1

$$\left\{ hl \to \frac{\text{Sqrt}\!\left[\dfrac{l^2-m^2}{-1+l^2}\right] R}{2},\quad hm \to \frac{\text{Sqrt}\!\left[\dfrac{l^2-m^2}{1-m^2}\right] R}{2},\right.$$

$$\left. hp \to \frac{\text{Sqrt}\!\left[(-1+l^2)\,(1-m^2)\right] R}{2}\right\}$$

and that relative to the Cartesian axes

$$\left\{ el \to \left\{ l\,\text{Sqrt}\!\left[\frac{1-m^2}{l^2-m^2}\right]\text{Cos}[p],\; l\,\text{Sqrt}\!\left[\frac{1-m^2}{l^2-m^2}\right]\text{Sin}[p],\; m\,\text{Sqrt}\!\left[\frac{-1+l^2}{l^2-m^2}\right]\right\},\right.$$

$$em \to$$

$$\left\{-\!\left(m\,\text{Sqrt}\!\left[\frac{-1+l^2}{l^2-m^2}\right]\text{Cos}[p]\right),\; -\!\left(m\,\text{Sqrt}\!\left[\frac{-1+l^2}{l^2-m^2}\right]\text{Sin}[p]\right),\; l\,\text{Sqrt}\!\left[\frac{1-m^2}{l^2-m^2}\right]\right\},$$

$$\left. ep \to \{-\text{Sin}[p],\, \text{Cos}[p],\, 0\}\right\}$$

Verify that the unit vectors form an orthogonal triad and that the element of volume is **R^3/8 (l^2 - m^2) Dt[l] Dt[m] Dt[p]**.

(d) Compute and simplify the derivatives of the unit vectors with respect to **l, m** and **p** and then relate them to the unit vectors. Create a set of replacement rules **deRep** for the derivatives in terms of the unit vectors, as we did in Exercise 20, and show that

$$\left\{ Dt[el, l] \to \frac{em\, m\,\text{Sqrt}\!\left[\dfrac{1-m^2}{-1+l^2}\right]}{l^2-m^2},\quad Dt[el, m] \to \frac{em\, l\,\text{Sqrt}\!\left[\dfrac{-1+l^2}{1-m^2}\right]}{l^2-m^2},\right.$$

$$\text{Dt[el, p]} \to \text{ep l Sqrt}[\frac{1-m^2}{1^2-m^2}], \quad \text{Dt[em, l]} \to -(\frac{\text{el m Sqrt}[\frac{1-m^2}{-1+1^2}]}{1^2-m^2}),$$

$$\text{Dt[em, m]} \to -(\frac{\text{el l Sqrt}[\frac{-1+1^2}{1-m^2}]}{1^2-m^2}), \quad \text{Dt[em, p]} \to -(\text{ep m Sqrt}[\frac{-1+1^2}{1^2-m^2}]),$$

$$\text{Dt[ep, l]} \to 0, \quad \text{Dt[ep, m]} \to 0,$$

$$\text{Dt[ep, p]} \to \text{em m Sqrt}[\frac{-1+1^2}{1^2-m^2}] - \text{el l Sqrt}[\frac{1-m^2}{1^2-m^2}]\}$$

Derive the divergence for a general vector field **fvecE = {fl, fm, fp}** by writing **fvecE** explicitly in terms of the unit vectors, computing derivatives with respect to **l**, **m** and **p** using **deRep**, and then computing the dot product, as in Problem 4. Finally, show that **laplacian[f]==**

$$(4 (2 l \text{ Dt}[f, l] + (-1 + 1^2) \text{ Dt}[f, \{l, 2\}]) +$$

$$4 (-2 m \text{ Dt}[f, m] + (1 - m^2) \text{ Dt}[f, \{m, 2\}])) /$$

$$((1^2 - m^2) R^2) + \frac{4 \text{ Dt}[f, \{p, 2\}]}{(-1 + 1^2) (1 - m^2) R^2}$$

(e) These coordinates are useful for describing two-center molecular Coulomb symmetries. Transform the function used in Problem 2 to elliptical coordinates and show that **v = -4 l/(R (1^2 - m^2))**. Show that **laplacian[v]** vanishes except when **l = 1** and **m = ±1**. Explain (cf. Exercise 9).

Problem 10

Consider a function of the form **f[R, rvec]** where **rvec = {x, y, z}** is expressed in terms of elliptical coordinates and thus depends parametrically on **R** (see Problem 9). Show that if **rvec** is held fixed then **Dt[f[R, rvec], R]==**

$$-(\frac{1 (-1 + 1^2) \text{ Dt}[f, l] + m (1 - m^2) \text{ Dt}[f, m]}{(1^2 - m^2) R}) + \text{Dt}[f, R]$$

where now **f = f[R, l, m, p]**.

Bibliography

[1] Abramowitz, M., and I. A. Stegun (eds.), *Handbook of Mathematical Functions*, National Bureau of Standards, Washington, D. C., 1964. Also available as a Dover reprint.

[2] Aharonov, Y., and L. Susskind, Phys. Rev. **158**, 1237 (1967).

[3] Alber, G., and P. Zoller, Phys. Rep. **199**, 233 (1991).

[4] Arfken, G., *Mathematical Methods for Physicists*, Academic Press, Orlando, 1985.

[5] Beckmann, C. E., and J. M. Feagin, Phys. Rev. A**36**, 4531 (1987).

[6] Bethe, H. A., and E. E. Salpeter, *Quantum Mechanics of One- and Two-Electron Atoms*, Springer-Verlag, New York, 1957.

[7] Biedenharn, L.C., and J. D. Louck, *The Racah-Wigner Algebra in Quantum Theory*, Encyclopedia of Mathematics and Its Applications, vol. 9, Addison-Wesley, Massachusetts, 1981.

[8] Blatt, J.M., and V. F. Weisskopf, *Theoretical Nuclear Physics*, Springer-Verlag, New York, 1979. Also available as a Dover reprint.

[9] Boas, M. L., *Mathematical Methods in the Physical Sciences*, 2nd ed., Wiley, New York, 1983.

[10] Bohm, D., *Quantum Theory*, Prentice-Hall, New York, 1951. Also available as a Dover reprint.

[11] Brandt, S., and H. D. Dahmen, *The Picture Book of Quantum Mechanics*, Wiley, New York, 1985.

[12] Brink, D. M., and G. R. Satchler, *Angular Momentum*, 2nd ed., Clarendon Press, Oxford, 1968.

[13] Caves, C. M., Phys. Rev. D**26**, 1817 (1982).

[14] Caves, C. M., K. S. Thorne, R. W. P. Drever, V. D. Sandberg, and M. Zimmerman, Rev. Mod. Phys. **52**, 341 (1980).

[15] Cohen-Tannoudji, C., J. Dupont-Roc, and G. Grynberg, *Photons and Atoms: Introduction to Quantum Electrodynamics*, Wiley, New York, 1989.

[16] Dirac, P. A. M., *The Principles of Quantum Mechanics*, 4th ed., Clarendon Press, Oxford, 1958.

[17] Engel, V., H. Metiu, R. Almeida, R. A. Marcus, and A. H. Zewail, Chem. Phys. Lett. **152**, 1 (1988).

[18] Engel, V., R. Schinke, S. Hennig, and H. Metiu, J. Chem. Phys. **92**, 1 (1990).

[19] Fano, U., and A. R. P. Rau, *Atomic Collisions and Spectra*, Academic Press, New York, 1986.

[20] Feagin, J. M., and J. S. Briggs, Phys. Rev. Lett. **57**, 984 (1986); Phys. Rev. A**37**, 4599 (1988).

[21] Fearn, H., and W. E. Lamb, Phys. Rev. A**46**, 1199 (1992).

[22] Fearn, H., and W. E. Lamb, Phys. Rev. A**48**, 2505 (1993).

[23] Fleck, J. A., J. R. Morris, and M. D. Feit, Appl. Phys. **10**, 129 (1976).

[24] Friedrich, H., *Theoretical Atomic Physics*, Springer-Verlag, New York, 1991.

[25] Goldstein, H., *Classical Mechanics*, 2nd ed., Addison-Wesley, Massachusetts, 1980.

[26] Goldstein, H., Am. J. Phys. **43**, 737 (1975); Am. J. Phys. **44**, 1123 (1976).

[27] Gottfried, K., *Quantum Mechanics*, Benjamin, Massachusetts, 1974.

[28] Gradshteyn, I. S., and I. M. Ryzhik, *Table of Integrals, Series, and Products*, corrected ed., Academic Press, New York, 1980.

[29] Griffiths, D. J., *Introduction to Electrodynamics*, 2nd ed., Prentice-Hall, Englewood Cliffs, 1989.

[30] Gutzwiller, M. C., *Chaos in Classical and Quantum Mechanics*, Springer-Verlag, New York, 1990.

[31] Hamming, R. W., *Numerical Methods for Scientists and Engineers*, 2nd ed., McGraw-Hill, New York, 1973. Also available as a Dover reprint.

[32] Heintz, W. H., Am. J. Phys. **42**, 1078 (1974).

[33] Home, D., and M. A. B. Whitaker, Phys. Rev. A**48**, 2502 (1993).

[34] Jackson, J. D., *Classical Electrodynamics*, 2nd ed., Wiley, New York, 1975.

[35] Johnson, B. R., J. Chem. Phys. **67**, 4086 (1977).

[36] Kalos, M. H., and P. A. Whitlock, *Monte Carlo Methods*, Wiley, New York, 1986.

[37] Koonin, S. E., *Computational Physics*, Addison-Wesley, Redwood City, 1986.

[38] Kosloff, R., J. Phys. Chem. **92**, 2087 (1988).

[39] Kulander, K. (ed.), *Time-Dependent Methods for Quantum Dynamics*, Comput. Phys. Commun. **63**, (1991).

[40] Lanczos, C., *Applied Analysis*, Prentice-Hall, Englewood Cliffs, 1956. Also available as a Dover reprint.

[41] Landau, L. D., and E. M. Lifschitz, *Quantum Mechanics: Non- Relativistic Theory*, 3rd ed., Pergamon, Oxford, 1977.

[42] Leforestier, C., R. H. Bisseling, C. Cerjan, M. D. Feit, R. Friesner, A. Guldberg, A. Hammerich, G. Jolicard, W. Karrlein, H.-D. Meyer, N. Lipkin, O. Roncero, and R. Kosloff, J. Comput. Phys. **94**, 59 (1991).

[43] Maeder, R., *Programming in Mathematica*, 2nd ed., Addison- Wesley, Redwood City, 1991.

[44] Marion, J. B., and S. T. Thornton, *Classical Dynamics*, Harcourt Brace Jovanovich, Orlando, 1970.

[45] Merzbacher, E., *Quantum Mechanics*, 2nd ed., Wiley, New York, 1970.

[46] Misner, C. W., K. S. Thorne, and J. A. Wheeler, *Gravitation*, Freeman, San Francisco, 1973.

[47] Morrison, M. A., T. L. Estle, and N. F. Lane, *Quantum States of Atoms, Molecules, and Solids*, Prentice-Hall, Englewood Cliffs, 1976.

[48] Morse, P. M., Phys. Rev. **34**, 57 (1929).

[49] Morse, P. M., and H. Feshbach, *Methods of Theoretical Physics*, 2 vols., McGraw-Hill, New York, 1953.

[50] Park, D., *Introduction to the Quantum Theory*, 3rd ed., McGraw- Hill, New York, 1992.

[51] Pauli, W., Z. Physik 36, 336 (1926). Translated in B. L. Van der Waerden (ed.), *Sources of Quantum Mechanics*, Dover, New York, 1968.

[52] Pauling, L. C., and B. E. Wilson, *Introduction to Quantum Mechanics with Applications to Chemistry*, McGraw-Hill, New York, 1935.

[53] Peierls, R., *More Surprises in Theoretical Physics*, Princeton University Press, Princeton, 1991.

[54] Press, W. H., B. P. Flannery, S. A. Teukolsky, and W. T. Vetterling, *Numerical Recipes*, Cambridge University Press, New York, 1986.

[55] Rau, A. R. P., Science **258**, 1444 (1992).

[56] Redmond, P. J., Phys. Rev. **133**, B1352 (1964).

[57] Rost, J. M., and J. S. Briggs, J. Phys. B**24**, 4293 (1991).

[58] Saxon, D. S., *Elementary Quantum Mechanics*, Holden-Day, San Francisco, (1968).

[59] Schiff, L. I., *Quantum Mechanics*, 3rd ed., McGraw-Hill, New York, 1968.

[60] Schrödinger, E., Naturwiss. **28,** 664 (1926).

[61] Thompson, W. J., *Computing for Scientists and Engineers*, Wiley, New York, 1992.

[62] Wolfram, S., *Mathematica: A System for Doing Mathematics by Computer*, 2nd ed., Addison-Wesley, Redwood City, 1991.

Index